AF677895

Industrial AI: Applications in Fault Detection, Diagnosis, and Prognosis

Industrial AI: Applications in Fault Detection, Diagnosis, and Prognosis

Guest Editors

Janet Lin
Liangwei Zhang
Haidong Shao

MDPI

Basel • Beijing • Wuhan • Barcelona • Belgrade • Novi Sad • Cluj • Manchester

Guest Editors
Janet Lin
Department of Civil, Environmental and Natural Resources Engineering
Luleå University of Technology
Luleå
Sweden

Liangwei Zhang
Department of Industrial Engineering
Dongguan University of Technology
Dongguan
China

Haidong Shao
College of Mechanical and Vehicle Engineering
Hunan University
Changsha
China

Editorial Office
MDPI AG
Grosspeteranlage 5
4052 Basel, Switzerland

This is a reprint of the Special Issue, published open access by the journal *Applied Sciences* (ISSN 2076-3417), freely accessible at: https://www.mdpi.com/journal/applsci/special_issues/Industrial_AI_Fault_Detection_Diagnosis_Prognosis.

For citation purposes, cite each article independently as indicated on the article page online and as indicated below:

Lastname, A.A.; Lastname, B.B. Article Title. *Journal Name* **Year**, *Volume Number*, Page Range.

ISBN 978-3-7258-5691-6 (Hbk)
ISBN 978-3-7258-5692-3 (PDF)
https://doi.org/10.3390/books978-3-7258-5692-3

© 2025 by the authors. Articles in this book are Open Access and distributed under the Creative Commons Attribution (CC BY) license. The book as a whole is distributed by MDPI under the terms and conditions of the Creative Commons Attribution-NonCommercial-NoDerivs (CC BY-NC-ND) license (https://creativecommons.org/licenses/by-nc-nd/4.0/).

Contents

About the Editors

Janet Lin

Jing (Janet) Lin (Senior Member, IEEE) is a renowned expert in reliability-focused engineering, with a distinguished career spanning academia, industry leadership, and global collaborations. She has been a Professor at the School of Innovation, Design, and Engineering at Mälardalen University (MDU), Sweden, since 2022, and an Associate Professor and Docent at Luleå University of Technology (LTU) since 2015.

Dr. Lin currently serves as the Vice President of the IEEE Reliability Society (RS), focusing on Membership (2023–2024) and Publications (2025 onward). She successfully established the IEEE RS Sweden and Norway Joint Section Chapter in 2021 and has served as its Chair since its inception. In 2024, she was honored with the "Annual Reliability Engineer of the Year" award, recognizing her outstanding contributions to the field.

Her research focuses on reliability-focused innovation, design, operation, and maintenance engineering, with expertise spanning Asset Management, RAM4S, e-Maintenance, Prognostics and Health Management (PHM), and Operational Research. Her work has practical applications in diverse industries, including transportation systems, energy system, advanced manufacturing, electronics and lighting, heavy industries, consumer goods and food, etc. She is a pioneer researcher in Dependability-Centered Asset Management (DCAM).

She earned her PhD in Management Science from Nanjing University of Science & Technology in 2008 and served as a Senior Asset Management Consultant at SKF Ltd., Co. from 2008 to 2011.

Liangwei Zhang

Liangwei Zhang (Senior Member, IEEE) received the Ph.D. degree in Operation and Maintenance Engineering from Luleå University of Technology, Sweden, in 2017. He also holds an M.Sc. degree in Management Science and Engineering from Nanjing University of Science and Technology, China, in 2009.

Dr. Zhang is currently an Associate Professor and the Head of the Department of Industrial Engineering at Dongguan University of Technology, China. Prior to his academic career, he worked as a consultant in Asset Management Service (AMS) at SKF China from 2009 to 2013.

Dr. Zhang is a Senior Member of IEEE and serves as an Editorial Board Member of the *International Journal of Reliability and Safety*. He is also a certified Project Management Professional (PMP®) of the Project Management Institute (PMI). In addition, he is an active member of several professional societies, including the Systems Reliability Engineering Committee of the Systems Engineering Society of China, the Dynamic Testing Committee of the Chinese Society of Vibration Engineering, and the Reliability Engineering Division of the Chinese Mechanical Engineering Society.

His research interests include machine learning, fault detection, and prognostics and health management (PHM).

Haidong Shao

Haidong Shao (Senior Member, IEEE) received his B.S. degree in Electrical Engineering and Automation and a Ph.D. in Vehicle Operation Engineering from Northwestern Polytechnical University, Xi'an, China, in 2013 and 2018, respectively. He is currently an Associate Professor at the College of Mechanical and Vehicle Engineering at Hunan University, Changsha, China. From 2019 to 2021, he was a Postdoctoral Fellow with the Division of Operation and Maintenance Engineering, Luleå University of Technology, Luleå, Sweden. His current research interests include prognostics and health management, intelligent diagnosis and maintenance, multi-source information fusion, and industrial internet. He has served as the Associate Editor of *IEEE Transactions on Industrial Informatics*, *IEEE Transactions on Systems, Man, and Cybernetics: Systems*, and *IEEE Transactions on Intelligent Transportation Systems*.

Preface

This Reprint, titled *Industrial AI: Applications in Fault Detection, Diagnosis, and Prognosis*, brings together a curated collection of 18 peer-reviewed papers originally published in *Applied Sciences*. This Reprint explores the convergence of artificial intelligence techniques with industrial systems, specifically targeting fault detection, fault diagnosis, and prognostics and health management (PHM).

The scope of this work spans multiple industrial domains—ranging from photovoltaic and wind power systems to bearings, elevators, electric vehicles, and semiconductor manufacturing—reflecting the versatility and growing impact of AI in modern maintenance and monitoring practices. The contributions included here demonstrate recent advances in machine learning, deep learning, hybrid modeling, time series analysis, and domain adaptation, as well as methods that promote model interpretability, transferability, and edge deployment.

The motivation behind compiling this Reprint was to document and share state-of-the-art developments at the intersection of AI and industrial reliability engineering. As Guest Editors, we aimed to create a valuable resource for academic researchers, industry professionals, and students who are seeking both theoretical insights and practical implementations of intelligent diagnostics and prognostics systems.

This Reprint is addressed to those engaged in designing, analyzing, and applying intelligent maintenance strategies, and who are committed to advancing the dependability, efficiency, and sustainability of industrial operations in the context of Industry 4.0 and beyond.

Janet Lin, Liangwei Zhang, and Haidong Shao
Guest Editors

applied sciences

MDPI

Article

A Generative Adversarial Network-Based Fault Detection Approach for Photovoltaic Panel

Fangfang Lu [1,2], Ran Niu [1], Zhihao Zhang [1], Lingling Guo [3] and Jingjing Chen [4,5,*]

1 College of Computer Science and Technology, Shanghai University of Electric Power, Shanghai 200090, China; lufangfang@shiep.edu.cn (F.L.); taitian12@yeah.net (R.N.); zhangzhihao@mail.shiep.edu.cn (Z.Z.)
2 Department of Electronic Engineering, School of Electronic Information and Electrical Engineering, Shanghai Jiao Tong University, Shanghai 200240, China
3 College of Chemical Engineering, Zhejiang University of Technology, Hangzhou 310014, China; lguo@zjut.edu.cn
4 Zhijiang College, Zhejiang University of Technology, Shaoxing 312000, China
5 School of Economics, Fudan University, Shanghai 200010, China
* Correspondence: jingjchen@fudan.edu.cn

Abstract: Photovoltaic (PV) panels are widely adopted and set up on residential rooftops and photovoltaic power plants. However, long-term exposure to ultraviolet rays, high temperature and humid environments accelerates the oxidation of PV panels, which finally results in functional failure. The traditional fault detection approach for photovoltaic panels mainly relies on manual inspection, which is inefficient. Lately, machine vision-based approaches for fault detection have emerged, but lack of negative samples usually results in low accuracy and hinders the wide adoption of machine vision-based approaches. To address this issue, we proposed a semi-supervised anomaly detection model based on the generative adversarial network. The proposed model uses the generator network to learn the data distribution of the normal PV panel dataset during training. When abnormal PV panel data are put into the model in the test phase, the reconstructed image generated by the model does not equal the input image. Since the abnormal PV panel data do not obey the data distribution learned by the generator, the difference between the original image and its reconstructed image exceeds the given threshold. So, the model can filter out the fault PV panel by checking the error value between the original image and its reconstructed image. The model adopts Gradient Centralization and SmoothL1 loss function to improve its generalization performance. Meanwhile, we use the convolutional block attention module (CBAM) to make the model pay more attention to the defective area and greatly improve the performance of the model. In this paper, the photovoltaic panels dataset is collected from a PV power plant located in Zhejiang, China. We compare the proposed approach with state-of-the-art semi-supervised and unsupervised approaches (i.e., AnoGAN (Anomaly Detection with Generative Adversarial Networks), Zhao's method, GANomaly, and f-AnoGAN), and the result indicates that the Area Under Curve (AUC) increases by 0.06, 0.052, 0.041 and 0.035, respectively, significantly improving the accuracy of photovoltaic panel fault detection.

Keywords: photovoltaic power generation; abnormal detection; generator adversarial networks; deep convolutional; gradient centralization

1. Introduction

Traditional thermal power generation uses heat energy generated by combustible materials (i.e., coal) and converts it into electric energy through power generation plants. The combustion of coal produces carbon dioxide, which is the main source of carbon emissions. Due to the rapid increase in carbon emissions, the concentration of global CO_2 continues to rise, and the caused greenhouse effect results in global warming, which poses a great threat to the survival of human beings. Since 1992, the Intergovernmental Panel on Climate Change (IPCC) has signed a series of agreements, including the Paris Agreement.

The overall goal of the Paris Agreement was to achieve a peak in global greenhouse gas emissions around 2020, and net zero greenhouse gas emissions in the second half of this century. Photovoltaic power generation is a technology that directly converts light energy into electric energy by using the photovoltaic effect at the semiconductor interface. Abundant solar radiant energy is an important energy that is inexhaustible, pollution-free, cheap and free to be used by human beings. Therefore, solar energy has become the focus of attention because of its unique advantages [1,2]. In the global promotion of clean and low-carbon energy transition, the adoption of photovoltaic (PV) panels in clean energy is increasing year by year [3]. According to International Energy Agency (IEA) statistics, since 2018, the cumulative installed capacity of global PV power generation has reached 480GW and is expected to reach 1721GW in 2030 [4].

In order to increase the exposure time of PV panels, large-scale PV power plants are located in areas without apparent shade, such as the roof of buildings, plains, hills and over fish ponds, as shown in Figure 1. Long-term exposure to ultraviolet rays, high temperature and humid environments accelerate the oxidation of PV panels, which results in functional failure [5]. Severe sealant delamination increases reflection, decreases irradiance, sinks moisture in the module, and accelerates panel oxidation [6]. The accumulation of dust directly reduces light transmittance and affects the efficiency of the PV power generation system, sometimes up to about 50% or, even worse, 80% [7]. Although snail trials have no direct effect on power generation efficiency, invisible cell cracks usually reduce their power output [8]. Therefore, it is necessary to inspect the status of photovoltaic panels regularly to find and replace defective photovoltaic panels in time. However, most of the photovoltaic panels are installed on roofs, fish ponds and hillsides with a wide area, so manual inspection cannot be carried out. Recently, several artificial intelligence algorithms (e.g., deep convolutional neural networks) were introduced to facilitate the process of PV panel fault detection [9–13]. Unfortunately, such fully supervised learning-based methods have been proven unsuitable for large-scale PV panel fault detection due to the lack of defective samples [14–16]. At present, semi-supervised learning-based approaches [17–20] are being explored by many researchers.

Figure 1. The fishery light complementary large-scale PV power plant.

To solve these issues, by leveraging adversarial auto-encoders and conditional Generative Adversarial Networks (GAN), we propose a semi-supervised anomaly detection approach (named ppFDetector) to facilitate the process of fault detection in PV panels. The main contributions in this paper can be summarized thus:

(1) A novel model-based semi-supervised generative adversarial network is proposed. Compared with the fully supervised learning model, the semi-supervised anomaly detection model does not require a large number of negative samples, which solves the problem of the anomaly detection model's inability to be trained without negative samples.

(2) The original optimization function is implemented with gradient centralization (GC) to regularize weight and output space to prevent the model from overfitting.
(3) Convolutional block attention module (CBAM) is used to make the model pay more attention to the defective area, and improve the performance of the model.
(4) SmoothL1loss is used to define the loss, which can combine the advantages of L1Loss and L2Loss to speed up the training of the model.

The remainder of this paper is organized as follows: Section 2 describes related literatures. Section 3 details the technical implementation of the ppFDetector. Experimental results using the ppFDetector on the PV panels dataset compared with state-of-the-art anomaly detection models are presented in Section 4. We finally conclude the paper in Section 5.

2. Related Work

2.1. Generative Adversarial Networks

The network structure of GAN is composed of a generation network and a discriminant network. The generator receives the random variable and generates fake sample data. The purpose of the generator is to try to make the generated samples the same as the real samples. The input of the discriminator consists of two parts, real data and data generated by the generator. The output is usually a probability value, which represents the probability that the input is a true distribution. If the input comes from real data, then output 1; otherwise, output 0. At the same time, the output of the discriminator will be fed back to guide's training. Ideally, it cannot distinguish whether the input data are real or generated, i.e., each time the output probability is 1/2 (equivalent to random guessing), the data generated by the generator have reached real. At this point, the model is optimal. In practical applications, the generation network and discriminant network are usually implemented by deep neural networks. As the number of network layers increases, it is difficult for GAN networks to converge. In order to solve this problem, a wide variety of GAN variant models have emerged. Deep Convolutional Generative Adversarial Networks (DCGAN) [21] is the first major improvement on GAN architecture. Its generator is an upsampling process, i.e., deconvolution. The discriminator is a common convolutional network. It is more stable in training and produces higher quality samples.

With the development of deep learning, generative adversarial networks can also perform image anonymization to protect training data. Ashish Undirwade and Sujit Das [22] proposed applying optimized noise to the latent space representation (LSR) of the original image, and then converting the input original image into a new synthetic image through generative adversarial networks.

At present, GAN is a promising network model that can be applied in many fields, such as image super-resolution, image translation and style conversion. Karnewar [23] proposed a Multi-Scale Gradient Generative Adversarial Network (MSG-GAN), which is a simple but effective technique for addressing this by allowing the flow of gradients from the discriminator to the generator at multiple scales. This technology provides a stable method for high-resolution image synthesis, and can replace the commonly used progressive growth technology.

2.2. PV Panel Fault Detection

PV panel fault detection roughly goes through the following phases: an initial phase, the phase of physical approaches and the phase of intelligent detection.

In the initial phase, manual inspection is the most primitive method for detecting PV panel faults. The worker uses an image of the PV panel taken by a camera to check whether there is a defect. Traditional manual inspection evaluates a single PV panel just by sight, which is usually high cost, high error rate and low efficiency [24], and has severe subjective inclines for different defects.

In the phase of physical approaches, Sawyer DE et al. [25] used laser scanning technology to aid the fault-detection process. It displays the continuity of resistance in the crystalline silicon of the forward conversion of laser scanning. If there exists a crack, there will be a discontinuity of resistance. Tsuzuki K et al. [26] proposed the use of sound waves for fault detection, which makes PV panels vibrate to generate sound waves. Based on a comparative analysis with sound waves generated by non-defective solar cells, the fault detection can be realized. Physical detection [25–27] has high detection accuracy for specific types of PV panel faults, but it cannot effectively detect other faults.

In the intelligent phase of fault detection, the intelligent PV panel fault detection method based on machine vision is now being gradually adopted by users. By machine vision detection, Bastari et al. [28] proposed a method for texture analysis of dark defect areas in electroluminescence images to classify defective and non-defective cells. Quarter et al. [29] first reported the UAV inspection system to perform non-destructive inspection of large-scale PV power plants. However, the effect of fault diagnosis cannot meet the requirements of practical use, and many researchers are further studying fault diagnosis methods. Li et al. [30] first extracted features by calculating the first derivative of the Gaussian function on the collected images, and then performed defect detection through feature matching. This method verifies the effectiveness of the fault analysis algorithm based on image processing to a certain extent, but strictly requires good quality of the collected images. However, due to various reasons such as the wind effect, the obtained image resolution is low, and the performance of this method is significantly reduced. In addition, traditional pattern recognition algorithms often fail to extract fault features with acceptable complexity from the acquired aerial images.

With the evolution of deep convolutional neural networks, the introduction of artificial intelligence algorithms into the intelligent patrol system [9] can improve its robustness and reliability [10]. Li et al. [11] proposed an intelligent diagnosis method of aerial PV module image defects based on the deep convolutional neural network (CNN). This method uses CNN to classify multiple deep features and diverse states. Compared with traditional methods, it can flexibly and reliably solve the problem of PV module images with low quality and distortions. Tang [12] proposed using a generative adversarial network combined with data augmentation to increase training samples; this achieves high classification accuracy, but cannot precisely locate defects. Zhao [13] presented a deep learning-based automatic detection of multitype defects to fulfill inspection requirements of the production line. These studies used CNN to solve PV panel fault detection, and the models were based on supervised learning. However, due to the lack of abnormal panel samples, supervised learning is not suitable for this kind of problem. Schlegl et al. [15] proposed an unsupervised anomaly detection method based on the generative adversarial network (AnoGAN) only using standard panel samples; it uses continuously iterative optimization to find a certain point in latent space. However, due to the need for iterative optimization, the AnoGAN algorithm is bound to consume a lot of time. Zhao et al. [16] proposed using GAN and auto-encoder to reconstruct the defect image, and then use Local Binary Pattern (LBP) to extract the local contrast of the image to detect the defect. However, for images with complex backgrounds, Zhao's method cannot reconstruct and repair defective images well for further fault detection. Also proposed was the f-AnoGAN [17], which can quickly map a picture to a certain point in hidden space and then use wasserstein GAN (WGAN) for anomaly detection. However, its quantitative evaluation of segmentation accuracy of anomaly detection only serves as a coarse indication.

3. ppFDetector Solution

In this paper, positive samples represent normal PV panel images without defect, and negative samples represent abnormal PV panel images with defect. As the diagram in Figure 2 shows, the workflow of ppFDetector is composed of GAN model training and anomaly detection strategy. During GAN model training, the generator network is designed to learn the data distribution of positive samples in the dataset of PV panel, and the discriminator network is designed for adversarial training, so as to reduce the error between the reconstructed image and its input original image. Then the generator can capture the training data distribution within both the normal PV panel image and its latent space vector. In the anomaly detection strategy, the test PV panel image (including positive samples and negative samples) enters the trained encoder–decoder to generate the latent space vector z and its reconstructed image. Next, the reconstructed image enters the encoder to generate its latent space vector $\widetilde{z}$, and the error $M(z, \widetilde{z})$ between latent space vectors z and $\widetilde{z}$ is computed. If $M(z, \widetilde{z}) > \theta$, the proposed model judges the test image as an abnormal PV panel, otherwise it is normal. The details of the model are described as follows.

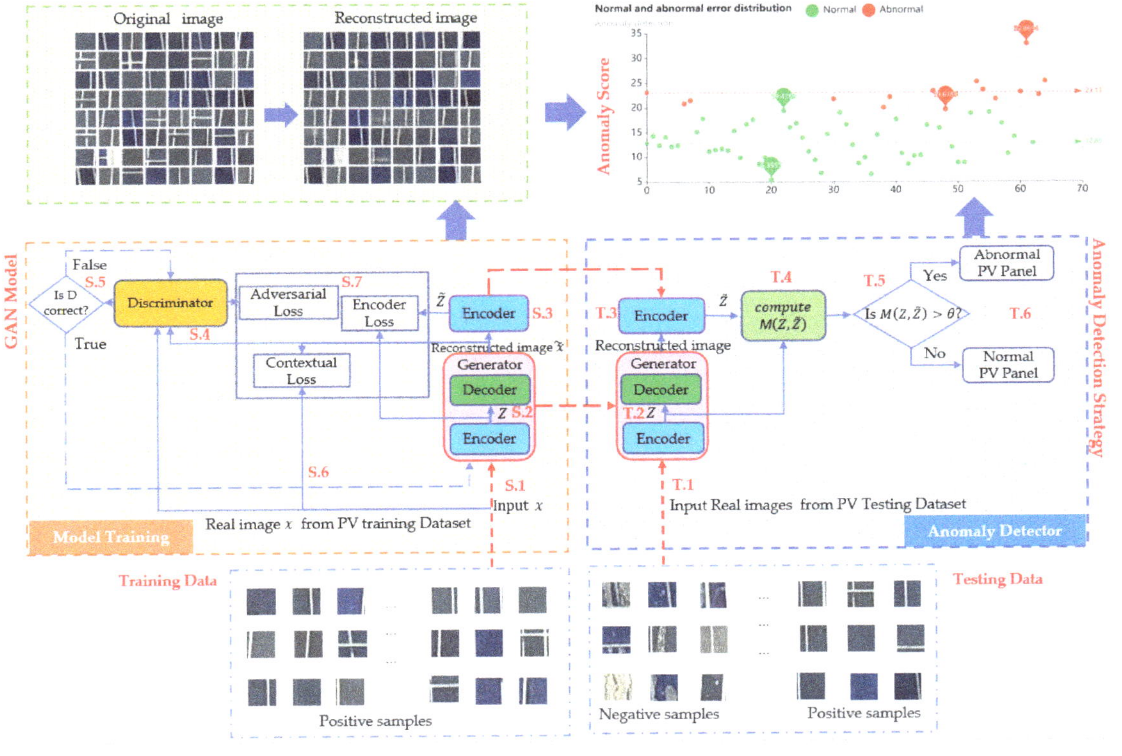

Figure 2. The integrated photovoltaic panel fault detection workflow of ppFDetector.

3.1. GAN Model

The network structure of GAN model is composed of a generator network G and a discriminator network D. The generator and the discriminator are constructed by iterative confrontation training. Considering that the color and shape patterns of the PV panel image are usually single and regular, the generator network in GAN can better learn the data distribution of the PV panel image set and generate a more ideal reconstructed image.

3.1.1. Generator Network

As shown in Figure 3, the generator network is mainly composed of encoder network G_E and decoder network G_D. The details of the encoder parameters and decoder parameters are shown in Tables 1 and 2, respectively. Firstly, the generator network G reads the input PV panel image x and passes it into the encoder network G_E. After x is passed through the convolutional layer, batch normalization layer, the leaky linear rectification activation function layer and CBAM layer, and scaled down several times, x is compressed into a latent space vector $z = G_E(x)$. The latent space vector z can theoretically be assumed as the best representation of x with the smallest dimension. Next, the decoder network G_D enlarges z through the transposed convolution layer, batch normalization layer and linear rectification activation function layer. After upscaling, the latent space vector z is reconstructed into the reconstructed image $\tilde{x}$ corresponding to the original PV panel image x, i.e., $\tilde{x} = G(x)$.

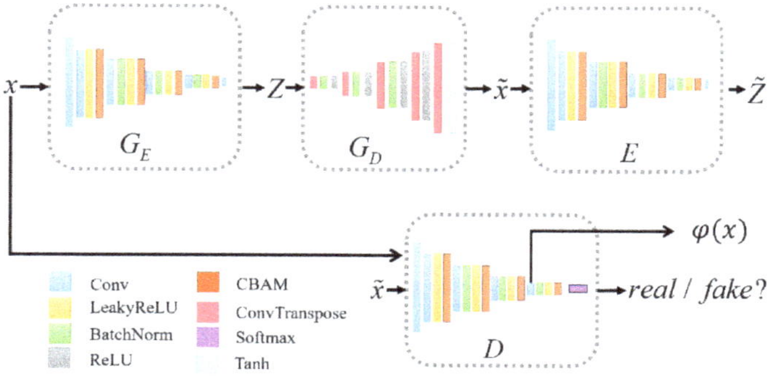

Figure 3. Pipeline of the generative adversarial network for anomaly detection.

Table 1. Details of the encoder parameters.

Layer 1	Input	Filter Number	Kernel	Stride	Padding	Batch Normalization	Activation Function	Attention Mechanism	Output
1	(32,32,3)	64	(4,4)	2	Same (1)	Batch Norm	LeakyReLU (0.2)	CBAM	(16,16,64)
2	(16,16,64)	128	(4,4)	2	Same (1)	Batch Norm	LeakyReLU (0.2)	CBAM	(8,8,128)
3	(8,8,128)	256	(4,4)	2	Same (1)	Batch Norm	LeakyReLU (0.2)	CBAM	(4,4,256)
4	(4,4,256)	100	(4,4)	1	Valid (0)	None	None	None	(1,1,100)

Table 2. Details of the decoder parameters.

Layer 1	Input	Filter Number	Kernel	Stride	Padding	Batch Normalization	Activation Function	Output
1	(1,1,100)	256	(4,4)	1	Valid (0)	Batch Norm	ReLU	(4,4,256)
2	(4,4,256)	128	(4,4)	2	Same (1)	Batch Norm	ReLU	(8,8,128)
3	(8,8,128)	64	(4,4)	2	Same (1)	Batch Norm	ReLU	(16,16,64)
4	(16,16,64)	3	(4,4)	2	Same (1)	None	tanh	(32,32,3)

3.1.2. Encoder Subnet

The structure of the encoder subnet E consists of the convolutional layer, batch normalization layer, leaky linear rectification activation function layer and CBAM layer. Through the G_{E-D}, the reconstructed image data $\tilde{x}$ is output. Put $\tilde{x}$ into E to obtain the encoded latent space vector $\tilde{z} = E(\tilde{x})$ of the reconstructed image, and the fault detection of PV panels can be realized by capturing the error between z and $\tilde{z}$.

3.1.3. Discriminator Network

In ppFDetector, the original PV panel image x and its reconstructed image $\tilde{x}$ are put into the discriminator network D. After the transposed convolution layer, the batch normalization layer and the leaky linear rectification activation function layer, the input image is downscaled by several times. After flattening, the compressed vector enters the fully connected layer to achieve the purpose of binary classification (by checking whether the input is the original PV panel image or the reconstructed one).

3.2. Loss Functions in GAN Model Training

In order to mathematicalize the above assumptions, three loss functions are defined and combined to establish the objective function of the model. Each loss function optimizes a single subnet. At the same time, this paper uses SmoothL1Loss instead of the combination of L1Loss and L2Loss. The formula of SmoothL1Loss is as follows:

$$smooth_{L1}(t) = \begin{cases} 0.5t^2 & |t| < 1 \\ |t| - 0.5 & otherwise \end{cases} \quad (1)$$

From the above formula, when the absolute value of t is less than 1, SmoothL1Loss is L2Loss. Because L2Loss will square the error, a smaller loss will be obtained, which is conducive to model convergence. Otherwise, SmoothL1Loss is the translation of L1Loss. L1Loss is not sensitive to outliers, and the magnitude of the gradient can be controlled so that it is not easy to explode during training. So SmoothL1Loss has the advantages of both L2Loss and L1Loss.

3.2.1. Adversarial Loss

The generator network G needs to update the training parameters according to the classification accuracy of discriminator network D (to determine whether the input is the original PV panel image or the reconstructed image). This section only refers to the generator part, but meanwhile, the discriminator is also trained and its weights are updated. In this paper, we update the generator network G based on the features of the intermediate layer of the discriminator denoted as $\varphi(x)$, as shown in Figure 3. Assuming that the independent variables x and $\widetilde{x} = G(x)$ both obey the positive sample data distribution P_x, the adversarial loss function L_{adv} is the distance between the original PV panel image x and its reconstructed image $\widetilde{x}$ using $smooth_{L1}$ distance, it is defined as follows:

$$L_{adv} = E_{x \sim P_x} ||\varphi(x) - \varphi(G(x))||_{smooth_{L1}} \quad (2)$$

where P_x is the data distribution of positive normal PV panel images; $\varphi(x)$ is the output feature of the input PV panel image x in the intermediate layer of the discriminator; and $G(x)$ is the reconstructed image of x in the generator. This loss is only activated with negative samples in the discriminator.

3.2.2. Contextual Loss

By optimizing the adversarial loss function L_{adv} in Equation (2), the generator G is able to produce images that are realistic enough to prevent discriminator D from discriminating between real and fake. However, for the reconstruction task, adversarial loss only allows the generator G to produce a reconstruction image $\widetilde{x}$ that obeys positive sample data distribution P_x. But the reconstruction image $\widetilde{x}$ may not correspond to its original PV panel image x, and thus cannot achieve our purpose of reconstruction of the PV panel.

In this paper, we define the context loss L_{con} by calculating the $smooth_{L1}$ distance between the reconstructed image $\widetilde{x}$ and its original PV panel image x, and the equation is as follows:

$$L_{con} = E_{x \sim P_x} ||x - G(x)||_{smooth_{L1}} \quad (3)$$

where $G(x)$ is the reconstructed image in generator when input PV panel image x.

3.2.3. Encoder Loss

The two losses defined in Equations (2) and (3) can help the generator produce reconstruction images that are both realistic and relevant to the original PV panel image. We also use the latent space vector z to replace the original PV panel image x as the data for detecting anomalies. In order to reduce the difference between latent space vectors z and $\tilde{z}$, this paper defines the encoder loss L_{enc}, and its function equation is as follows:

$$L_{enc} = E_{x \sim P_x} ||G_E(x) - E(G(x))||_{smooth_{L1}} \tag{4}$$

where $G_E(x)$ is the latent space vector z output from encoder subnet of generator G_E when the input is PV panel image x. $E(G(x))$ is the reconstructed latent space vector $\tilde{z}$ when input PV panel image x firstly passes through the generator G and then passes through encoder subnet E.

Up to now, the objective loss function L of the GAN model can be defined as follows:

$$L = L_{adv} + \lambda L_{con} + L_{enc} \tag{5}$$

where λ is a parameter for adjusting the sharpness of the reconstructed image.

3.3. Gradient Centralizaton

Since the gradient of neural network becomes more and more difficult to descend as the number of network layers increases, problems such as gradient disappearance, gradient explosion, and inability to converge will occur. So, optimization techniques are crucial for efficient training of neural networks.

In this paper, we merge the idea of gradient centralization (GC) [31] to optimize the gradient directly. Using weight space and output feature space regularization to avoid overfitting, this will also improve the generalization performance of neural networks. As shown in Figure 4a, W denotes the weight, L denotes the loss function, $\nabla_W L$ denotes the weight gradient, and $\Phi_{GC}(\nabla_W L)$ denotes the central gradient. As shown in Figure 4a, replace $\nabla_W L$ by $\Phi_{GC}(\nabla_W L)$ to implement embedding GC to the existing network optimizer. The GC is calculated as the mean of each column of the gradient matrix/tensor, and transfers the center of each column to zero mean as shown in Figure 4b.

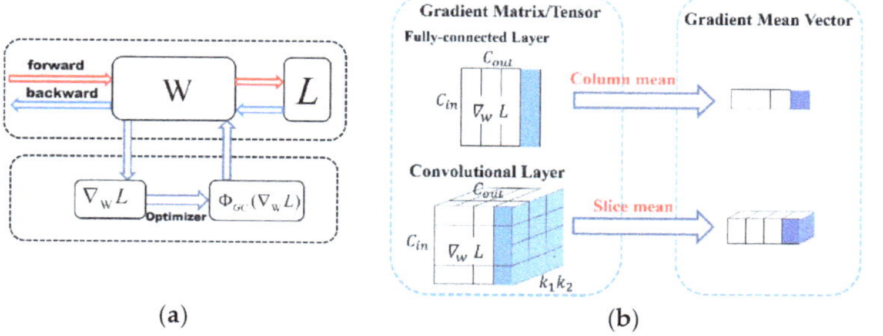

Figure 4. Schematic diagram of gradient centralization. (**a**) Schematic diagram of using GC; (**b**) GC operation of gradient matrix/weight tensor.

3.3.1. GC Formula

As for the fully connected layer or convolutional layer, we assume that the gradients have been obtained by back propagation. As for the weight vector W_i, its gradient is $\nabla_{W_i} L (i = 1, 2, \ldots, N)$, the subscript i represents the column i vector in the gradient matrix. The operator of GC, which is denoted by Φ_{GC}, can be calculated as follows:

$$\Phi_{GC}(\nabla_{W_i} L) = \nabla_{W_i} L - \mu_{\nabla_{w_i} L} \tag{6}$$

where $\mu_{\nabla_{W_i} L} = \frac{1}{M} \sum_{j=1}^{M} \nabla_{W_{i,j}} L$. As shown in Equation (6), we only need to calculate the mean of the column vector of weight matrix, then remove the mean from each column vector.

3.3.2. Properties of GC

The following is a theoretical analysis of how GC can improve the generalization ability of the model.

(1) Weight space regularization: The projection of the weight gradient is able to constrain the weight space in a hyperplane, as shown in Figure 5, in which P is a projection matrix of hyperplane with normal vector e, and $P\nabla_W L$ is projection gradient. Firstly, projecting gradient $\nabla_W L$ to the hyperplane is determined by $e^T(W - W^t) = 0$, in which W^t is the weight vector of the t times iteration, then weight W is updated along $-P\nabla_{W^t} L$ direction. It can be concluded that $e^T W^{t+1} = e^T W^t = \cdots = e^T W^0$, i.e., $e^T W$ is a constant during the training, so GC regularizes the solution space of W, thus decreasing the possibility of overfitting.

(2) Output space regularization: After the introduction of GC, a constant intensity change in an input feature causes a change in the output activation, which is unrelated to the current weight vector. If the mean value of the initial weight vector converges to 0, then the output activation is insensitive to changes in the interference intensity of the input features, so the output feature space is more stable to the changes in the training samples.

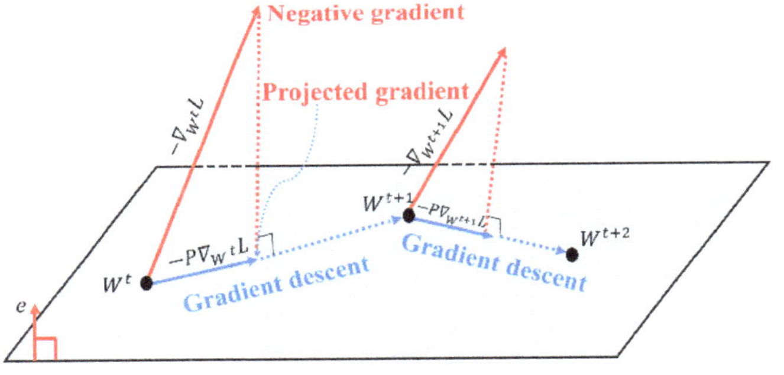

Figure 5. The geometrical interpretation of GC.

In this paper, the ppFDetector model uses Gradient Centralization to regularize the weight space and output space in order to avoid overfitting.

3.4. Convolutional Block Attention Module

Introducing an attention mechanism into a convolutional neural network can improve the performance of the model in classification tasks. Woo [32] proposed convolutional block attention module (CBAM). Considering that the traditional CNN network only uses convolution operation to extract different features in the images, we use the CBAM module to emphasize the meaningful features of channel and spatial axis. As shown in Figure 6, we use the channel attention module and the spatial attention module in turn, whose functions are to learn 'what' in the channel axis and 'where' in the spatial axis respectively.

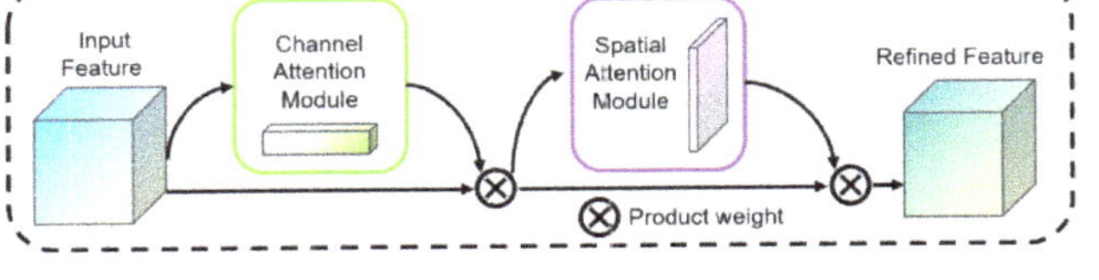

Figure 6. Module of CBAM.

3.4.1. Channel Attention

CBAM uses both average-pooled and max-pooled features, then feeds them sequentially into a weight-sharing Multilayer Perceptron (MLP), and finally uses element-wise summation to combine the output feature vectors. Channel attention mainly focuses on what features of the input image are meaningful. The channel attention module of CBAM is shown in Figure 7.

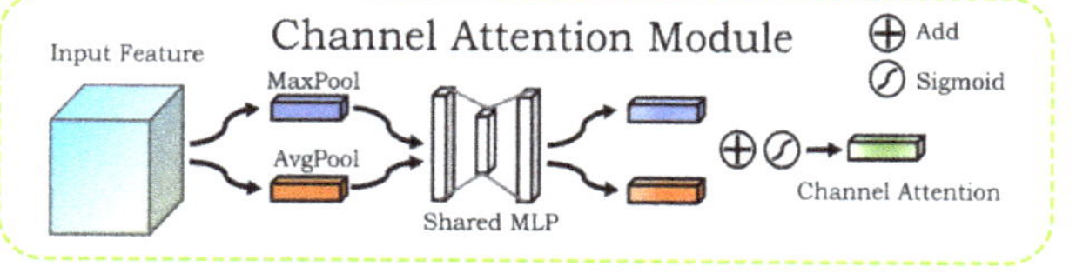

Figure 7. Channel attention module.

3.4.2. Spatial Attention

Adding the spatial attention module makes up for the lack of channel attention to a certain extent because spatial attention mainly focuses on which part of the input image has more effective information. In order to calculate spatial attention, a feature map is obtained through max pooling and average pooling, spliced into a 2D feature map, sent to the standard 7×7 convolution for parameter learning, and finally a 1D weight feature map is obtained. The spatial attention module of CBAM is shown in Figure 8.

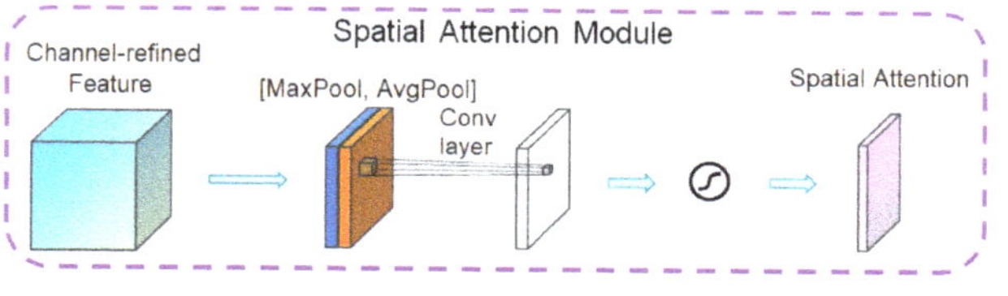

Figure 8. Spatial attention module.

In this paper, CBAM module is added to make the ppFDetector model pay more attention to the defect area, which can improve the performance of the proposed model.

3.5. Anomaly Detection Strategy

In the anomaly detection strategy, when the normal PV panel image is put into the trained model, the generator first encodes it into latent space vector and then decodes it into the reconstructed image. Because it obeys data distribution learned by the generator, the error between the input image and its reconstructed image is smaller than the threshold defined by the anomaly model. However, when the abnormal PV panel image is put into the model, the reconstructed image generated by the model is not equal to the input image. Since it doesn't obey the data distribution learned by the generator, the error between the input image and its reconstructed image is bigger than the threshold. Therefore, the model detects abnormal PV panel by the error between the input image and its reconstructed image. However, the input image x is a three-channel color image that is stored in a matrix. If the error between the input image and its reconstructed image is directly calculated by the image matrix, the complexity of space and time is unacceptable. The latent space vector z can theoretically be assumed as the best representation of x with the smallest dimension. In order to reduce the time and space complexity of the proposed method, we used the error between latent space vectors z and $\widetilde{z}$ instead of the error between the input image and its reconstructed image.

In this paper, the error of the latent space vector denoted as $M(z,\widetilde{z})$ can be calculated by the mean of the differences between the latent space vectors z and $\widetilde{z}$. The relevant expressions are as follows:

$$M(z,\widetilde{z}) = \frac{1}{N}\sum\nolimits_{i=1}^{N}|z(i) - \widetilde{z}(i)|, \quad (7)$$

where N is the length of vector z.

Finally, if $M(z,\widetilde{z}) > \theta$, the proposed model judges the test image as an abnormal PV panel, otherwise, it is normal.

4. Experiment

4.1. Dataset

The dataset is collected from a PV power plant located in Zhejiang province. The original image resolution is 3840 × 2048. In order to speed up the training process, we split the original images into a set of small images (i.e., 32 × 32 images as shown in Figure 9a). By segmenting the original image, 32,000 small images are obtained, of which 25,600 small images are included under the training dataset and 6400 under test set. An additional 3200 abnormal images are obtained from the negative samples by splitting defective images, as shown in Figure 9b.

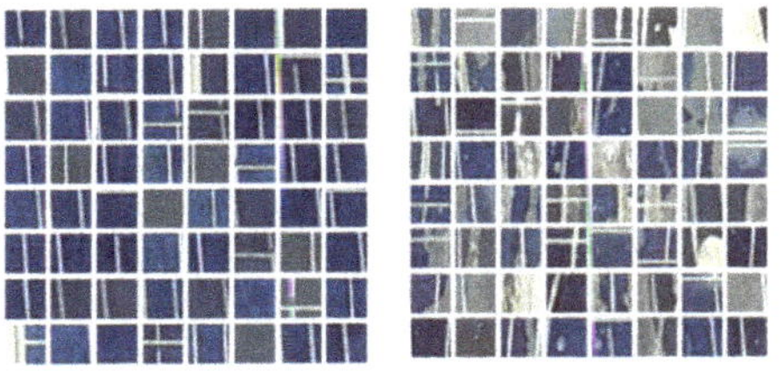

(a) (b)

Figure 9. PV Segmentation images: (**a**) Positive sample; (**b**) Negative sample.

4.2. Model Construction

The construction of the generator and the discriminator in the generative confrontation network is a key task. The generator includes an encoder and a decoder, and the discriminator is an encoder-like structure. The details of the encoder parameters and decoder parameters are shown in Tables 1 and 2, respectively. The input image size of the encoder is 32 × 32 × 3, and a 4 × 4 convolution kernel is used. The first three layers of convolution use edge padding. The core step size is 2, and the batch normalization layer [33], the LeakyReLU activation function layer [34] and CBAM layer [32] are added after the convolution layer. The last layer of convolution directly outputs a hidden space vector with a size of 1 × 1 × 100. The decoder is symmetrical to the encoder structure, the first layer of transposed convolution has no padding and step size, and the last three layers of transposed convolution padding is 1, and the step size is 2. After the first three transposed convolutional layers, the batch normalization layer and the ReLU activation function layer are added, and the output size is 32 × 32 × 3 of the reconstruction image.

In this paper, three experiments are performed on where the CBAM layer is embedded. Firstly, the CBAM layer is embedded after the first convolution block of the encoder network G_E. Secondly, the CBAM layer is embedded after both the first and second convolution blocks of G_E. Thirdly, the CBAM layer is embedded after the first, second and third convolution blocks, as shown in Figure 3. The final results of the three different CBAM embedding methods are shown in Figure 10. It can be seen that the third embedding effect is the best, so the proposed ppFDetector model chooses the third CBAM embedding method, and the AUC reaches the maximum value.

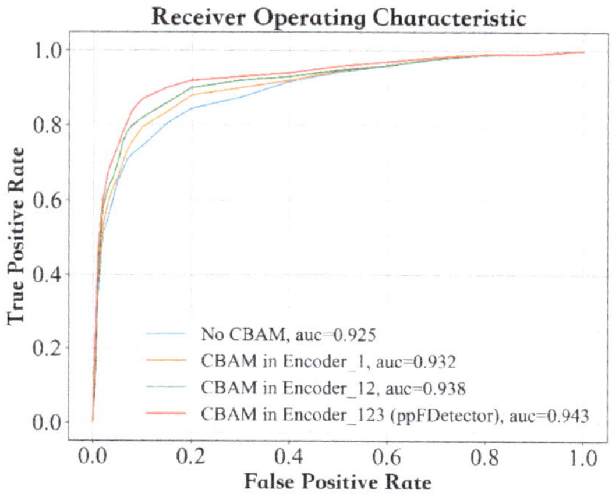

Figure 10. ROC curves of different CBAM embedding methods.

4.3. Model Training

As shown in Figure 2, in the model training, only positive PV panel image x enters the encoder–decoder (S.2) to generate the reconstructed image $\tilde{x}$. Then it is divided into two branches. In the first branch, the reconstructed image $\tilde{x}$ enters the discriminator through S.4, and then goes through S.5 to determine whether the result of the discriminator is correct or not. In the second branch, the reconstructed image $\tilde{x}$ enters the encoder (S.3) to generate the latent space vector $\tilde{z}$ of the reconstructed image. Then, the adversarial loss, encoder loss and contextual loss (S.7) are calculated and sent back to the generator. The generator and the discriminator are constructed by iterative confrontation training.

4.4. Model Validation

After the model training is completed, the model needs to be evaluated to ensure that efficient and accurate anomaly detection tasks are completed at minimum cost. Therefore, the model is evaluated from the following perspectives:

(1) We evaluate the likelihood of the original image and its reconstructed image of the generator. As shown in Figure 11, in the training process, as the number of iterations increases, the difference between the original image and the reconstructed image gradually shrinks.

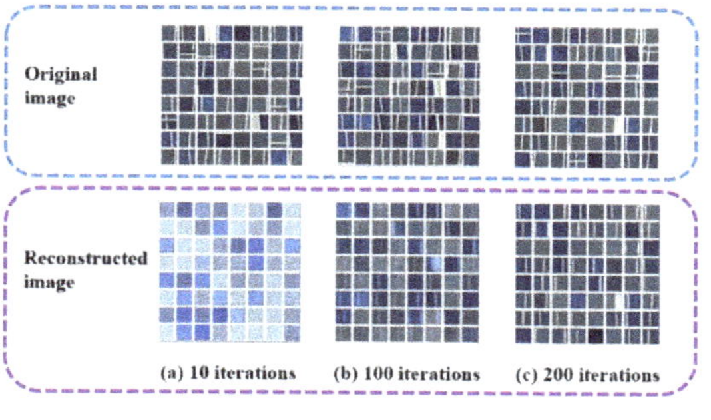

Figure 11. Original images and reconstructed images during training. (**a**) 10 iterations results; (**b**) 100 iterations results; (**c**) 200 iterations results.

(2) We evaluate the data distribution comparison between the original image and its reconstructed image for the positive sample and negative sample, respectively. Since

only positive sample dataset X is fed into the generator G_{E-D} during the training process, G_{E-D} only learns by the data distribution P_x of the normal PV panel images while the data distribution P_y of the abnormal PV panel images is unknown. When a normal PV panel image x is put into the generator, the reconstructed image $\widetilde{x}$ generated by the generator is equal to the input image x because it obeys the data distribution learned by the generator, then its data distribution $P_{\widetilde{x}}$ is infinitely close to P_x as shown in Figure 12a. When an abnormal PV panel image y is put into the generator, G_{E-D} still encodes it into z_y and then decodes it into $\widetilde{y}$ in a manner that obeys the data distribution P_x. Due to the difference between P_x and P_y, y and $\widetilde{y}$ are different, and its data distribution $P_{\widetilde{y}}$ is different from P_y as shown in Figure 12b.

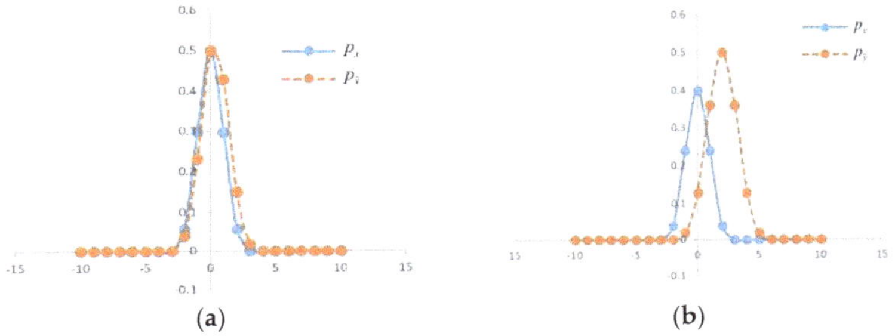

Figure 12. Data distribution comparison between the original image and its reconstructed image. (a) Positive sample; (b) Negative sample.

(3) We evaluate the error between the latent space vectors of the original image and its reconstructed image. As shown in Figure 13a, the $M(z,\widetilde{z})$ of the positive samples (in blue color) is distributed around 10, but the $M(z,\widetilde{z})$ of the negative samples (in red color) is mainly distributed between 30 and 70. It can also be seen in Figure 13b. So, the $M(z,\widetilde{z})$ of the positive samples and negative samples has a clear decision boundary.

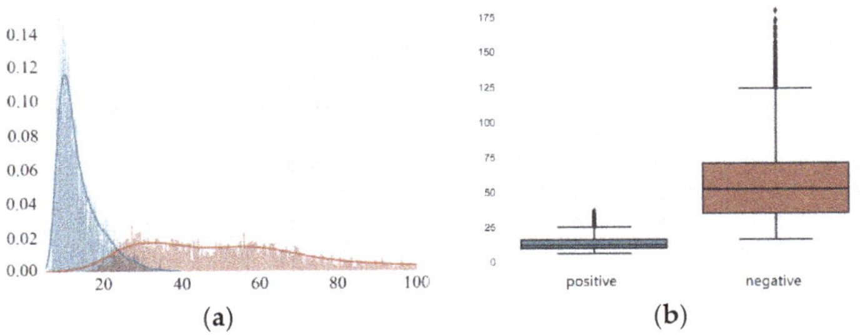

Figure 13. Plots of $M(z,\widetilde{z})$ of the positive and negative samples. (a) Histogram; (b) Boxplot.

4.5. Model Checking

As shown in Figure 2, in the anomaly detection process, the test PV panel image $x \in R^{256\times256}$ is first split into a set of small images (i.e., 32 × 32 images as shown in Figure 9a). The small image $x \in R^{32\times32}$ enters the encoder–decoder (T.2), to generate the latent space vector z and its reconstructed image. Next, the reconstructed image enters the encoder (T.3) to generate the latent space vector $\widetilde{z}$. Then the error $M(z,\widetilde{z})$ between latent space vectors z and $\widetilde{z}$ (T.4) is computed. If $M(z,\widetilde{z}) > \theta$, the test small image is judged as an abnormal PV panel, Otherwise, it is normal. According to the above method, each small image is judged whether it is abnormal or not in turn. Finally, we merge the 64 small image $x \in R^{32\times32}$ back into the large original image $x \in R^{256\times256}$. As shown in Figure 14, a is the test image, and c is a line chart of the latent vector z of the test image and its reconstructed image. The horizontal axis represents the sequence number of the image block, and the total is 64. The vertical axis represents the value of latent vector z. Figure 14d is a scatter plot of the error $M(z, \widetilde{z})$ of the test image, the green points are normal, but the red points are abnormal. It can be clearly seen that there is an obvious decision boundary between the normal point and the abnormal point. When we get the outliers in $M(z, \widetilde{z})$, and select the corresponding feature map as abnormal, the output result is shown in Figure 14b.

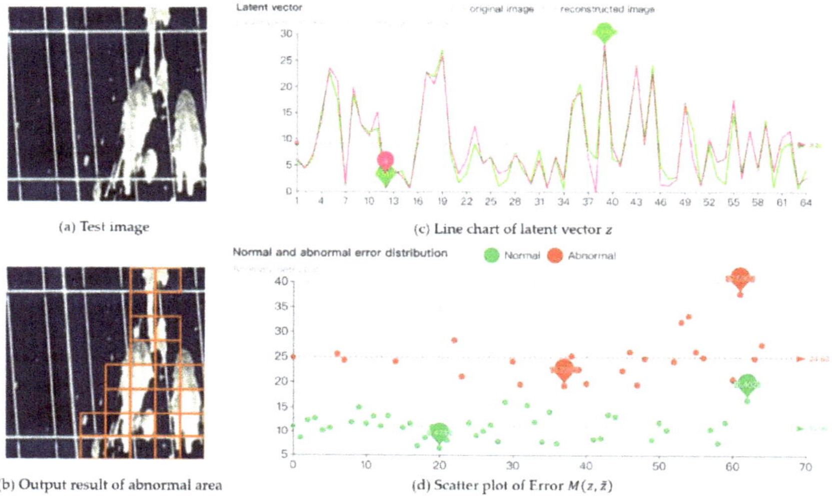

Figure 14. Output results of abnormal detection. (**a**) Test image; (**b**) Output result of abnormal area; (**c**) Line chart of latent vector z; (**d**) Scatter plot of Error $M(z, \widetilde{z})$.

4.6. Model Evaluation

In order to verify the effectiveness of the proposed method, this paper compares the ppFDetector with AnoGAN [17], Zhao's method [18], f-AnoGAN [19], GANomaly [20], and Pre-trained Vgg16. Considering that the minimum input size of Vgg16 is 48 × 48, we need to enlarge the image from 32 × 32 to 48 × 48. For a fully supervised model, pre-trained Vgg16, 10, 100 and 200 negative samples are respectively appended to the training set to explore the influence of different numbers of negative samples on the metric of accuracy.

We take ROC (Receiver Operating Characteristic curve) curve, precision, accuracy, F1 and sensitivity scores as the performance metrics. Precision represents the proportion of examples classified as positive examples that are actually so. The higher the index, the better the performance. Accuracy is the ratio of the number of samples correctly classified by the classifier to the total number of samples for a given test dataset. And the higher the score, the better the model. F1 score can be regarded as a weighted average of the model precision and recall. Its maximum value is 1, and its minimum value is 0. The larger the value, the better the model. Sensitivity refers to the proportion of all positive examples identified to all positive examples. The higher the sensitivity rate, the higher the probability that the actual abnormal sample is predicted. As shown in Table 3, the performance of a fully supervised learning method (i.e., Pre-trained Vgg16) improves by increasing the proportion of negative samples in the training process. Although the precision and accuracy performance metrics of vgg16(200) are higher than ppFDetector, ppFDetector does not require a negative sample during training, and so ppFDetector outperforms the Pre-trained Vgg16 method with limited negative samples. Table 4 shows the performance comparison between the ppFDetector, semi-supervised learning methods (i.e., Zhao's method and GANomaly), and unsupervised learning methods (i.e., AnoGAN and f-AnoGAN). It shows that the ppFDetector only training with positive samples is generally competitive with state-of-the-art semi-supervised and unsupervised methods, and achieves higher accuracy, precision, F1 score and sensitivity.

Table 3. Performance comparison with fully supervised model.

Model	Precision	Accuracy	F1	Sensitivity
Pre-trained Vgg16 (10)	0.324	0.265	0.378	0.453
Pre-trained Vgg16 (100)	0.746	0.750	0.795	0.852
Pre-trained Vgg16 (200)	**0.930**	**0.960**	0.936	0.941
ppFDetector	0.929	0.945	**0.944**	**0.960**

Bold shows the best performance between the four models.

Table 4. Performance comparison with semi-supervised and unsupervised models.

Model	Precision	Accuracy	F1	Sensitivity
AnoGAN	0.878	0.749	0.852	0.827
Zhao's method	0.869	0.887	0.899	0.931
GANomaly	0.908	0.892	0.929	0.952
f-AnoGAN	0.912	0.930	0.930	0.950
ppFDetector	**0.929**	**0.945**	**0.944**	**0.960**

Bold shows the best performance between the four models.

Figure 15a illustrates the comparison of ROC curves of semi-supervised and unsupervised methods. The ppFDetector outperforms in PV panel fault detection. The AUC value of the ppFDetector is 0.943, which is 0.041 higher than GANomaly, and 0.052, 0.06 and 0.035 higher than Zhao's method, AnoGAN and f-AnoGAN, respectively. Figure 15b illustrates the ROC curves of the ppFDetector and fully supervised method (i.e., Pre-trained Vgg16) with a different number of negative samples in the training process. The ppFDetector outperforms Pre-trained Vgg16 with limited negative samples, i.e., 10 and 100. But when the Pre-trained Vgg16 trained with more than 200 negative samples, its performance is close to ppFDetector.

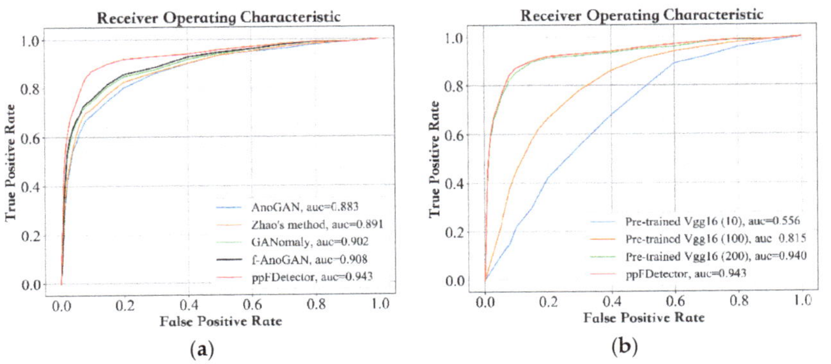

Figure 15. ROC curve comparison chart. (**a**) semi-supervised and unsupervised methods; (**b**) fully supervised methods.

4.7. Ablation Experiment

Figure 16a shows the comparison of the loss with GC and without GC in 50 rounds of training. The original network only takes Batch Normalization (BN) as optimization. As shown in Figure 16a, the training loss of BN+GC is reduced faster than BN, so GC can further speed up convergence of the model during training. Figure 16b compares SmoothL1Loss with the manual combination of L2loss and L1Loss. Using SmoothL1Loss is significantly lower than the one using the manual combination of L2loss and L1Loss.

This proves that SmoothL1Loss can make the model more accurate. The GC is applied on conditional GAN of ppFDetector to regularize weight space and output feature space to prevent overfitting, so as to improve the generalization performance in the deep convolution network. Figure 16c is a comparison of the loss with and without the CBAM module. CBAM first extracts features based on the attention of the spatial domain, then extracts features based on the attention of the channel domain, and finally achieves the optimal attention extraction effect, thereby improving the performance of the model.

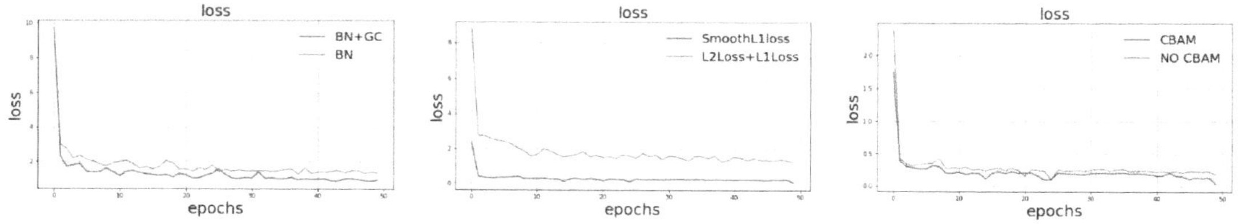

(a) (b) (c)

Figure 16. Loss comparison chart. (**a**) BN, BN+GC loss curve; (**b**) L2Loss+L1Loss, SmoothL1Loss loss curve. (**c**) NO CBAM, CBAM loss curve.

5. Conclusions

In this paper, with an aim to detect the fault of PV panels in a fast and accurate manner, a novel PV panel fault detection approach-ppFDetector is proposed. The ppFDetector does not require a large number of negative samples. It employs the generative adversarial network to extract image features to approximate the image data distribution of the positive samples. Thus, the model can generate a reconstructed image extremely similar to its positive samples. By calculating the differences between the original image and its reconstructed image, ppFDetector can accurately determine whether the PV panel is abnormal. This anomaly detection model is with strong transplantation. As there are a large number of positive samples, binary classifications can be performed after the model is trained; i.e., ppFDetector can solve the problem that models cannot train without negative samples. Moreover, ppFDetector accelerates the model training speed while increasing the generalization performance of the model through the application of gradient centralized. The ppFDetector also uses the CBAM and SmoothL1 loss function to further improve the accuracy and robustness of PV panel fault detection, which can replace manual inspection. In the future, we will work towards expanding the abnormal types that ppFDetector can detect by collecting abnormal datasets with various types of defects (such as dust cover, snail pattern and rust), and employ the CNN to classify the types of the defect after detecting abnormal areas.

Author Contributions: Conceptualization J.C.; Methodology, F.L. and R.N.; Software, R.N. and Z.Z.; Validation, R.N. and Z.Z.; Writing—original draft, F.L. and R.N.; Writing—review and editing, L.G. and J.C. All authors have read and agreed to the published version of the manuscript.

Funding: This work is supported by the Ministry of Education Humanities Social Sciences Research Project (No. 20YJC760101), the Educational Commission of Zhejiang Province of China (No. Y202147553), Zhejiang Provincial Natural Science Foundation of China (No. LGF20C050001) and the National Nature Science Foundation of China (No. 61805213).

Institutional Review Board Statement: Not applicable.

Informed Consent Statement: Not applicable.

Data Availability Statement: https://github.com/RobotvisionLab/ppFDetector.git (accessed on 3 February 2022).

Acknowledgments: We acknowledge a photovoltaic power plant located in Zhejiang province for providing data for our experiment.

Conflicts of Interest: The authors declare no conflict of interest.

References

1. Fan, Z.; Yang, Z.; Yang, L. Daylight performance assessment of atrium skylight with integrated semi-transparent photovoltaic for different climate zones in China. *Build. Environ.* **2021**, *190*, 107299. [CrossRef]
2. Xuan, Q.; Li, G.; Jiang, B.; Zhao, X.; Zhao, B.; Ji, J.; Pei, G. Analysis and quantification of effects of the diffuse solar irradiance on the daylighting performance of the concentrating photovoltaic/daylighting system. *Build. Environ.* **2021**, *193*, 107654. [CrossRef]
3. Liu, Y.; Zhang, R.; Huang, Z.; Cheng, Z.; López-Vicente, M.; Ma, X.; Wu, G. Solar photovoltaic panels significantly promote vegetation recovery by modifying the soil surface microhabitats in an arid sandy ecosystem. *Land Degrad. Dev.* **2019**, *30*, 2041–2047. [CrossRef]
4. Du, W.Z.; Shen, H.Y. Information Technology: Investigators from Huazhong University of Science and Technology Target Information Technology (Defect Image Sample Generation with GAN for Improving Defect Recognition). *Inf. Technol. Newsweekly* **2020**, *27*, 24–33.
5. Mohammed, S.; Boumediene, B.; Miloud, B. Assessment of PV modules degradation based on performances and visual inspection in Algerian Sahara. *Int. J. Renew. Energy Res.* **2016**, *6*, 106–116.
6. Kentges, M. *Review of Failures of Photovoltaic Modules*; IEA-Photovoltaic Power Systems Programme T13-01; SUPSI, 2014; pp. 1–140. Available online: https://iea-pvps.org/wp-content/uploads/2020/01/IEA-PVPS_T13-01_2014_Review_of_Failures_of_Photovoltaic_Modules_Final.pdf (accessed on 3 February 2022).
7. Tabanjat, A.; Becherif, M.; Hissel, D. Reconfiguration solution for shaded PV panels using switching control. *Renew. Energy* **2015**, *82*, 4–13. [CrossRef]
8. Dolara, A.; Lazaroiu, G.; Leva, S.; Manzolini, G. Snail trails and cell micro-cracks impact on PV module maximum power and energy production. *IEEE J. Photovolt.* **2016**, *6*, 1269–1277. [CrossRef]
9. Yang, T.; Zhao, L.Y.; Wang, C.S. Overview of application of Artificial Intelligence in Power System and Integrated Energy System. *Autom. Electr. Power Syst.* **2019**, *43*, 2–14.
10. Zhou, N.C.; Liao, J.Q.; Wang, Q.G. Analysis and Prospect of Deep Learning Application in Smart Grid. *Autom. Electr. Power Syst.* **2019**, *43*, 180–191.
11. Li, X.; Yang, Q.; Yan, W. Intelligent fault pattern recognition of aerial photovoltaic module images based on deep learning technique. *Syst. Cybern. Inform.* **2018**, *16*, 67–71.
12. Tang, W.; Yang, Q.; Xiong, K.; Yan, W. Deep learning based automatic defect identification of photovoltaic module using electroluminescence images. *Sol. Energy* **2020**, *201*, 453–460. [CrossRef]
13. Zhao, Y.; Zhan, K.; Wang, Z.; Shen, W. Deep learning-based automatic detection of multitype defects in photovoltaic modules and application in real production line. *Prog. Photovolt. Res. Appl.* **2021**, *29*, 471–484. [CrossRef]
14. Mittal, S.; Dutta, M.K.; Issac, A. Non-destructive image processing based system for assessment of rice quality and defects for classification according to inferred commercial value. *Measurement* **2019**, *148*, 1792–1797. [CrossRef]
15. Singh, A.P.; Kumar, A.; Akshayveer, O.P.S. Performance enhancement strategies of a hybrid solar chimney power plant integrated with photovoltaic panel. *Energy Convers. Manag.* **2020**, *218*, 113020. [CrossRef]
16. Zhang, E.; Li, B.; Li, P.; Chen, Y. A Deep Learning Based Printing Defect Classification Method with Imbalanced Samples. *Symmetry* **2019**, *11*, 1440. [CrossRef]
17. Schlegl, T.; Seebock, P.; Waldstein, S.M.; Schmidt-Erfurth, U.; Langs, G. Unsupervised anomaly detection with generative adversarial networks to guide marker discovery. In *International Conference on Information Processing in Medical Imaging*; Lecture Notes in Computer Science, 10265 LNCS; Springer: Cham, Switzerland, 2017; pp. 146–147.
18. Zhao, Z.; Li, B.; Dong, R.; Zhao, P. A surface defect detection method based on positive samples. In Proceedings of the Pacific Rim International Conference on Artificial Intelligence (PRICAI), Nanjing, China, 28–31 August 2018; pp. 473–481.
19. Schlegl, T.; Seeböck, P.; Waldstein, S.M.; Langs, G.; Schmidt-Erfurth, U. f-AnoGAN: Fast unsupervised anomaly detection with generative adversarial networks. *Med. Image Anal.* **2019**, *54*, 30–44. [CrossRef]
20. Akcay, S.; Abarghouei, A.A.; Breckon, T.P. Ganomaly: Semi-supervised anomaly detection via adversarial training. In Proceedings of the ACCV, Perth, Australia, 2–6 December 2018; pp. 622–637.
21. Radford, A.; Metz, L.; Chintala, S. Unsupervised Representation Learning with Deep Convolutional Generative Adversarial Networks. *arXiv* **2015**, arXiv:1511.06434.
22. Undirwade, A.; Das, S. Image anonymization using deep convolutional generative adversarial network. In *Machine Learning Algorithms and Applications*; Srinivas, M., Sucharitha, G., Matta, A., Chatterjee, P., Eds.; Crc Press: Boca Raton, FL, USA, 2021. [CrossRef]
23. Karnewar, A.; Wang, O. MSG-GAN: Multi-Scale Gradients for Generative Adversarial Networks. In Proceedings of the IEEE Conference on Computer Vision and Pattern Recognition, Seattle, WA, USA, 13–19 June 2020; pp. 7796–7805.
24. Tatapudi, S.; Libby, C.; Raupp, C. Defect and safety inspection of 6 PV technologies from 56000 modules representing 257000 modules in 4 climatic regions of the United States. In Proceedings of the 43rd IEEE Photovoltaic Specialists Conference, Portland, OR, USA, 5–10 June 2016; pp. 1747–1751.
25. Sawyer, D.; Kessler, H.K. The solar cell is scanned by laser for displaying the working characteristics of the cell and detecting the defects of the cell. *J. Electron. Equip.* **1980**, *27*, 864–872.

26. Tsuzuki, K.; Murakami, T.; Yoshino, T.; Takeyama, Y.; Shimizu, K. Inspection Method and Production Method of Solar Cell Modules. U.S. Patent 6,271,462, 7 August 2001.
27. Tsai, D.M.; Wu, S.C.; Li, W.C. Defect detection of solar cells in electroluminescence images using Fourier image reconstruction. *Sol. Energy Mater. Sol. Cells* **2012**, *99*, 250–262. [CrossRef]
28. Bastari, A.; Bruni, A.; Cristalli, C. Classification of silicon solar cells using electroluminescence texture analysis. In Proceedings of the IEEE International Symposium on Industrial Electronics, Bari, Italy, 4–7 July 2010; pp. 1722–1727.
29. Quater, P.; Grimaccia, F.; Leva, S. Light unmanned aerial vehicles (UAVs) for cooperative inspection of PV plants. *IEEE J. Photovolt.* **2014**, *4*, 1107–1113. [CrossRef]
30. Li, X.; Yang, Q.; Chen, Z. Visible defects detection based on UAV-based inspection in large-scale photovoltaic systems. *IET Renew. Power Gener.* **2017**, *11*, 1234–1244. [CrossRef]
31. Yong, H.; Huang, J.; Hua, X.; Zhang, L. Gradient Centralization: A New Optimization Technique for Deep Neural Networks. In Proceedings of the ECCV, Cambridge, UK, 14–18 April 2020; pp. 635–652.
32. Woo, S.; Park, J.; Lee, J.Y.; Kweon, I.S. CBAM: Convolutional Block Attention Module. In Proceedings of the ECCV, Munich, Germany, 8–14 September 2018; pp. 3–19.
33. Ioffe, S.; Szegedy, C. Batch normalization: Accelerating deep network training by reducing internal covariate shift. In Proceedings of the International Conference on International Conference on Machine Learning (ICICML), Lille, France, 6–11 July 2015; pp. 234–238.
34. Xu, B.; Wang, N.; Chen, T. Empirical Evaluation of Rectified Activations in Convolutional Network. CoRR. 2015, pp. 14–23. Available online: http://arxiv.org/abs/1505.00853 (accessed on 3 February 2022).

applied sciences

MDPI

Article

Reverse Knowledge Distillation with Two Teachers for Industrial Defect Detection

Mingjing Pei [1,2,3,4], Ningzhong Liu [1,3,4,*], Pan Gao [1,3,4,*] and Han Sun [1,3,4]

1 College of Computer Science and Technology, Nanjing University of Aeronautics and Astronautics, Nanjing 211106, China; peimj@wxc.edu.cn (M.P.)
2 School of Electronics and Information Engineering, West Anhui University, Lu'an 237012, China
3 MIIT Key Laboratory of Pattern Analysis and Machine Intelligence, Nanjing 211106, China
4 Collaborative Innovation Center of Novel Software Technology and Industrialization, Nanjing 210093, China
* Correspondence: lnz_nuaa@163.com (N.L.); pan.gao@nuaa.edu.cn (P.G.)

Abstract: Industrial defect detection plays an important role in smart manufacturing and is widely used in various scenarios such as smart inspection and product quality control. Currently, although utilizing a framework for knowledge distillation to identify industrial defects has achieved great progress, it is still a significant challenge task to extract better image features and prevent overfitting for student networks. In this study, a reverse knowledge distillation framework with two teachers is designed. First, for the teacher network, two teachers with different architectures are used to extract the diverse features of the images from multiple models. Second, considering the different contributions of channels and different teacher networks, the attention mechanism and iterative attention feature fusion idea are introduced. Finally, to prevent overfitting, the student network is designed with a network architecture that is inconsistent with the teacher network. Extensive experiments were conducted on Mvtec and BTAD datasets, which are industrial defect detection datasets. On the Mvtec dataset, the average accuracy values of image-level and pixel-level ROC achieved 99.43% and 97.87% , respectively. On the BTAD dataset, the average accuracy values of image-level and pixel-level ROC reached 94% and 98% , respectively. The performance on both datasets is significantly improved, demonstrating the effectiveness of our method.

Keywords: knowledge distillation; industrial defect detection; anomaly detection; deep learning

1. Introduction

Industrial defect detection is one of the most important technologies to ensure the quality of products and maintain the stability of production, aiming at detecting defects in various industrial products. Recently, with the emergence of new technologies in the fields of computer vision and deep learning [1,2], vision-based industrial defect detection technology has been developed significantly. Industrial defect detection has become one of the important basic research and technology in intelligent manufacturing. It is widely used in unmanned quality inspection [3], intelligent inspection [4], video surveillance [5], etc. At present, the main difficulties and challenges of industrial defect detection lie in two aspects. One is the about data, such as the lack of abnormal samples, unpredictable defect patterns, more types of defects, and complex background of defect images. The other is that industrial vision defect detection requires methods with high accuracy, even to detect subtle defects.

Facing the difficulties and challenges in industrial defect detection, many approaches are currently available in the literature, which can be roughly categorized into three categories. The first category is currently designed for detection of synthetic abnormal samples, such as Cutpaste [6], and NSA [7]. The main purpose is to use the neural network to classify normal samples and synthetic abnormal samples for binary classification and then perform abnormality detection by the Gaussian probability density function. These methods require image pre-processing operations before model training, thus increasing the model complexity. Meanwhile, there is still a gap between the synthesized anomalous samples and the real ones. So, how to synthesize good anomaly samples is another challenge. The second category is based on the comparison at the pixel level of the image. The idea is to reconstruct a normal image that most closely resembles the input image, and then the difference between the reconstructed image and the input image is lie in the defective region. Such methods often use self-encoding models and generative models, including auto-encoder (AE) [8,9], generative adversarial networks (GAN) [10,11], etc. However, the pixels of the reconstructed image may not be aligned with the input image, and the reconstruction process may also change the style of the image, which can lead to detection errors and degrade the performance of the detection. In addition, the pixel level is more susceptible to noise interference, resulting in poor robustness of detection. In this regard, researchers turn to focus on feature space. Naturally, the third category is based on the feature similarity [12,13]. It is generally considered that the feature vectors of normal regions are more similar or closer, while the feature vectors of anomalous regions are different. The core objective of such methods is to find distinguishable feature embeddings and reduce the interference of irrelevant features. The features extracted by CNN contain information about the local perceptual field, instead of considering only individual pixels. Meanwhile, it does not require very strict spatial alignment, increasing the tolerance to noise interference. Some teacher-student network frameworks are proposed based on feature similarity [14,15]. The main idea is that the teacher network and the student network will produce different feature representations in the anomaly region since the training uses normal samples and the student network has not seen anomalous samples. Their framework has two main shortcomings. One is that they all use a single-teacher network, which may be deficient in extracting features, and the other is that the teacher and student networks use the same architecture, which may produce overfitting.

To address the issues of the teacher–student network, a new network architecture was designed. Firstly, for a single teacher network, the extracted features may be not representative enough. To overcome this, a two teacher network with inconsistent architectures is proposed to extract features for diverse feature learning. This can extract more information from the image and facilitate the detection of defective areas of the image. Secondly, the feature maps of the two teacher networks show numerous channels, but the relative importance of each channel varies. Meanwhile, the contribution of each teacher to the anomaly detection results is different. So, the attention mechanism is introduced to solve these issues. This allows the model to focus more on the important features and improves accuracy of the model. Finally, if the student network and the teacher network architectures are consistent, it may produce overfitting. It means that the student network is also able to reconstruct the anomaly region well, which can have a significant impact on our test results. The performance of the student network needs to be lower than the teacher network. So, a simple student network was redesigned to improve our AUC results, and its structure was not consistent with that of the teacher network.

The main contributions are summarized as follows:

- A teacher–student network architecture with two teachers is adopted for industrial image anomaly detection and localization, extracting features of images from multiple models.
- For each channel of the teacher network, the attention mechanism is employed. In addition, the iterative attention feature fusion method for the feature fusion of the two teachers is used. For the teacher and student networks, inconsistent architectures are used to mitigate overfitting.
- A large number of experiments were conducted on Mvtec and BTAD datasets to demonstrate the effectiveness of our approach.

The rest of this paper is organized as follows:

We briefly review the related work in Section 2. In Section 3, the proposed teacher–student network architecture model with two teachers is detailed. The experimental results are presented in Section 4. We summarize our work in Section 5.

2. Related Work

Industrial defect anomaly detection and localization have been studied in recent years and a lot of literature has appeared. There are different divisions from different perspectives. We divide it into four components, image reconstruction and recovery, embedding-based, abnormal sample synthesis, and knowledge distillation.

2.1. Image Reconstruction and Recovery

The idea of image restoration methods is to assume that the parameters of the model are only trained from normal samples. The model can only reconstruct the normal samples well, while the defective areas of the abnormal samples will produce large reconstruction errors. Such methods include autoencoder (AE) [8,9], variational autoencoder (VAE) [16], and generative adversarial networks (GAN) [10,11]. However, the image reconstruction-based approach has the disadvantage that the model may still reconstruct the unseen defects completely. So, another popular approach was introduced to store normal features by introducing feature memory [17,18]. The researchers introduced memory to store the features of normal training samples, and in the inference phase, the sample features of the image to be tested are matched with the features in memory and reconstructed with them.

Image recovery methods treat defects as noise and consider it as a denoising process. The core idea of such methods is to add defects to a normal image and then train a network model to recover it to the corresponding original image. One of the core issues of such methods is how to design defective images for construction. For example, Ref. [19] superimposed Gaussian noise on the original image. Ref. [20] used attribute cancellation. Ref. [21] took into account the positional relationships in the image, borrowed from puzzle restoration [22] to disrupt the order of image blocks in the original image, and trained the network to restore them. Other methods are to overlay a mask on the original image and train the network to restore what is covered by the mask. Ref. [23] added random rectangular masks to normal samples to simulate real defects. Ref. [24] randomly selected superpixels of the image as a mask for training the model.

2.2. Embedding-Based

The purpose of feature similarity methods is to find feature embeddings that are distinguishable. It consists of deep one-class and feature distance metric methods. The deep one-class classification method [25] extracts the features of normal images through neural networks during training, and makes the extracted feature vectors of normal images as compact as possible, constructing a separation plane of normal feature boundaries in the feature space. During testing, the model extracts the features of the sample to be tested and maps them to the feature space. Then, it determines whether the test sample is abnormal according to the separation plane. Ref. [13] proposed the deep support vector data description (Deep SVDD) method, where the features of normal samples would be near a centroid. The centroid in Deep SVDD is replaced by a bias term in the neural network in FCDD [26]. FCDD directly locates the abnormal sites by the feature maps obtained from the fully convolutional network. Ref. [27] avoid the problem of model degradation by maintaining the semantic information of the features. Furthermore, Ref. [28] introduced multiscale to improve the detection capability. To improve the performance of one-class classification, self-supervised learning is another effective tool. The key to this approach is to design a self-supervised proxy task that facilitates defect detection by using normal samples to construct supervised information. Ref. [29] proposed a method based on RotNet [30], which applies a geometric transformation to the input data and trains a network that recognizes the transformation used. In addition, researchers have used self-supervised methods that create negative samples [6,31] to improve the discriminative power of classification models. Ref. [31] used images generated early in the training process.

The method of feature distance metric performs distance metric in feature space. Ref. [32] proposed SPADE based on the k-nearest neighbor (kNN) method, which first extracts the feature vectors of the training set using a CNN model pretrained on ImageNet [33] to build a database of normal samples. Then, it uses the kNN method to retrieve the k normal samples that are most semantically similar to the sample to be tested. The deep statistical model-based approach models the probability distribution of the features of the normal samples, eliminating the need to build a large normal sample database. Ref. [34] first extracted multiscale features of normal samples using a pre-training network and modeled each feature map as a multivariate Gaussian distribution. In the testing phase, the Marxian distance between the feature vector of the sample to be tested and the normal distribution is used to measure the presence of defects in the sample. Ref. [35] further modeled the multivariate Gaussian distribution on the image blocks, achieving pixel-level segmentation. Ref. [36] introduced the normalizing flow (NF) model to enhance flexibility.

2.3. Abnormal Sample Synthesis

Recently, many papers have used various methods to synthesize anomalous samples. The Cutpaste method [6] simulated anomalous samples by using a random patch from the image, and then the patch was pasted to any region of the image. In the object class, the patch may be pasted to the outside of the object, so Ref. [7] cut the patch from the surface of the object and pasted it to the object. Ref. [37] was to simulate various anomalies on images using data augmentation methods. Ref. [38] adopted two-dimensional Perlin noise to simulate the anomaly, but it was still a bit different from the actual anomaly.

2.4. Knowledge Distillation

The knowledge distillation network architecture consists of a pretrained teacher network and a trainable student network. In anomaly detection training, normal samples are used. If the input is a normal sample, the extracted features of the teacher network and the student network are closer together, and conversely, they are further apart. For example, US [14] used a teacher and multiple students to extract features of the samples. the US method utilized only the last layer of the network for distillation, while MKD [39] utilized multiple feature layers in the middle of the network for knowledge distillation. To improve the detection performance, Ref. [15] proposed an inverse distillation architecture, where the extracted features by the teacher are first compressed and then decoded by the student network. DeSTSeg [40] introduced the idea of segmentation into the knowledge distillation architecture, which also yields good results.

3. Methods

Given an input image, the task aims to detect anomalies while leveraging a reverse distillation framework, in which the student network imitates the behavior of the teacher network. The key is how to design a teacher and student network. Our framework is proposed in Figure 1. It consists of three modules: two different pretrained teacher encoders E, a multiple attentional feature fusion module, and a student decoder D. First, the two different pre-trained teachers extract multi-scale features in different layers and concatenate them separately. Second, we fuse features of the same layer and different branches using an attention mechanism and compress the obtained feature map into a low-dimensional feature vector. Finally, the student network decodes the low-dimensional feature vectors to obtain the feature maps, in which we can calculate the loss and generate multiscale similarity maps of each layer. In the testing stage, the anomaly map is generated by fusing these multiscale similarity maps.

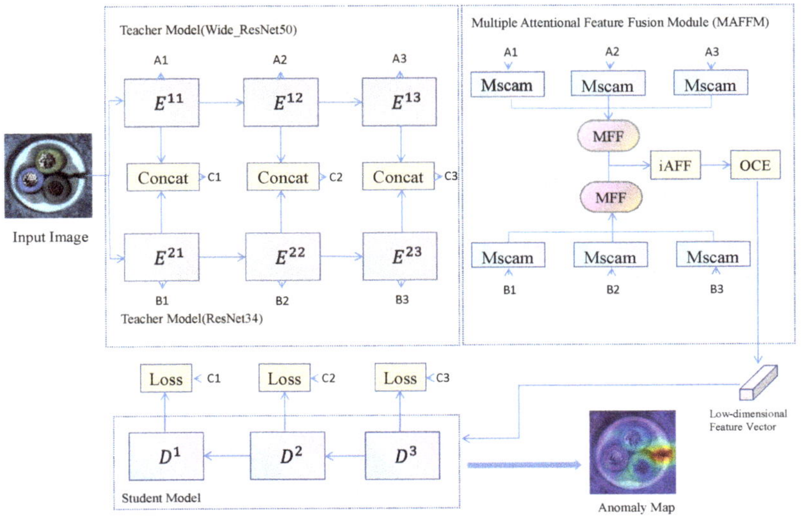

Figure 1. Overview of our proposed framework for anomaly detection and localization. A contains A1, A2, and A3. Similarly, A2 and A3 are the same as A1. In the same way, B and C have the same properties as A. Our model consists of two different pretrained teachers encoder E, a multiple attentional feature fusion module, and a student decoder D. A multiscale channel attention module (Mscam) is used to fuse features in different layers from E and ensemble high-level and low-level features from Mscam by multiscale feature (MFF) block. Iterative attentional feature fusion (iAFF) is adopted to fuse features from two MFF and map it onto a compact feature vector by a one-class embedding (OCE) block. By minimizing the similarity loss between the two teacher models, student D learns to imitate the behavior of concatenated features in training. The anomaly map is generated by fusing the multiscale similarity maps of the student decoder.

3.1. Teacher Model

Currently, many distillation network architectures focus on distillation from a single teacher to a single student for anomaly detection and localization [15]. However, in the human learning process, a student does not learn from only one teacher; instead, he may learn from different teachers or different sources of information. Multiple teacher models provide multiple interpretations of the student model's task by providing multiple streams of information, and the student model can use the teacher model's views of the target task to improve the model's performance. Thus, an ensemble of multiple teacher network predictions typically yields better performance than a single teacher network. Therefore, two teachers' distillation networks are adopted.

Due to the different teachers needed, ResNet34 and Wide_ResNet50 pretrained on ImageNet are used in our framework [41,42]. They are enabled to extract rich features from images. In them, the last block (conv5_x) is removed and the output feature maps are extracted by the remaining blocks (conv2_x, conv3_x, and conv4_x). In Wide_ResNet50, they are named E11, E12, and E13, respectively. Similarly, they are known as E21, E22, and E23 in ResNet34, respectively. Finally, we fuse E11 and E21, E12 and E22, and E13 and 23, respectively, and then use them as intermediate feature layers of the network.

Since it has only one student decoding network for calculating the loss, it needs to fuse the two teachers' feature channels and use a simple concatenation operation. For knowledge transfer, cosine similarity is used as our loss. The loss function is defined in Equation (1). In training, the size of the input image is $224 \times 224 \times 3$, so the size of the first layer (E11, E21) feature map is $64 \times 64 \times 320$, the second layer (E12, E22) feature map is 32 × 32 × 640, and the third layer (E13, E23) feature map is 16 × 16 × 1280. After fusion, the feature maps of each layer are 64 × 64 × 320, 32 × 32 × 640, and 16 × 16 × 1280, respectively.

In order to obtain a two-dimensional anomaly map M^k, the formula is calculated as follows.

$$M^k(h,w) = 1 - \frac{f_T^k(h,w) \cdot f_S^k(h,w)}{\|f_T^k(h,w)\| \cdot \|f_S^k(h,w)\|} \quad (1)$$

where f_T^k and f_S^k represent the feature vectors extracted from the teacher network and the student network, respectively. The h and w are defined as the height and width of the feature map, and k denotes the number of feature layers.

In the training process, the loss function L is represented as follows.

$$L = \sum_{k=1}^{3} \left\{ \frac{1}{H_k W_k} \sum_{h=1}^{H_k} \sum_{w=1}^{W_k} M^k(h,w) \right\} \quad (2)$$

where h, w and k have the same meaning as in Equation (1).

3.2. Multiple Attentional Feature Fusion Module

On the one hand, every feature layer in a teacher network has many channels, but every channel is not equally important. On the other hand, for the learned features of a multiteacher network, it may not be the best operation to fuse them by simple add and concatenation operations. Therefore, the attention mechanism and iterative attention feature fusion are used to design a module.

A combination module named MAFFM is designed. It consists of the Mscam module, iAFF module, MFF module, and OCE module. Details of the modules can be found in the reference [15,43]. The Mscam module uses two branches with different scales to extract channel attention weights. One branch uses Global Avg Pooling to extract the attention of global features, and the other branch directly uses point-wise convolution to extract the channel attention of local features. Therefore, it is used in every layer of two teachers. In the Mscam network structure, there is a channel average distribution coefficient r, for which we made ablation experiments. After the Mscam, the feature map of every layer is still 64 × 64 × 320, 32 × 32 × 640, and 16 × 16 × 1280, respectively. The MFF module is a multiscale fusion module, which is to fuse feature maps of different sizes into the same size. Thus, it can get a feature map of size 16 × 16, and the channel number is 3072. The iAFF module uses an additional layer of attentional feature fusion(AFF) module to generate better initial features. After iAFF module, it can obtain a feature map of size 16 × 16 × 3072. The purpose of the OCE block is to project the high-dimensional feature vector into the low-dimensional space, using the residual connection block of ResNet. The design of the module can be found in the literature [15]. Finally, it can get a low-dimensional feature vector with a size of 8 × 8 × 2048.

3.3. Student Model

In some distillation architectures for anomaly detection, the same architecture is used in some literature for both teacher and student networks [15]. Since neural networks are generalizable, the student network may be able to reconstruct the anomaly region features well, leading to reduced performance of the detection. Therefore, it needs to design a more suitable network. In addition, in order to match the feature channels of the two teacher intermediate layers, a student decoding network is designed to decode the feature maps from the low-dimensional features. The student decoding network is shown in Figure 2.

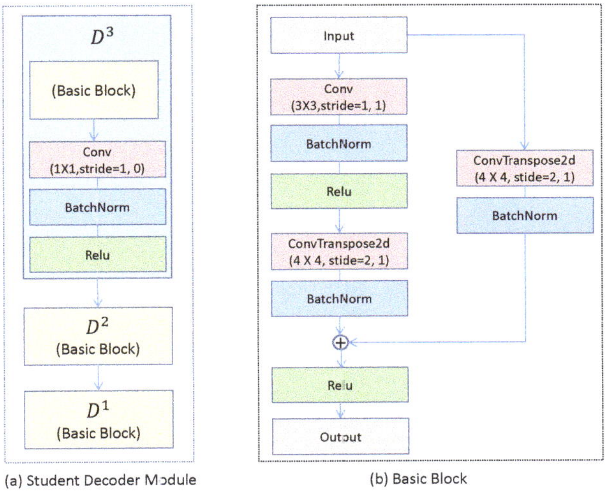

Figure 2. Student decoder module and basic block in our student decoding network.

Motivated by the residual connection block in ResNet [41], we adopt it to fuse different scales of feature maps in the basic block (see Figure 2b). In the basic block, one branch is composed of a 4 $\times$ 4 deconvolution(ConvTranspose2d) layer and a BatchNorm layer, which aims to better preserve the original features. The other branch mainly consists of a 3 $\times$ 3 convolution and a 4 $\times$ 4 deconvolution(ConvTranspose2d). The 3 $\times$ 3 convolution is followed by a BatchNorm layer and a Relu layer, changing the input channels to match the number of output channels of the teacher network. The 4 $\times$ 4 deconvolution(ConvTranspose2d) is to change the size of the feature map to match the size of the feature map of the teacher network. Finally, the two branches are summed and outputted by the Relu activation layer.

The student decoder module(see Figure 2a) is the same as the teacher network. There are three layers in total, and it is an inverse structure. The D3 module consists of a basic block and a 1 $\times$ 1 convolutional layer, followed by a BatchNorm layer and a Relu layer. The purpose of the 1 $\times$ 1 convolution is mainly to change the input channel to match the output channel of the corresponding teacher network. The D2 and D1 modules are both composed of the basic block.

3.4. Training and Inference

In the training, the model uses all normal images and no abnormal images. The student network tries to mimic the output of the teacher network so that the features learned by both networks are as similar as possible for the same image. The loss function is to calculate the similarity of the feature maps of the corresponding layers. The main purpose of the training is to reduce the loss of function.

In the inference stage, both image-level anomaly score and pixel-level anomaly score are considered for anomaly detection and localization. When a normal image is input, the teacher network and the student network learn feature representations on a normal region with high similarity. On the contrary, when the image is abnormal, the pretrained teacher network can generate discriminative feature representations in both normal and anomalous regions, but the student network cannot, causing low similarity. In the student network, it needs to calculate a similarity value in each pixel. To be consistent with the image size, we will restore the feature map to the same size as the original image by bilinear up-sampling operation. Specific details can be found in the references [15]. Finally, it can get the pixel-level anomaly scores map by Gaussian filtering. In the image-level anomaly score, the maximum value is used in the pixel anomaly map as the sample-level score.

4. Experiments

To verify the effectiveness of our proposed method, the MVTec AD and BTAD datasets are used to evaluate our performance. Additionally, the ablation study is conducted on the MVTec dataset to examine the impact of various blocks or parameters on the results.

4.1. Datasets

MVTec AD is an industrial inspection-focused dataset for comparing anomaly detection methods. There are 15 different object and texture categories. Each class is split into a training set and a testing set. There are 3629 training images of typical objects and 1725 test images with various types of anomalies. For the test image, pixel-level binary annotations for anomalous images in the test set are provided.

BTAD is a dataset of actual industrial anomalies. The datasets include 2830 images of 3 industrial products that display flaws on their bodies and their surfaces. Additionally, it is separated into 1800 training images and 741 test images. It contains 3 categories, two objects, and one texture. The test datasets only contained one type of anomaly.

4.2. Evaluation Protocol

There are two evaluations, which are image-level evaluation and pixel-level evaluation. They both use the operating characteristic curve (ROC) and area under the curve (AUC) as the evaluation metric, called ROC-AUC. The ROC curve is a graph that shows the performance of a classification model across all thresholds. The two parameters that control this curve are the true positive rate (TPR) and the false positive rate (FPR). The equations are defined in (2) and (3).

$$TPR = \frac{TP}{TP + FN} \tag{3}$$

$$FPR = \frac{FP}{FP + TN} \tag{4}$$

The AUC calculates the area under the ROC curve in two dimensions between (0,0) and (1,1). The network reports the ROCAU scores after being trained separately for each object class.

For more accurate evaluation of anomaly localization performance, the per-region-overlap (PRO) metric is used. Unlike for pixel-by-pixel measurements, the PRO score treats anomalous regions of any size equally.

4.3. Implementation Details

The NVIDIA GTX Geforce 2070s GPU and the Ubuntu 18.04 operating system are used for the experiments. In experiments, the pre-trained Wide_ResNet50 and ResNet34 are used as our backbone in two teacher networks, and the student network is designed by ourselves. All the image input resolutions in MVTec and BTAD are set to 256 × 256. To train our model, Adam is used as the optimizer with β parameter between 0.5 and 0.999. The learning rate uses the cosine annealing strategy, which the parameter period is 32. The initial value of the learning rate is 0.005. A total of 200 epochs are trained with a batch size of 4. For the student network, the Kaiming initialization method for network parameters is used. The random seed is set to 88. For additional details, it can refer to the open source code of the ADRD method.

4.4. Results

For comparisons with other work, we use the results mentioned in the original papers for compared methods. In image-level anomaly detection, we compare our model with the five kinds of existing models, including image restoration models, embedding-based models, abnormal sample synthesis models, flow-based models, and knowledge distillation models. The image restoration models contain AnoGAN [44], UniAD [45] and RIAD [46]. The embedding-based models consist of Psvdd [47] and PaDiM [35]. Abnormal sample synthesis models contain Cutpaste [6], NSA [7] and Draem [38]. Models for knowledge distillation involve US [14] and ADRD [15]. CFLOW-AD [48] is based on flow models. Our baseline is based on the ADRD model.

Table 1 shows the results of our image-level anomaly detection. From Table 1, it can see that our model outperformed other models in terms of the overall average value, achieving 99.43. In addition, from the perspective of texture and object class, the results of our method are also higher than other models, reaching 99.68 and 99.18, respectively. It shows that the knowledge distillation model is valuable in image anomaly detection.

Table 1. Image-level anomaly detection results on MVTec AD dataset. Methods achieved for the top one ROC-AUC (%) are highlighted in bold. (Note: added reference numbers and 2 latest methods in table, one of which is SOTA method).

Category		AnoGAN [44]	RIAD [45]	US [14]	Psvdd [46]	PaDiM [35]	Cutpaste [6]	NSA [7]	Draem [38]	CFLOW-AD [49]	UniAD [50]	ADRD [15]	Ours
Texture	Carpet	54	84.2	91.6	92.9	99.8	93.9	95.6	97	98.73	99.9	98.9	**100**
	Grid	58	99.6	81	94.6	96.7	**100**	99.9	99.9	99.6	98.5	**100**	99.8
	Leather	64	100	88.2	90.9	**100**	**100**	99.9	**100**	**100**	**100**	**100**	**100**
	Tile	50	98.7	99.1	97.8	98.1	94.6	100	**99.6**	99.88	99	99.3	99
	Wood	62	93	97.7	96.5	99.2	99.1	97.5	99.1	99.12	97.9	99.2	**99.6**
	Average	57.6	95.1	91.5	94.5	98.8	97.5	98.6	99.12	99.47	99.06	99.48	**99.68**
Object	Bottle	86	99.9	99.0	98.6	99.9	98.2	97.7	99.2	**100**	**100**	**100**	99.8
	Cable	78	81.9	86.2	90.3	92.7	81.2	94.5	91.8	97.59	97.6	95	**99.6**
	Capsule	84	88.4	86.1	76.7	91.3	98.2	95.2	**98.5**	97.68	85.3	96.3	98.3
	Hazelnut	87	83.3	93.1	92	92	98.3	94.7	**100**	99.98	99.9	99.9	**100**
	Metal nut	76	88.5	82.0	94	98.7	99.9	98.7	98.7	99.26	99	**100**	**100**
	Pill	87	83.8	87.9	86.1	93.3	94.9	99.2	**98.9**	96.82	88.3	96.6	97.4
	Screw	80	84.5	54.9	81.3	85.8	88.7	90.2	93.9	91.89	91.9	97	**98.7**
	Toothbrush	90	100	95.3	100	96.1	99.4	100	**100**	99.65	95	99.5	**100**
	Transistor	80	90.9	81.8	91.5	97.4	96.1	95.1	93.1	95.21	**100**	96.7	98.1
	Zipper	78	98.1	91.9	97.9	90.3	99.9	99.8	**100**	98.48	96.7	98.5	99.9
	Average	82.6	89.9	85.8	90.8	93.8	95.5	96.5	97.41	97.66	95.37	97.95	**99.18**
Average		74.27	91.7	87.7	92.1	95.5	96.1	97.2	98.27	99.27	96.6	98.72	**99.43**

In addition, it's obvious that our model gains further improvement compared with the ADRD baseline and outperforms other models. It demonstrates that the two teacher networks, the attention mechanism, and the inconsistent teacher–student network module have a positive effect. The US is similar to our method, but it does not use the inverse teacher–student network architecture, we believe that the student network may not be good at identifying anomalous regions in the inverse structure. Some categories have better results than ours, such as Grid, Capsule, and Pill. However, our results are almost the same as theirs.

For pixel-level anomaly detection, the comparison baselines include US [14], SPADE [32], PaDiM [35], Cutpaste [6], Draem [38], ADRD [15], NSA [7], CFLOW-AD [48] and UniAD [45]. Compared to the methods provided at the image-level anomaly detection, some models are not added to our table, and the data are not provided in the paper.

From Table 2, it can see that our method outperforms the other methods in pixel-level anomaly detection with 97.87, which proves that our method is effective. In the texture class and target class, it can see that the result of our method is the best in the target class, reaching 98.27, but the texture class is not. The Draem method process segments the simulated anomaly samples after simulating the anomaly samples. With this approach, it may have good results for pixel-level discrimination. But our method is directly comparing the similarity between regional feature vectors.

In Table 3, we show the PRO scores of various methods in the Mvtec dataset. As can be seen from the table, we achieved the best result of 94.66. We only compared three methods since other literature did not provide PRO scores.

We also contrast our model with a few others on the BTAD dataset, such as the autoencoder [50] and the VT-ADL method [49]. From Table 4, it can see that the performance of our method outperforms other models in image-level anomaly detection. Since the other methods do not give pixel-level anomaly detection scores, it can not compare them. Individually, the two categories 1 and 2 are better than our results in the AE + MSE + SSIM method, which mainly uses the autoencoder method. The MSE is the mean square error. The SSIM is the luminance, structure, and contrast of the image.

Table 2. Pixel-level anomaly detection results on MVTec AD dataset. Methods achieved for the top one ROC-AUC (%) are highlighted in bold. (Note: added reference numbers and 2 latest methods in table, one of which is SOTA method).

Category		US [14]	SPADE [32]	PaDiM [35]	Cutpaste [6]	Draem [38]	NSA [7]	CFLOW-AD [49]	UniAD [50]	ADRD [15]	Ours
Texture	Carpet	93.5	97.5	**99.1**	98.3	96.2	95.5	**99.23**	98	98.9	99.1
	Grid	89.9	93.7	97.3	97.5	**99.5**	99.2	96.89	94.6	99.3	99.2
	Leather	97.8	97.6	99.2	99.5	98.9	99.5	**99.61**	98.3	99.4	99.3
	Tile	92.5	87.4	94.1	90.5	**99.5**	99.3	97.71	91.8	95.6	95
	Wood	92.1	88.5	94.9	95.5	**97**	90.7	94.49	93.4	95.3	94.7
	Average	93.2	92.9	96.9	96.3	**98.2**	96.8	97.60	97.8	97.7	97.46
Object	Bottle	97.8	98.4	98.3	97.6	**99.3**	98.3	98.76	98.1	98.7	98.6
	Cable	91.9	97.2	96.7	90	95.4	96	97.64	96.8	97.4	**98.3**
	Capsule	96.8	99.0	98.5	97.4	94.1	97.6	**98.98**	97.9	98.7	98.8
	Hazelnut	98.2	99.1	98.2	97.3	**99.5**	97.6	98.82	98.8	98.9	99
	Metal nut	97.2	98.1	97.2	93.1	**98.7**	98.4	98.56	95.7	97.3	97.8
	Pill	96.5	96.5	95.7	95.7	97.6	98.5	98.95	95.1	98.2	97.7
	Screw	97.4	98.9	98.5	96.7	**99.7**	96.5	98.10	97.4	99.6	99.6
	Toothbrush	97.9	97.9	98.8	98.1	98.1	94.9	98.56	97.8	99.1	**99.1**
	Transistor	73.7	94.1	97.5	93.0	90.0	88	93.28	98.7	92.5	**94.7**
	Zipper	95.6	96.5	98.5	99.3	98.6	94.2	98.4	96.0	98.2	**99.1**
	Average	94.6	97.6	97.8	95.7	97.1	96	98.14	97	97.86	**98.27**
Average		93.9	96.5	97.5	96	97.5	96.3	**97.87**	97.4	97.78	**97.87**

Table 3. PRO scores of various methods on MVTec dataset.

Category		US [14]	PaDim [35]	ADRD [15]	Ours
Texture	Carpet	87.9	96.2	97	**97.4**
	Grid	95.2	94.6	**97.6**	97.5
	Leather	94.5	97.8	99.1	**99.2**
	Tile	**94.6**	86	90.6	89.7
	Wood	91.1	91.1	90.9	**92**
	Average	92.7	93.2	95.0	**95.16**
Object	Bottle	93.1	94.8	**96.6**	**96.6**
	Cable	81.8	88.8	91.0	**94.1**
	Capsule	**96.8**	93.5	95.8	96.5
	Hazelnut	**96.5**	92.6	95.5	95.8
	Metal nut	**94.2**	85.6	92.3	92.3
	Pill	96.1	92.7	**96.4**	95.4
	Screw	94.2	94.4	98.2	**98.5**
	Toothbrush	93.3	93.1	**94.5**	94.1
	Transistor	66.6	**84.5**	78.0	81
	Zipper	95.1	95.9	95.4	**97.3**
	Average	90.8	91.6	93.4	**94.16**
Average		91.4	92.1	93.9	**94.66**

Table 4. Image-level and Pixel-level anomaly detection results on the BTAD dataset. Methods achieved for the top one ROC-AUC (%) are highlighted in bold. "-" means that no pixel-level anomaly score is given in the method. (Note: added reference numbers).

BTAD	AE + MSE [47]	AE + MSE + SSIM [47]	VT-ADL [48]	Ours
0	0.49/-	0.53/-	0.99/-	1.00/0.99
1	0.92/-	0.96/-	0.94/-	0.95/0.96
2	0.95/-	0.89/-	0.77/-	0.87/0.97
Mean	0.78/-	0.79/-	0.9/-	0.94/0.98

Figure 3 shows the visualization results of our image anomaly detection. The red color indicates the region considered by the model to be a relatively high anomaly, the blue color indicates the region with a relatively low anomaly, and the green color region indicates the region with a slight anomaly. The first row represents the original image, the second row represents the mask image, and the third row is the visualization result. From Figure 3, it can see that the defective regions can be located, including the places where there are multiple anomalous regions. However, some areas are incorrectly localized. For example, most of the background areas of the screws and carpet show slight anomalies.

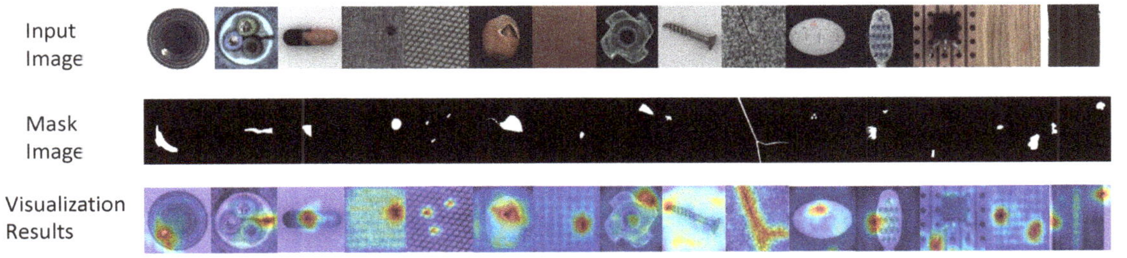

Figure 3. Image anomaly detection visualization results. The first row represents the original image, the second row shows the mask image, and the third row is the visualization result.

In Figure 4, failure cases are shown. From our visualization results, we have selected a few samples where errors occurred. First, the background may affect the results of our model predictions in the object class. In the first and second columns of Figure 4, it can see that there is a little contamination in the background, marked with red colored boxes, and the visualization results show that there is a high level of outliers in these areas. Second, the area of the anomalous region is not fully localized. In the third column of Figure 4, it can see that the display of the anomalous area is incomplete and only a part of it is shown.

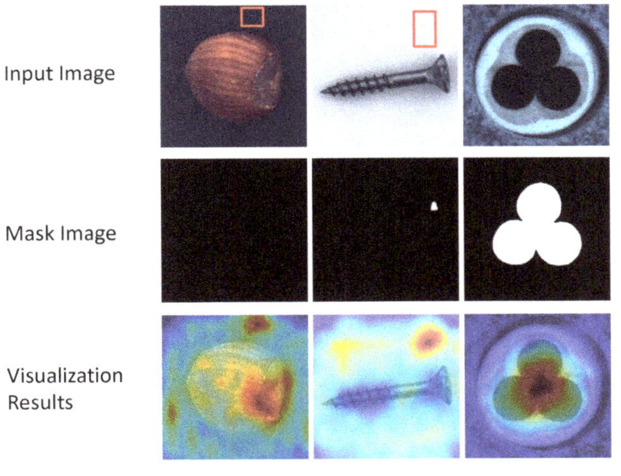

Figure 4. The case of failure in our method. Several samples were selected from our test dataset, the top is the input image, the middle is the mask image, and the bottom is the visualization result.

In Figure 5, we did the statistics of the abnormal scores. Normal samples are shown in blue and abnormal samples are shown in yellow. As can be seen from Figure 5, the model is able to distinguish the normal samples from the abnormal samples well. However, some categories cannot be distinguished well and may require a better model or more training epochs.

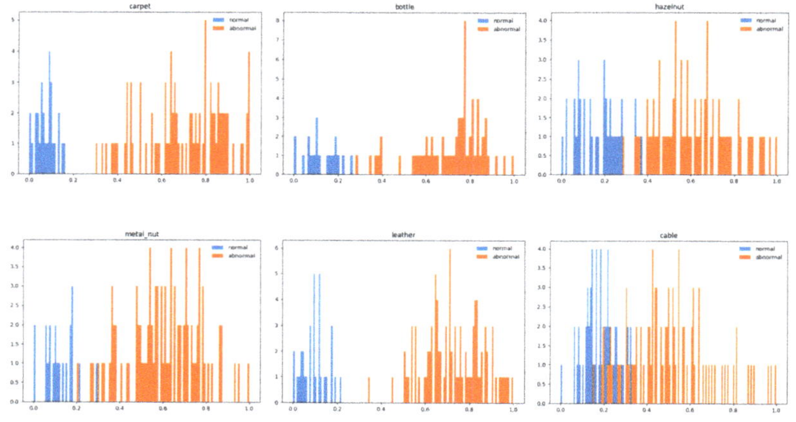

Figure 5. Histogram of sample abnormal scores. Blue indicates normal samples and yellow indicates abnormal samples. The horizontal axis indicates the abnormal scores and the vertical axis indicates the counts.

4.5. Ablation Study

Teacher networks are important for feature extraction of images, and in Table 5, we verified the effect of using pretraining and nonpretraining networks on the image-level and pixel-level anomaly detection results. From Table 5, it can be seen that using pretraining can improve the results by about 20 points, indicating that using pretraining can extract the features of the image well.

Table 5. Image-level and pixel-level anomaly detection results with teacher network using pretraining and nonpretraining.

Two Teachers Network	Unpretrained	Pretrained
Texture	86.42/77.52	99.68/97.46
Object	71.55/79.83	99.18/98.27
Average	78.99/78.68	99.43/97.87

In the attention mechanism module, we investigated the image feature channel division parameter r to see its effect on the anomaly detection results at the image-level and pixel-level, and the results are shown in Table 4. As can be seen from Table 6, the pixel-level results have a relatively small change in value as the r parameter changes. At the image-level anomaly detection results have some variation as the r parameter changes. It indicates that the division of image feature channels has some influence on the image-level anomaly detection results.

Table 6. Image-level and pixel-level anomaly detection results for the attention mechanism module with channel division parameter r.

Div	2	4	6	8
Texture	99.68/97.3	99.96/97.46	98.46/97.40	99.68/97.46
Object	97.58/97.99	99.12/98.20	99.15/98.27	99.18/98.27
Average	98.63/97.65	99.44/97.83	98.81/97.84	99.43/97.87

In Table 7, the baseline represents the adoption of the original network, including one teacher and student network. New Student Network represents the replacement of the original student network with a new student network, which is designed by ourselves. Two Teachers Network means replacing a single teacher network with two teacher networks. Mscam represents the addition of two teacher network feature channels to the attention mechanism module separately. Iaff represents the feature fusion of the two-teacher network feature channels again.

Table 7. Ablation study of designed networks and modules for image-level and pixel-level anomaly detection results.

Baseline	New Sutdent	Two Teachers	Mscam	Iaff	Value
√					98.72/97.78
√	√				98.90/97.83
√	√	√			99.18/97.78
√	√	√	√		99.29/97.75
√	√	√	√	√	99.43/97.87

Both image-level and pixel-level anomaly detection results are improved by using different architectures for the teacher and student networks. This suggests that the powerful performance student network may cause overfitting and recover the anomalies as well.

Better feature representation can be extracted using two teacher networks. In Table 6, it can observe that the two teacher network achieves good results in image-level anomaly detection, but the pixel-level anomaly detection results are almost unchanged.

The use of an attention mechanism and feature fusion module enables better extraction of useful features. As can be seen from Table 6, the image-level and pixel-level anomaly detection results achieve a good result. The attention mechanism can extract better feature channels according to their importance, and the feature fusion module can assign higher weights to the teacher network that provides valuable features.

5. Conclusions

We propose to use reverse knowledge distillation with two teachers for industrial defect detection. In the teacher–student network, a two-teacher network with inconsistent architectures is employed to extract features for diverse feature learning. To focus on the important features and improves the accuracy of the model, an attention mechanism is adopted in the teacher network and an iterative attentional feature fusion is used for the feature fusion of the two teacher networks. In the student network, the student network is designed to prevent overfitting. Extensive experiments are conducted on Mvtec and BTAD datasets. From the results, both image-level and pixel-level anomaly scores are improved. However, the pixel-level performance improvement is not high, and it needs to be investigated in the future.

Author Contributions: Conceptualization, N.L. and M.P.; methodology, N.L. and H.S.; software, M.P.; validation, M.P.; formal analysis, N.L.; investigation, M.P.; resources, N.L.; data curation, M.P.; writing—original draft preparation, M.P.; writing—review and editing, N.L. and P.G.; visualization, M.P.; supervision, N.L.; project administration, N.L.; funding acquisition, N.L. All authors have read and agreed to the published version of the manuscript.

Funding: This research is supported in part by the Natural Science Foundation of Jiangsu Province of China (BK20222012), Guangxi Science and Technology Project (AB22080026/2021AB22167), National Natural Science Foundation of China (No. 61375021) and the Natural Science Key Project of Anhui Provincial Education Department (No. KJ2020A0636, No. KJ2021A0937, No. 2022AH051683, No. 2022AH051670, No. 2022AH051669).

Institutional Review Board Statement: Not applicable.

Informed Consent Statement: Not applicable.

Data Availability Statement: The data that support the findings of this study are available from the first author, Mingjing Pei, upon reasonable request.

Conflicts of Interest: The authors declare no conflict of interest.

References

1. Chai, J.; Zeng, H.; Li, A.; Ngai, E.W. Deep learning in computer vision: A critical review of emerging techniques and application scenarios. *Mach. Learn. Appl.* **2021**, *6*, 100134. [CrossRef]
2. Chen, Z.; Lu, Z.; Gao, H.; Zhang, Y.; Zhao, J.; Hong, D.; Zhang, B. Global to Local: A Hierarchical Detection Algorithm for Hyperspectral Image Target Detection. *IEEE Trans. Geosci. Remote Sens.* **2022**, *60*, 5544915. [CrossRef]
3. Neubauer, K.; Bullard, E.; Blunt, R. *Collection of Data with Unmanned Aerial Systems (UAS) for Bridge Inspection and Construction Inspection*; Technical Report; Federal Highway Administration, Office of Infrastructure: Washington, DC, USA, 2021.
4. Kuric, I.; Klarák, J.; Bulej, V.; Sága, M.; Kandera, M.; Hajdučík, A.; Tucki, K. Approach to Automated Visual Inspection of Objects Based on Artificial Intelligence. *Appl. Sci.* **2022**, *12*, 864. [CrossRef]
5. Sharma, R.; Sungheetha, A. An efficient dimension reduction based fusion of CNN and SVM model for detection of abnormal incident in video surveillance. *J. Soft Comput. Paradig. (JSCP)* **2021**, *3*, 55–69. [CrossRef]
6. Li, C.L.; Sohn, K.; Yoon, J.; Pfister, T. Cutpaste: Self-supervised learning for anomaly detection and localization. In Proceedings of the IEEE/CVF Conference on Computer Vision and Pattern Recognition, Online, 19–25 June 2021; pp. 9664–9674.
7. Schlüter, H.M.; Tan, J.; Hou, B.; Kainz, B. Self-supervised out-of-distribution detection and localization with natural synthetic anomalies (nsa). *arXiv* **2021**, arXiv:2109.15222.
8. Gong, D.; Liu, L.; Le, V.; Saha, B.; Mansour, M.R.; Venkatesh, S.; Hengel, A.v.d. Memorizing normality to detect anomaly: Memory-augmented deep autoencoder for unsupervised anomaly detection. In Proceedings of the IEEE/CVF International Conference on Computer Vision, Seoul, Republic of Korea, 27 October–2 November 2019; pp. 1705–1714.
9. Zhou, K.; Xiao, Y.; Yang, J.; Cheng, J.; Liu, W.; Luo, W.; Gu, Z.; Liu, J.; Gao, S. Encoding structure-texture relation with p-net for anomaly detection in retinal images. In Proceedings of the European Conference on Computer Vision, Glasgow, UK, 23–28 August 2020; pp. 360–377.
10. Sabokrou, M.; Khalooei, M.; Fathy, M.; Adeli, E. Adversarially learned one-class classifier for novelty detection. In Proceedings of the IEEE/CVF International Conference on Computer Vision, Salt Lake City, UT, USA, 18–23 June 2018; pp. 3379–3388.
11. Schlegl, T.; Seeböck, P.; Waldstein, S.M.; Langs, G.; Schmidt-Erfurth, U. f-AnoGAN: Fast unsupervised anomaly detection with generative adversarial networks. *Med. Image Anal.* **2019**, *54*, 30–44. [CrossRef]
12. Wang, S.; Wu, L.; Cui, L.; Shen, Y. Glancing at the patch: Anomaly localization with global and local feature comparison. In Proceedings of the IEEE/CVF Conference on Computer Vision and Pattern Recognition, Online, 19–25 June 2021; pp. 254–263.
13. Ruff, L.; Vandermeulen, R.; Goernitz, N.; Deecke, L.; Siddiqui, S.A.; Binder, A.; Müller, E.; Kloft, M. Deep one-class classification. In Proceedings of the International Conference on Machine Learning, PMLR, Stockholm Sweden, 10–15 July 2018; pp. 4393–4402.
14. Wang, G.; Han, S.; Ding, E.; Huang, D. Student-teacher feature pyramid matching for unsupervised anomaly detection. *arXiv* **2021**, arXiv:2103.04257.
15. Deng, H.; Li, X. Anomaly Detection via Reverse Distillation from One-Class Embedding. In Proceedings of the IEEE/CVF Conference on Computer Vision and Pattern Recognition, Seattle, WA, USA, 19–20 June 2022; pp. 9737–9746.
16. Zhi-Han, Y. Training Latent Variable Models with Auto-encoding Variational Bayes: A Tutorial. *arXiv* **2022**, arXiv:2208.07818.
17. Hou, J.; Zhang, Y.; Zhong, Q.; Xie, D.; Pu, S.; Zhou, H. Divide-and-assemble: Learning block-wise memory for unsupervised anomaly detection. In Proceedings of the IEEE/CVF International Conference on Computer Vision, Online, 11–17 October 2021; pp. 8791–8800.
18. Park, H.; Noh, J.; Ham, B. Learning memory-guided normality for anomaly detection. In Proceedings of the IEEE/CVF Conference on Computer Vision and Pattern Recognition, Seattle, WA, USA, 13–19 June 2020; pp. 14372–14381.
19. Zhang, X.; Mu, J. Deep anomaly detection with self-supervised learning and adversarial training. *Pattern Recognit.* **2022**, *121*, 108234. [CrossRef]
20. Ye, F.; Huang, C.; Cao, J.; Li, M.; Zhang, Y.; Lu, C. Attribute restoration framework for anomaly detection. *IEEE Trans. Multimed.* **2020**, *24*, 116–127. [CrossRef]
21. Salehi, M.; Eftekhar, A.; Sadjadi, N.; Rohban, M.H.; Rabiee, H.R. Puzzle-ae: Novelty detection in images through solving puzzles. *arXiv* **2020**, arXiv:2008.12959.
22. Noroozi, M.; Favaro, P. Unsupervised learning of visual representations by solving jigsaw puzzles. In Proceedings of the European Conference on Computer Vision, Amsterdam, The Netherlands, 11–14 October 2016; pp. 69–84.
23. Haselmann, M.; Gruber, D.P.; Tabatabai, P. Anomaly detection using deep learning based image completion. In Proceedings of the 2018 17th IEEE International Conference on Machine Learning and Applications (ICMLA), Orlando, FL, USA, 17–20 December 2018; pp. 1237–1242.
24. Li, Z.; Li, N.; Jiang, K.; Ma, Z.; Wei, X.; Hong, X.; Gong, Y. Superpixel masking and inpainting for self-supervised anomaly detection. In Proceedings of the BMVC, Online, 7–10 September 2020.
25. Sohn, K.; Li, C.L.; Yoon, J.; Jin, M.; Pfister, T. Learning and evaluating representations for deep one-class classification. *arXiv* **2020**, arXiv:2011.02578.
26. Liznerski, P.; Ruff, L.; Vandermeulen, R.A.; Franks, B.J.; Kloft, M.; Müller, K.R. Explainable deep one-class classification. *arXiv* **2020**, arXiv:2007.01760.
27. Wu, P.; Liu, J.; Shen, F. A deep one-class neural network for anomalous event detection in complex scenes. *IEEE Trans. Neural Netw. Learn. Syst.* **2019**, *31*, 2609–2622. [CrossRef]

28. Massoli, F.V.; Falchi, F.; Kantarci, A.; Akti, Ş.; Ekenel, H.K.; Amato, G. MOCCA: Multilayer One-Class Classification for Anomaly Detection. *IEEE Trans. Neural Net. Learn. Syst.* **2021**, *33*, 2313–2323. [CrossRef]
29. Golan, I.; El-Yaniv, R. Deep anomaly detection using geometric transformations. *Adv. Neural Inf. Process. Syst.* **2018**, *31*.
30. Komodakis, N.; Gidaris, S. Unsupervised representation learning by predicting image rotations. In Proceedings of the International Conference on Learning Representations (ICLR), Vancouver, BC, Canada, 30 April–3 May 2018.
31. Pourreza, M.; Mohammadi, B.; Khaki, M.; Bouindour, S.; Snoussi, H.; Sabokrou, M. G2D: Generate to detect anomaly. In Proceedings of the IEEE/CVF Winter Conference on Applications of Computer Vision, Online, 5–9 January 2021; pp. 2003–2012.
32. Cohen, N.; Hoshen, Y. Sub-image anomaly detection with deep pyramid correspondences. *arXiv* **2020**, arXiv:2005.02357.
33. Krizhevsky, A.; Sutskever, I.; Hinton, G.E. Imagenet classification with deep convolutional neural networks. *Commun. ACM* **2017**, *60*, 84–90. [CrossRef]
34. Rippel, O.; Mertens, P.; Merhof, D. Modeling the distribution of normal data in pre-trained deep features for anomaly detection. In Proceedings of the 2020 25th International Conference on Pattern Recognition (ICPR), Milan, Italy, 10–15 January 2021; pp. 6726–6733.
35. Defard, T.; Setkov, A.; Loesch, A.; Audigier, R. Padim: A patch distribution modeling framework for anomaly detection and localization. In Proceedings of the International Conference on Pattern Recognition, Online, 10–15 January 2021; pp. 475–489.
36. Rudolph, M.; Wandt, B.; Rosenhahn, B. Same same but differnet: Semi-supervised defect detection with normalizing flows. In Proceedings of the IEEE/CVF Winter Conference on Applications of Computer Vision, Waikoloa, HI, USA, 3–8 January 2021; pp. 1907–1916.
37. Yoa, S.; Lee, S.; Kim, C.; Kim, H.J. Self-supervised learning for anomaly detection with dynamic local augmentation. *IEEE Access* **2021**, *9*, 147201–147211. [CrossRef]
38. Zavrtanik, V.; Kristan, M.; Skočaj, D. Draem-a discriminatively trained reconstruction embedding for surface anomaly detection. In Proceedings of the IEEE/CVF International Conference on Computer Vision, Online, 11–17 October 2021; pp. 8330–8339.
39. Salehi, M.; Sadjadi, N.; Baselizadeh, S.; Rohban, M.H.; Rabiee, H.R. Multiresolution knowledge distillation for anomaly detection. In Proceedings of the IEEE/CVF Conference on Computer Vision and Pattern Recognition, Online, 19–25 June 2021; pp. 14902–14912.
40. Zhang, X.; Li, S.; Li, X.; Huang, P.; Shan, J.; Chen, T. DeSTSeg: Segmentation Guided Denoising Student-Teacher for Anomaly Detection. *arXiv* **2022**, arXiv:2211.11317.
41. He, K.; Zhang, X.; Ren, S.; Sun, J. Deep residual learning for image recognition. In Proceedings of the IEEE Conference on Computer Vision and Pattern Recognition, Las Vegas, NV, USA, 26 June–1 July 2016; pp. 770–778.
42. Zagoruyko, S.; Komodakis, N. Wide residual networks. In Proceedings of the British Machine Vision Conference (BMVC), York, UK, 19–22 September 2016.
43. Dai, Y.; Gieseke, F.; Oehmcke, S.; Wu, Y.; Barnard, K. Attentional feature fusion. In Proceedings of the IEEE/CVF Winter Conference on Applications of Computer Vision, Waikola, HI, USA, 5–9 January 2021; pp. 3560–3569.
44. Deecke, L.; Vandermeulen, R.; Ruff, L.; Mandt, S.; Kloft, M. Image anomaly detection with generative adversarial networks. In Proceedings of the Joint European Conference on Machine Learning and Knowledge Discovery in Databases, Wurzburg, Germany, 16–20 September 2019; pp. 3–17.
45. You, Z.; Cui, L.; Shen, Y.; Yang, K.; Lu, X.; Zheng, Y.; Le, X. A unified model for multi-class anomaly detection. In Proceedings of the Neural Information Processing Systems, New Orleans, LA, USA, 2–4 December 2022.
46. Zavrtanik, V.; Kristan, M.; Skočaj, D. Reconstruction by inpainting for visual anomaly detection. *Pattern Recognit.* **2021**, *112*, 107706. [CrossRef]
47. Yi, J.; Yoon, S. Patch Svdd: Patch-Level Svdd for Anomaly Detection and Segmentation. In *Lecture Notes in Computer Science*; Springer: Berlin/Heidelberg, Germany, 2020; pp. 375–390.
48. Gudovskiy, D.; Ishizaka, S.; Kozuka, K. Cflow-ad: Real-time unsupervised anomaly detection with localization via conditional normalizing flows. In Proceedings of the IEEE/CVF Winter Conference on Applications of Computer Vision (WACV), Waikoloa, HI, USA, 3–8 January 2022; pp. 98–107.
49. Mishra, P.; Verk, R.; Fornasier, D.; Piciarelli, C.; Foresti, G.L. VT-ADL: A vision transformer network for image anomaly detection and localization. In Proceedings of the 2021 IEEE 30th International Symposium on Industrial Electronics (ISIE), Kyoto, Japan, 20–23 June 2021; pp. 1–6.
50. Guo, X.; Liu, X.; Zhu, E.; Yin, J. Deep clustering with convolutional autoencoders. In Proceedings of the International Conference on Neural Information Processing, Guangzhou, China, 14–18 November 2017; pp. 373–382.

Disclaimer/Publisher's Note: The statements, opinions and data contained in all publications are solely those of the individual author(s) and contributor(s) and not of MDPI and/or the editor(s). MDPI and/or the editor(s) disclaim responsibility for any injury to people or property resulting from any ideas, methods, instructions or products referred to in the content.

applied sciences

MDPI

Article

Rolling Bearing Health Indicator Extraction and RUL Prediction Based on Multi-Scale Convolutional Autoencoder

Zijian Ye [1,*], Qiang Zhang [2], Siyu Shao [2], Tianlin Niu [2] and Yuwei Zhao [2]

[1] Graduate School, Air Force Engineering University, Xi'an 710043, China
[2] Air Defense and Anti-Missile Academy, Air Force Engineering University, Xi'an 710043, China; zhangq292@126.com (Q.Z.); cathygx.sy@gmail.com (S.S.); ntl2368@163.com (T.N.); ywzhao1987@126.com (Y.Z.)
* Correspondence: zijianye2016@163.com

Abstract: Rolling bearings are some of the most crucial components in rotating machinery systems. Rolling bearing failure may cause substantial economic losses and even endanger operator lives. Therefore, the accurate remaining useful life (RUL) prediction of rolling bearings is of tremendous research importance. Health indicator (HI) construction is the critical step in the data-driven RUL prediction approach. However, existing HI construction methods often require extraction of time-frequency domain features using prior knowledge while artificially determining the failure threshold and do not make full use of sensor information. To address the above issues, this paper proposes an end-to-end HI construction method called a multi-scale convolutional autoencoder (MSCAE) and uses LSTM neural networks for RUL prediction. MSCAE consists of three convolutional autoencoders with different convolutional kernel sizes in parallel, which can fully exploit the global and local information of the vibration signals. First, the raw vibration data and labels are input into MSCAE, and then, MSCAE is trained by minimizing the composite loss function. After that, the vibration data of the test bearings are fed into the trained MSCAE to extract HI. Finally, RUL prediction is performed using the LSTM neural network. The superiority of the HI extracted by MSCAE was verified using the PHM2012 challenge dataset. Compared to state-of-the-art HI construction methods, RUL prediction using MSCAE-extracted HI has the highest prediction accuracy.

Keywords: MSCAE; RUL; rolling bearing; health indicator

1. Introduction

Rolling bearings are the joints of rotating machinery and play an essential role in industrial production, intelligent manufacturing, and transportation [1,2]. Bearing mounting methods, lubricant quality, bearing operating environment, load and speed can all affect bearing vibration and life [3–5]. Once the rolling bearing fails, it will cause the machinery and equipment to stop working, causing substantial economic loss or even threatening the operator's life [6]. Therefore, timely analysis of their working conditions and prediction of their remaining useful life (RUL) is of great research importance [7]. The Prognostic and Health Management (PHM) refers to the use of sensor monitoring data to predict, monitor, and manage the health status of a system through models and algorithms [8–12]. RUL prediction is a crucial technique in PHM and is defined as the time interval between the current moment and when a system or internal component fails [10,11].

Currently, RUL prediction methods are mainly divided into model-based and data-driven methods [13]. The model-based method uses a mathematical approach to construct a physical model of the degradation trend of mechanical components [14]. However, modern mechanical systems have dramatically increased in complexity, have highly coupled internal components, and often operate in severe environments with heavy loads, variable operating conditions, and multiple noise levels [15]. Therefore, building accurate degradation models is a challenging task, limiting the development of model-based methods.

The data-driven method is based on machine learning algorithms to construct a mapping relationship between the monitoring data from the sensors and the RUL [16]. Deep learning is currently one of the up-to-date research directions in machine learning. With its powerful data processing and feature extraction capabilities, deep learning is widely used in computer vision, natural language processing, and fault diagnosis [17]. With the advantages of deep learning, more and more researchers are applying it in the RUL prediction model. Wang et al. [18] proposed a multi-scale convolutional attention network. The multi-sensor data are first fused, and then, feature extraction is performed using a multi-scale convolution module with a self-attentive mechanism. Finally, regression analysis is performed on the high-level representation using a dynamic dense layer. The validity of the model was verified using milling tool life data. Ma et al. [19] proposed a convolutional long and short-term memory model for bearing RUL prediction. Unlike the traditional method that directly connects the convolutional neural network (CNN) and LSTM, this method performs convolutional operations on all the state transitions in the LSTM. Cao et al. [20] proposed a temporal convolution prediction framework with a residual self-attention mechanism. The marginal spectrum of the vibration signal is first extracted; then, it is used as the input of the temporal convolution network, and finally, the residual self-attention mechanism is introduced to achieve end-to-end remaining lifetime prediction. The validity of the proposed method was verified with the PHM2012 challenge dataset and the XJTU-SY dataset. Yao et al. [21] proposed a prediction method that combines a one-dimensional convolutional neural network with simple recurrent units.

The extraction of health indicators (HI) is a crucial step in the RUL prediction process. HI mainly reflect the degradation of mechanical equipment. Therefore, the quality of HI will directly affect the RUL prediction accuracy. Qiu et al. [22] used self-organizing mapping (SOM) to fuse the extracted features to construct the HI of rolling bearings. Qin et al. [23] used the root mean square (RMS) error as the HI of a rolling bearing and predicted the future HI sequence using a gated dual attention unit neural network to determine the magnitude of the rolling bearing RUL. Chen et al. [24] proposed a simple HI construction scheme that uses the ratio of the current moment's RUL value to the initial RUL as the HI, reducing the need for expensive prior knowledge. Five bandpass energy values of the spectrum are used as features. The mapping relationship between the features and HI is constructed directly using a circular autoencoder (AE) based on the attention mechanism. Finally, the value of the RUL is calculated using linear regression. Guo et al. [25] used CNN to extract the bearings' HI and used outlier region correction techniques to detect and correct the outlier points in the obtained HI.

There are three main problems with the above methods. (1) In the HI construction process, it is often necessary to analyze the vibration signals in the time-frequency domain and extract the features in the time domain, frequency domain or time-frequency domain as the input to the model. However, different combinations of features may produce different results. In the literature [26,27], 14 specially designed features and 14 commonly used features were used for HI construction, respectively, and the trend, monotonicity, and scale similarity of the two HI differed. This approach requires a large amount of expert prior knowledge and does not allow automatic HI extraction. (2) The temporal information of the vibration signals is not fully utilized, and usually, only a single size time window is used for feature extraction of the vibration signals, which cannot utilize the local and the global information comprehensively. The bearing vibration signal is a vector containing time information, and the lack of adequate consideration of the time scale can affect the effectiveness of HI. (3) Usually, when using HI for RUL prediction, the failure threshold of HI needs to be determined. In the literature [23], the RMS maximum value of 1.903 at bearing failure was chosen as the failure threshold. In the literature [28], the failure threshold for HI was determined to be 0.17 by calculating the average of the last five local minimum HI points of the tested motor. These methods often carry a particular element of subjectivity and conjecture, leading to errors in RUL predictions.

To solve the problems mentioned above, a novel deep learning framework, called a multi-scale convolutional autoencoder (MSCAE), is proposed in this paper for the automatic extraction of rolling bearing HI. MSCAE is obtained by fusing multiple convolutional autoencoders with different convolutional kernel sizes. Convolutional autoencoders with small convolutional kernels can perform feature extraction for locally degraded features. In contrast, convolutional autoencoders with large convolutional kernels can perform feature extraction considering the degenerate trend of the whole sequence. Combining both of the advantages can solve the problem of underutilization of temporal information. Meanwhile, the construction of labels using quadratic functions and the training of MSCAE using composite loss functions can realize end-to-end HI extraction and automate HI extraction. After obtaining the HI of MSCAE, the LSTM is used for bearing RUL prediction. Unlike other methods, HI obtained by the proposed method does not require the additional step of failure threshold determination, and the failure threshold is directly set to 0. The main contributions of this paper are listed below.

(1) A novel autoencoder-based HI construction method, called MSCAE, is proposed. Compared with the traditional HI construction method, it can effectively use local and global temporal information to obtain a more robust HI and, at the same time, realize the automatic HI extraction.
(2) The HI obtained by MSCAE is compared with single-scale convolutional autoencoders with different convolutional kernel sizes, and the superiority of the proposed method is verified by a comprehensive evaluation index consisting of monotonicity, correlation and robustness.
(3) The HI obtained by MSCAE does not require artificial determination of the failure threshold, which is directly set to 0 and can be directly used for RUL prediction.

The remaining parts of this paper are organized as follows. Section 2 introduces the relevant theoretical knowledge covered in this paper. Section 3 presents the proposed HI construction method. Section 4 conducts relevant experiments and comparisons to verify the validity of the proposed method. Section 5 concludes this paper.

2. Related Theory

2.1. AE

AE was first proposed in the literature [29] as an unsupervised learning method and is often used for feature dimensionality reduction. AE consists of two parts: the encoder and the decoder. The encoder reduces the dimensionality of the input signals and extracts the high-level representations. The decoder takes the output results of the encoder as the input and reconstructs the input signals. AE uses a back-propagation algorithm to update the internal parameters to minimize the reconstruction error. The structure of a typical AE is shown in Figure 1.

Assume that the input of the encoder is $\boldsymbol{x} = [x_1, x_2, \ldots, x_l]$. l is the length of the input data. Then, the output $\boldsymbol{s}$ of the encoder can be expressed as:

$$\boldsymbol{s} = f_e(W_e\boldsymbol{x} + b_e) \tag{1}$$

where W_e and b_e denote the weight and bias, respectively, and f_e denotes the encoder activation function. Then, the the decoder output $\hat{\boldsymbol{x}} = [\hat{x}_1, \hat{x}_2, \ldots, \hat{x}_l]$ can be expressed as:

$$\hat{\boldsymbol{x}} = f_d(W_d\boldsymbol{s} + b_d) \tag{2}$$

where W_d and b_d denote the weight and bias, respectively, and f_d denotes the decoder activation function. The autoencoder optimizes the internal parameters by minimizing the reconfiguration error. The reconfiguration error L_{AE} can be expressed as:

$$L_{AE} = \frac{1}{l}\sum_{i=1}^{l}(x_i - \hat{x}_i)^2 \tag{3}$$

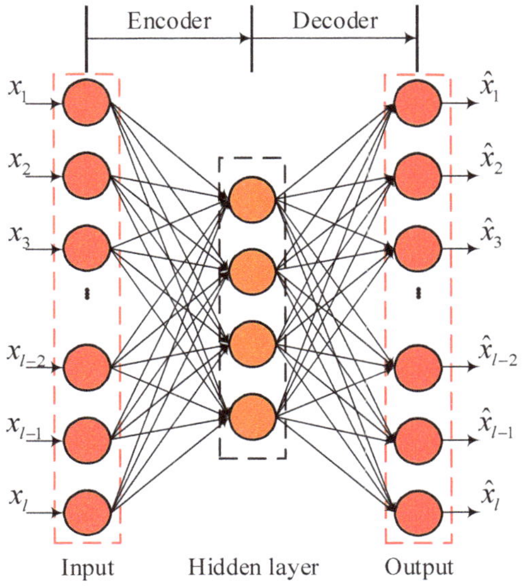

Figure 1. The structure of AE.

2.2. LSTM Neural Network

Data-driven methods for predicting the bearing RUL are usually based on monitoring data collected by sensors, such as temperature and vibration. There is a strong time dependence on these data. The recurrent neural network (RNN) model is characterized by taking sequence data as input and performing operations in the evolutionary direction of the sequence. Therefore, RNN can recognize time series and suit to solve RUL prediction problems. However, traditional RNN suffers from long-term dependency problems, and when training on long sequence data, gradient disappearance or gradient explosion may occur, resulting in ineffective training of the model. To solve this problem, Hochreiter and Schmidhuber proposed the LSTM model [30]. The structure of LSTM is shown in Figure 2. There are two information pathways in LSTM: one for storing long-term memory and one for short-term processing information and adding valid information to the other information pathway. In addition, three gating units, the input gate, forget gate and output gate, are added to the LSTM to control the information flow of input data in the LSTM unit. The formula of LSTM can be expressed as:

$$f_t = \sigma\left(W_f x_t + V_f h_{t-1} + b_f\right) \tag{4}$$

$$i_t = \sigma(W_i x_t + V_i h_{t-1} + b_i) \tag{5}$$

$$\tilde{c}_t = tanh(W_c x_t + V_c h_{t-1} + b_c) \tag{6}$$

$$o_t = \sigma(W_o x_t + V_o h_{t-1} + b_o) \tag{7}$$

$$c_t = f_t \odot c_{t-1} + i_t \odot \tilde{c}_t \tag{8}$$

$$h_t = o_t \odot tanh(c_t) \tag{9}$$

where x_t represents the input at moment t. h_t and c_t represent the hidden state and the cell state at the moment t, respectively. f_t is the output of the forget gate. i_t is the output of the input gate. o_t is the output of the output gate. $W_f, V_f, W_i, V_i, W_c, V_c, W_o$, and V_o represent the weight matrixes. b_f, b_i, b_c, and b_o represent the bias matrixes. σ and $tanh$ denote the sigmoid and hyperbolic tangent activation functions, respectively. $\odot$ denotes point-by-point multiplication.

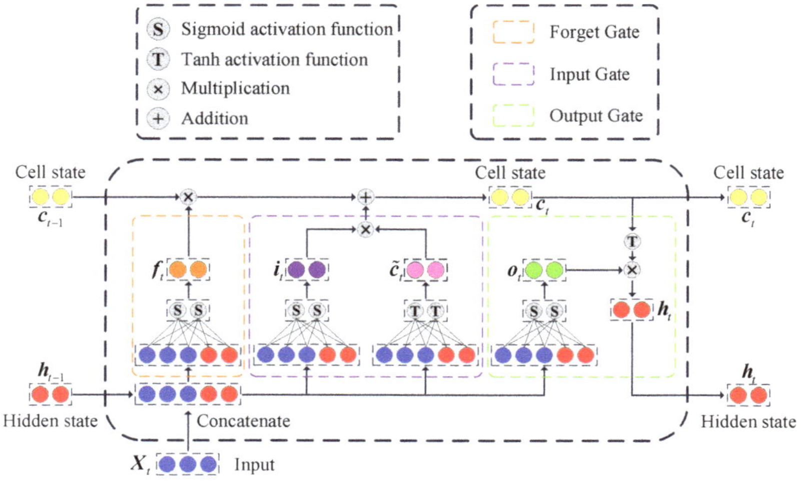

Figure 2. The structure of LSTM.

3. The Proposed Method

3.1. MSCAE

Based on the AE introduced in Section 2.1, the structure of the convolutional autoencoder is obtained by replacing the matrix operations in it with convolutional operations. Usually, features are extracted using convolutional kernels of the same size in the convolutional autoencoder. The literature [18] verified the effectiveness of the multi-scale convolutional network. Compared to traditional CNN, the use of parallel convolutional paths with different convolutional kernel sizes enables multi-scale learning, allowing the model to extract features from different time scales, ensuring the integrity of the representations and making full use of both global and local information. In this paper, we adopt the idea of multi-scale convolution to construct MSCAE to extract a more robust HI based on the full utilization of global and local degradation information, and the automatic extraction of HI can be realized. The structure of the proposed MSCAE is shown in Figure 3.

As shown in Figure 3, the proposed MSCAE uses three convolutional paths for feature extraction of the input data. It is worth noting that the three convolutional paths use one-dimensional convolution with different convolutional kernel sizes, aiming to make full use of the global and local information of the input data for more effective feature extraction.

In the encoding stage, three pathways perform parallel convolution and pooling operations on the input data. Suppose $E^{i}_{n,m}$ denotes the n-th channel of convolutional layer data in the m-th encode block in the i-th$(i = 1,2,3)$ pathway and N^{e}_{m} is the number of convolutional layer channels in the m-th encode block. Then, the one-dimensional convolution operation can be expressed as:

$$Z^{i,e}_{k,m+1} = f_r\left(\sum_{n=1}^{N^e_m} E^{i}_{n,m} * w^{i,e}_{k,n,m} + b^{i,e}_{m}\right) \tag{10}$$

where $*$ is the one-dimensional convolution operation. $w^{i,e}_{k,n,m}$ denotes the weight of the k-th convolutional kernel of the convolutional layer in the m-th encode block. $b^{i,e}_{m}$ is bias. f_r adopts the ReLU activation function. $Z^{i,e}_{k,m+1}$ denotes the k-th channel data of the convolutional layer operation result in the m-th encode block of the i-th pathway.

After performing a convolution operation, a downsampling operation is performed on the convolution result using maximum pooling to reduce the size of the data. The pooling result $I^{i}_{k,m+1}$ for the k-th channel data of the pooling layer in the m-th encoding block of the i-th pathway can be expressed as:

$$E^{i}_{k,m+1} = P\left(Z^{i,e}_{k,m+1}, p_m, s_m\right) \tag{11}$$

where P denotes the maximum pooling operation, p_m is the size of the pooling window in the m-th convolution block, and s_m is the stride. After the input data are extracted by L encoding blocks of features, the Flatten layer is used to turn the high-level representations into one-dimensional data, which is input to the fully connected block for HI extraction. It is worth noting that the fully connected block of MSCAE proposed in this paper uses only one fully connected layer in both encoding and decoding parts, and the number of hidden neurons is the same. Then, the extracted HI can be expressed as:

$$\hat{y} = \sigma\left(W_{fe}X_e + b_{fe}\right) \tag{12}$$

where $\hat{y}$ denotes the extracted HI, W_{fe} and b_{fe} denote the encoded partial weight and bias, respectively, and σ denotes the Sigmoid activation function.

In the decoding stage, there are also three parallel paths. The decoding part in the fully connected block performs a dimensional expansion operation on the extracted HI, and the results obtained can be expressed as:

$$X_d = f_r\left(W_{fd}\hat{y} + b_{fd}\right) \tag{13}$$

where X_d denotes the output of the decoding part in the fully connected block, W_{fd} and b_{fd} denote the weight and bias of the decoding part in the fully connected block, respectively, and f_r is the ReLU activation function. After obtaining X_d, it is dimensionally changed using the Reshape layer, and the transformed shape is the same as the shape of the input data in the Flatten layer. L decoding blocks reconstruct the results of the Reshape layer, i.e., three parallel upsampling and convolution pathways. Suppose $D^{i}_{n,m}$ denotes the n-th channel data of the upsampling layer in the m-th encoding block of the i-th$(i = 1,2,3)$ pathway. Then, the operation of the upsampling layer can be expressed as:

$$Z^{i,d}_{n,m} = U\left(D^{i}_{n,m}, u_m, l_m\right) \tag{14}$$

where $Z^{i,d}_{n,m}$ is the result of the upsampling layer, U denotes the upsampling operation, and u_m and l_m represent the upsampling window size and stirde of the m-th decoding block, respectively. In a decoding block, the upsampling layer is connected after the convolution layer, and the convolution result of the m-th decoding block can be expressed as:

$$D^{i}_{k,m+1} = f_r\left(\sum_{n=1}^{N^d_m} Z^{i,d}_{n,m} * w^{i,d}_{k,n,m} + b^{i,d}_m\right) \tag{15}$$

where $D^{i}_{k,m+1}$ is the k-th channel data of the output of the m-th decoding block, N^d_m denotes the total number of channels of the input data of the m-th decoding block, $w^{i,d}_{k,n,m}$ denotes the weight of the k-th convolutional kernel in the m-th decoding block, $b^{i,d}_m$ is the bias, and f_r is the ReLU activation function. The reconstruction of the input data is obtained after processing by L decoding blocks. Assume that the i-th input data of MSCAE is $\boldsymbol{x_i} = [x_{i,1}, x_{i,2}, \ldots, x_{i,l}]$, and the corresponding real RUL label is y_i, $i = 1,2,\ldots B$. l is the length of the input data, and B is the total life of the bearing. Then, the HI extracted by MSCAE is $\hat{y}_i$, and the reconstruction of the input data is $\hat{\boldsymbol{x}}_i = [\hat{x}_{i,1}, \hat{x}_{i,2}, \ldots \hat{x}_{i,l}]$. In this paper, we use the composite loss function to evaluate the HI extraction capability of MSCAE. The composite loss consists of two parts: one is the error between the input data and the reconstructed data, and the other is the error between the HI and the true RUL. The composite loss function is calculated as follows.

$$J_\theta = \frac{1}{2}\sum_{i=1}^{B} \|x_i - \hat{x}_i\|^2 + \frac{\nu}{2}\sum_{i=1}^{B} \|y_i - \hat{y}_i\|^2 \tag{16}$$

where θ denotes the internal parameters of MSCAE and ν is a scaling factor to adjust the weight between the two errors. Compared with the traditional AE, the composite loss function constructed by MSCAE combines the supervised learning approach with the unsupervised learning approach for model training, which fully utilizes the degradation information of the bearings and enhances the extraction capability of the model HI. In the training phase of the model, the vibration data and labels of the training bearings are used to optimize the internal parameter θ using the back-propagation algorithm to minimize the composite loss. In the testing phase of the model, the vibration data of the testing bearing serve as the input, and the HI is obtained using the encoder part of the trained MSCAE.

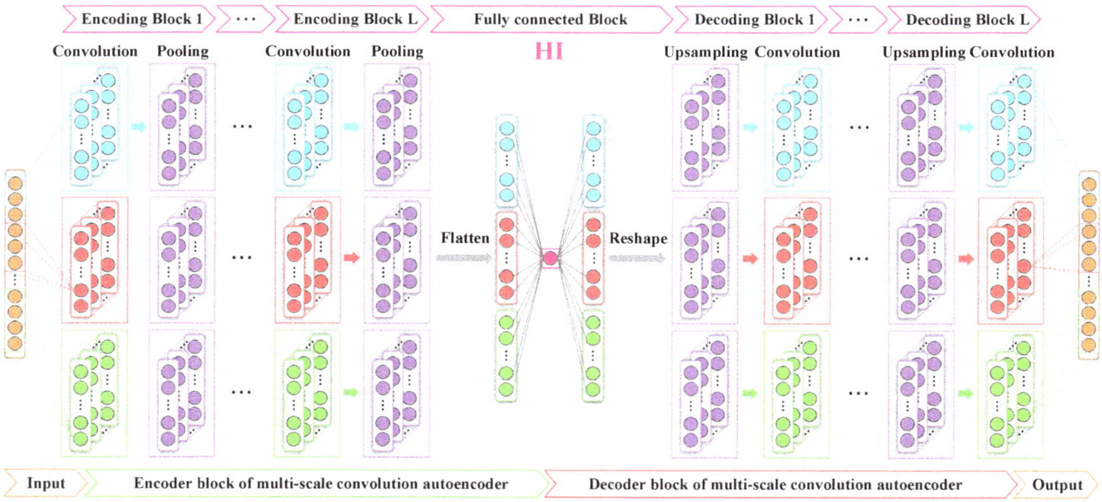

Figure 3. The structure of MSCAE.

3.2. HI Extraction and RUL Prediction Process

3.2.1. Construction Method of Degradation Labels

After obtaining the vibration data of the bearings, it is necessary to divide them into training and testing sets. Since the obtained original data do not have corresponding labels, it is necessary to construct the degradation labels corresponding to the vibration data. There are mainly two conventional methods for constructing degradation labels: linear degradation and segmental smoothing, and Figure 4 shows the differences between the two methods.

The degenerate label construction equation of Figure 4a can be expressed as

$$y_i = \frac{B - t_i}{B} \tag{17}$$

where B denotes the total life of the bearing, t_i denotes the current time the bearing has been in operation, and y_i denotes the current degradation level of the bearing. The degenerate label construction equation of Figure 4b can be expressed as:

$$y_i = \begin{cases} 1 & if \quad t_i \le t_h \\ 1 - \frac{t_i - t_h}{B - t_h} & else \end{cases} \tag{18}$$

where t_h indicates the degradation starting threshold; when t_i is less than or equal to t_h, it is determined that the bearing has not started to degrade and the degradation label is always 1. When t_h exceeds the threshold, it starts to show a linear degradation trend.

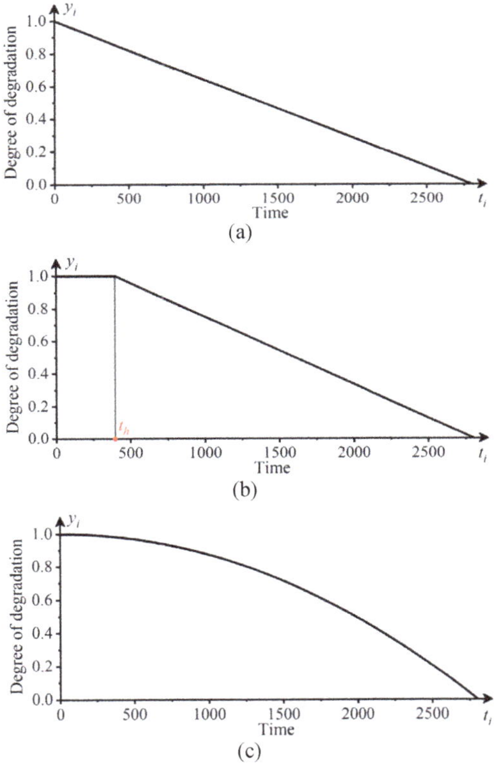

Figure 4. Three degradation label construction methods. (**a**) Linear degradation method. (**b**) Piecewise smoothing method. (**c**) Quadratic function method.

The first two methods of describing bearing degradation trends shown in Figure 4 have their own limitations compared to the true degradation of the bearing. The true bearing degradation does not exhibit a linear characteristic but rather degrades faster and faster as the operating time of the bearing increases. Therefore, the linear degradation method does not satisfy this operating characteristic. For the segmented smoothing method, the threshold t_h at the degradation onset moment is usually determined artificially. However, t_h is usually not the same for different operating conditions, and there are different bearings t_h for the same operating conditions, which limits the application of the segmented smoothing method. The literature [31] proposed a method for constructing a degenerate trend based on a quadratic function, as shown in Figure 4c, and the specific expression can be expressed as:

$$y_i = 1 - \frac{t_i^2}{B^2} \tag{19}$$

The quadratic function-based method overcomes the shortcomings of the above two methods, and its constructed labels are more satisfying to the real degradation of the bearings. With the increase of the operation time, the degradation of the bearings is gradually accelerated, which is shown in Figure 4c as the slope of the curve gradually increases. Moreover, it is a convenient and effective method without artificially specifying the degradation threshold, so this paper adopts the quadratic function method to construct the degradation labels of bearings.

3.2.2. Overall Framework Flow

The overall framework flow of MSCAE-based HI extraction and RUL prediction proposed in this paper is shown in Figure 5. The framework is divided into two main phases, offline training and online testing. There are three main steps in the offline training phase, and the process is as follows.

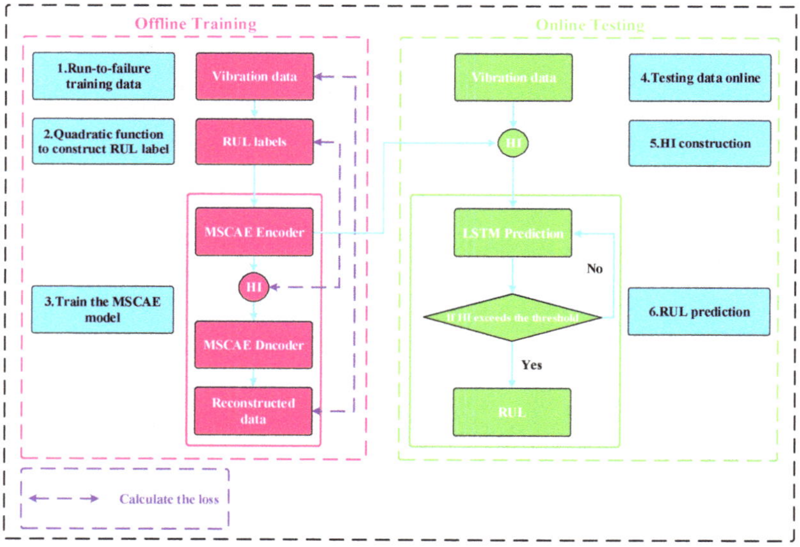

Figure 5. Flow chart of HI extraction and RUL prediction.

1. Vibration data from training bearing operation to failure are obtained using vibration sensors. Suppose the vibration data are $X = [\boldsymbol{x_1}, \boldsymbol{x_2}, \ldots, \boldsymbol{x_B}]$, $\boldsymbol{x_i} = [x_{i,1}, x_{i,2}, \ldots, x_{i,l}]$, $\boldsymbol{x_i}$ represents the vibration data at the moment t_i, $i = 1, 2, \ldots B$, B is the total time of bearing operation, and l is the length of the data.
2. The degenerate labels for each moment of the vibration data are calculated using Equation (19), and the labels $\boldsymbol{y} = [y_1, y_2, \ldots y_B]$ are obtained by the calculation.
3. Training the proposed MSCAE model. First, the vibration datum X of the training bearing is input to the model, and the output of the MSCAE encoding part is the extracted HI, which is denoted as $\hat{\boldsymbol{y}} = [\hat{y}_1, \hat{y}_2, \ldots \hat{y}_B]$. After that, HI is used as an input to the decoding part of MSCAE to obtain the reconstructed data $\hat{X} = [\hat{\boldsymbol{x}}_1, \hat{\boldsymbol{x}}_2, \ldots, \hat{\boldsymbol{x}}_B]$ of the vibration data X. Finally, the composite loss function is calculated using Equation (16), and the internal parameters θ of the model are updated using the back-propagation algorithm. After the MSCAE is trained offline, it moves to the online testing phase.

4. The vibration data from test bearing operation to failure are obtained, which are defined as $X' = [x'_1, x'_2, \dots x'_T]$, and T is the total operating time of the testing bearing.
5. The vibration signal X' of the testing bearing is input to the encoder of the MSCAE trained in step 3 to obtain in HI $\boldsymbol{H} = [h_1, h_2, \dots h_T]$ of the test bearing.
6. After obtaining the HI of the testing bearing, the first N points of $\boldsymbol{H}$ are taken to construct the training data for training the LSTM model. The training matrix can be expressed as follows.

$$\boldsymbol{V} = \begin{bmatrix} h_1 & h_2 & \cdots & h_{N-M} \\ h_2 & h_3 & \cdots & h_{N-M+1} \\ \vdots & \vdots & \ddots & \vdots \\ h_{M+1} & h_{M+2} & \cdots & h_N \end{bmatrix} = \begin{bmatrix} \boldsymbol{v_1} \\ \boldsymbol{v_2} \\ \vdots \\ \boldsymbol{v_{M+1}} \end{bmatrix} \tag{20}$$

where M is the neurons number of the LSTM output layer, $\boldsymbol{v_i} = [h_i, h_{i+1}, \dots, h_{i+N-M-1}]$, $(i = [1, 2, \dots, M+1])$. The LSTM is trained by taking the first M vectors of the training matrix $\boldsymbol{V}$ as the input of the LSTM and the last vector $\boldsymbol{v_{M+1}}$ as the output. Assuming that the mapping function of the trained LSTM is denoted as f, the last M vectors of the matrix $\boldsymbol{V}$ are passed as input to the trained LSTM to obtain the first prediction result $\boldsymbol{\bar{v}_{M+2}}$, and the specific expression is:

$$\boldsymbol{\bar{v}_{M+2}} = f([\boldsymbol{v_2}, \boldsymbol{v_3}, \cdots, \boldsymbol{v_{M+1}}]) \tag{21}$$

Then, the matrix $\boldsymbol{V}$ can be updated as follows:

$$\boldsymbol{V} = [\boldsymbol{v_1}, \boldsymbol{v_2}, \cdots, \boldsymbol{v_{M+1}}, \boldsymbol{\bar{v}_{M+2}}]^T \tag{22}$$

The above method allows the prediction of HI vectors to be performed continuously and the matrix $\boldsymbol{V}$ to be continuously updated. Thus, $\boldsymbol{\bar{v}_k}$ can be expressed as:

$$\boldsymbol{\bar{v}_k} = f\left(\left[\boldsymbol{v_{M-k}}, \boldsymbol{v_{M+1}}, \dots, \boldsymbol{\bar{v}_{M+2}}, \dots, \boldsymbol{\bar{v}_{M+(k-M-1)}}\right]\right) \tag{23}$$

Meanwhile, the matrix $\boldsymbol{V}$ is updated as follows:

$$\boldsymbol{V} = [\boldsymbol{v_1}, \dots, \boldsymbol{v_{M+1}}, \boldsymbol{\bar{v}_{M+2}}, \dots, \boldsymbol{\bar{v}_k}]^T \tag{24}$$

where $\boldsymbol{\bar{v}_k} = \left[h_k, \dots, h_N, \bar{h}_{N+1}, \dots, \bar{h}_{k+N-M-1}\right]$. When $\bar{h}_{k+N-M-1}$ is less than the threshold 0, the prediction step is stopped, and the RUL of the bearing is obtained as $(k - M) \times T_s$, and T_s is the sampling interval of the vibration sensor. If the prediction result is not less than the threshold, the iterative prediction is continued until the predicted value is less than the threshold and the corresponding RUL is obtained.

4. Experimental Analysis

4.1. Data Introduction

The experimental data in this paper were obtained from the PHM Challenge [32] organized by the Institute of Electrical and Electronics Engineers in 2012, and the data were obtained from the PRONOSTIA experimental bench, as shown in Figure 6. The experimental bench consists of three main parts: the rotation part, the degradation generation part and the sensor part. The rotating part consists of an asynchronous motor with a gearbox and two shafts to provide the working environment for the test bearing. The degradation generating part can apply radial load to the testing bearing to simulate the actual working operation with load and accelerate the degradation, which can complete the whole process of bearing operation to failure in a short time. The sensor part consists of three sensors, two of which are vibration sensors positioned 90° apart, measuring the magnitude of horizontal and radial vibrations, respectively, and the other is a temperature sensor, measuring the temperature of the bearing during operation.

The PHM2012 challenge data give the full cycle life data of the bearings from operation to failure for three operating conditions. The three operating conditions are 1800 rpm and 4000 N; 1650 rpm and 4200 N; and 1500 rpm and 5000 N. There are 7, 7, 3 test bearings for each of the three operating conditions. The PRONOSTIA test bench has a sampling interval of 10 s, a sampling duration of 0.1 s, and a sampling frequency of 25.6 kHz, indicating that every 10 s, the sensor can collect 2560 data points.

In order to verify the effectiveness of the MSCAE proposed in this paper, the data set needs to be divided into training data and testing data. The division method in this paper is shown in Table 1. As can be seen from Table 1, all bearings under three operating conditions are used in this paper to validate the proposed method. The training set contains 5, 6, 2 bearings, and the testing set contains 2, 1, 1 bearings.

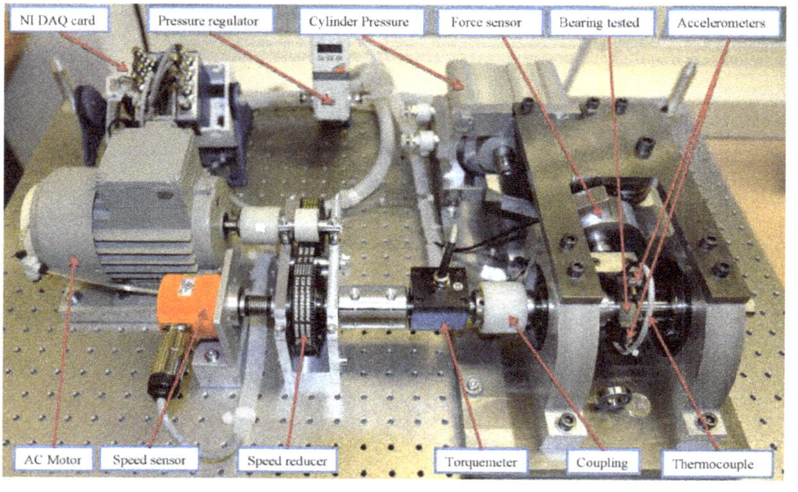

Figure 6. PRONOSTIA test bench.

Table 1. Training set and test set division.

	Training Set	Test Set
Condition 1	Bearing1_2 Bearing1_4 Bearing1_5 Bearing1_6 Bearing1_7	Bearing1_1 Bearing1_3
Condition 2	Bearing2_1 Bearing2_2 Bearing2_3 Bearing2_4 Bearing2_5 Bearing2_7	Bearing2_6
Condition 3	Bearing3_1 Bearing3_2	Bearing3_3

4.2. Evaluation Metrics

To verify the effectiveness of the proposed MSCAE to extract HI in this paper, three evaluation metrics, monotonicity, correlation and robustness, were used to quantify the performance of HI [31]. It is worth noting that the range of all three metrics is within [0, 1].

- Monotonicity: It aims to assess the tendency of HI to increase monotonically or decrease monotonically as the running time increases. The stronger the monotonicity of HI, the closer it is to 1. The specific formula for monotonicity can be expressed as:

$$Mon(H) = \left| \frac{\text{No. of } dH > 0}{T-1} - \frac{\text{No. of } dH < 0}{T-1} \right| \tag{25}$$

where dH denotes the first-order derivative between two HI values and T denotes the number of HI and also the number of sensor samples.

- Correlation: It aims to measure the correlation between HI and runtime. The more correlated the two are, the closer the value of correlation is to 1, and vice versa. The formula for correlation can be expressed as:

$$Corr(H) = \frac{\left|\sum_{i=1}^{T} (h_i - \bar{H})(t_i - \bar{T})\right|}{\sqrt{\sum_{i=1}^{T} (h_i - \bar{H})^2 \sum_{i=1}^{T} (t_i - \bar{T})^2}} \tag{26}$$

where $\bar{H} = \sum_{i=1}^{T} h_i / T$ and $\bar{T} = \sum_{i=1}^{T} t_i / T$.

- Robustness: It aims to measure the ability of HI to resist outlier interference; the stronger its ability, the closer the robustness is to 1, and vice versa. The extracted HI can be seen as a superposition of the average trend and noise, whereby H can be expressed as:

$$H(t_i) = H_T(t_i) + H_R(t_i) \tag{27}$$

where $H_T(t_i)$ denotes the average trend of HI at the moment T, and $H_R(t_i)$ denotes the noise disturbance of HI at the moment T. Then, the robustness is calculated by the formula:

$$Rob(H) = \frac{1}{T} \sum_{i=1}^{T} \exp\left(-\left|\frac{H_R(t_i)}{H(t_i)}\right|\right) \tag{28}$$

In order to comprehensively evaluate the advantages and disadvantages of the extracting HI, a Composite Indicator (CI) containing the above three indicators is proposed, which is defined as:

$$\text{CI} = \frac{1}{3}(\text{Mon} + \text{Corr} + \text{Rob}) \tag{29}$$

4.3. The Validity of MSCAE

The specific structure of the MSCAE proposed in this paper is shown in Figure 7. As can be seen from Figure 7, MSCAE consists of three encoding blocks, three decoding blocks and one fully connected block. The number of convolutional kernels in the three coding blocks is 8, 16, 4, and the size of convolutional kernels in the three pathways is $3 \times 1, 7 \times 1$ and 11×1, respectively, and the size and stride of maximum pooling are 8. The number of convolutional kernels in the three decoding blocks is 16, 8, 1, and the size of convolutional kernels in the three pathways is the same as the decoder, and the size and stride of upsampling are 8. The Sigmoid activation function is used in the fully connected block when the second layer obtains HI. The activation function is not used when reconstructed data are obtained. All the remaining layers use the ReLU activation function, and each layer uses the BatchNormalization (BN) layer to improve the model generalization. The detailed hyperparameter settings of the model are shown in Table 2. Comparative experiments are conducted in this paper to choose the scale factor v of the composite loss function. In this paper, five values of 0.2, 0.4, 0.6, 0.8 and 1.0 are selected for comparison and validation. Ten experiments are conducted on four test bearings using these five values, respectively, and the final box line diagram is obtained, as shown in Figure 8. From Figure 8, it can be seen that v is taken as 0.6, 0.6 and 0.4, respectively, as the best choice for the three working conditions.

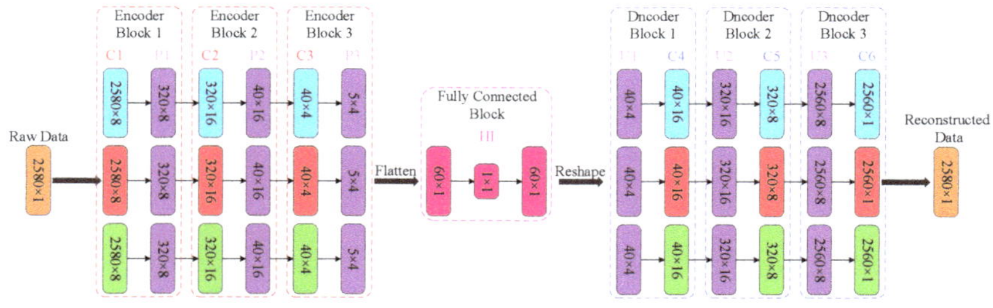

Figure 7. The details of MSCAE.

Table 2. Hyperparameter setting of the model.

Hyperparameter	Value
Batch size	128
Epoch	100
Learning rate lr	Adam (0.0001)
Number of Encoding Block	3
Number of Decoding Block	3
Kernel size	$[3 \times 1], [7 \times 1], [11 \times 1]$
Pooling size	8
Upsampling size	8
Number of fully connected layer units	$[60, 1, 60]$

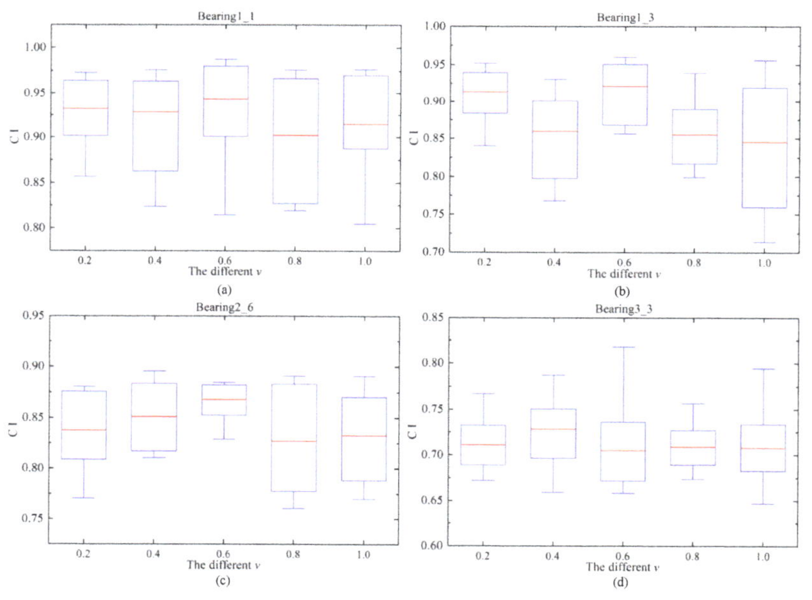

Figure 8. The results of the selection of different scale factors ν. (**a**) The result of Bearing1_1. (**b**) The result of Bearing1_3. (**c**) The result of Bearing2_6. (**d**) The result of Bearing3_3.

After training the MSCAE using the training bearing data, the MSCAE was tested using testing bearings, and the HI extracted by four testing bearings is shown in Figure 9. In Figure 9, the green color indicates that the value of HI is closer to 1, and the blue color is closer to 0. The red curve is the HI trend curve fitted using polynomials. Four bearings in Figure 9 have a clear downward trend in HI, and by the time sampling stops, the value of HI is close to 0.

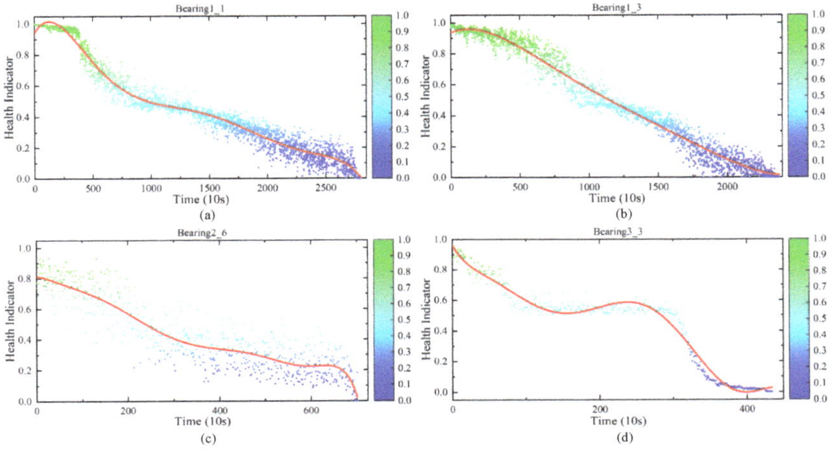

Figure 9. The HI of four testing bearings. (**a**) The HI of Bearing1_1. (**b**) The HI of Bearing1_3. (**c**) The HI of Bearing2_6. (**d**) The HI of Bearing3_3.

In order to verify the superiority of MSCAE, this paper sets up a comparison between MSCAE and three convolutional autoencoders (CAE) with constant convolutional kernel sizes of 3×1, 7×1 and 11×1, respectively. To ensure the completeness of the experiments, the structure of the CAE is the same as the structure of a pathway in MSCAE. Moreover, when the model is trained, the hyperparameters are set the same as MSCAE. The obtained results are shown in Figure 10. From Figure 10, it can be seen that MSCAE exceeds the other three models in CI for the tested bearings under each operating condition. It shows that MSCAE can combine the advantages of different convolutional kernel sizes to identify different time scale information and make full use of the local and global information of the original vibration signal to extract a more effective HI. To further verify the superiority of MSCAE, four models were used to extract the HI of Bearing2_6 bearing for analysis, as shown in Figure 11. When the convolution kernel size is set to 3, the final HI trend increases, which is against the degradation law of the real bearing. Moreover, when the convolution kernel size is 7 and 11, the final HI trend decreases to 0. However, both undergo abrupt changes when the bearing is damaged, i.e., the HI value directly decreases to 0. The abrupt change is most obvious when the convolution kernel size is 7, and this phenomenon affects the prediction of RUL. Therefore, it is desired to obtain an HI that can satisfy the real degradation trend of the bearing, and the degradation process of the HI is relatively flat, which is conducive to the continuous prediction of RUL. It can be found that the HI extracted by MSCAE proposed in this paper can meet the above requirements.

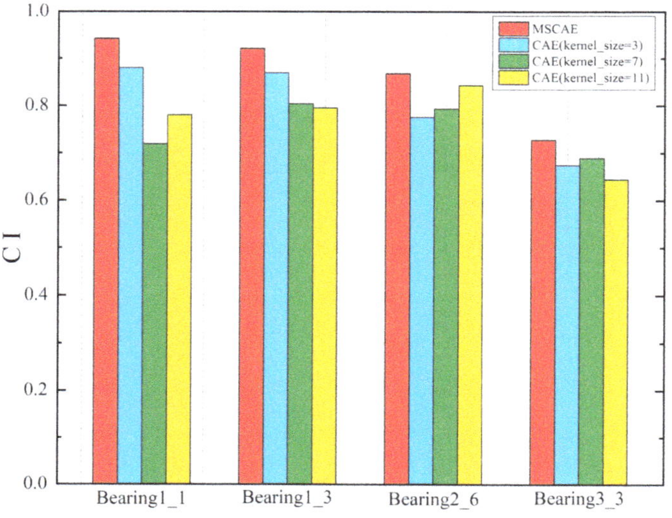

Figure 10. The HI of four testing bearings.

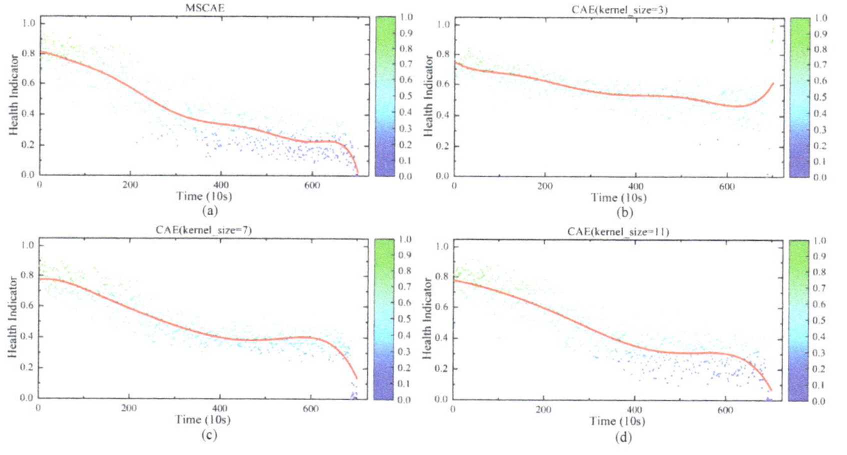

Figure 11. Four models to extract HI of Bearing2_6. (**a**) MSCAE. (**b**) CAE (kernel_size = 3). (**c**) CAE (kernel_size = 7). (**d**) CAE (kernel_size = 11).

4.4. RUL Prediction

After extracting the HI of the testing bearing using MSCAE, it can be found that the extracted HI has obvious degradation characteristics with time, which indicates that the HI is a time series. Therefore, LSTM is used to perform the RUL prediction of bearings. For Bearing1_1, assuming that the last 100 HI are unknown and the first 2703 HI are known, the known 2703 HI points are used as training data for the LSTM to predict the unknown HI points, and when the predicted value of the LSTM is less than the threshold 0, it means that the predicted current moment bearing has been damaged, and the time interval between the current moment and the starting moment of the prediction is the predicted RUL value. Similarly, for Bearing1_3, assuming that the last 100 HI points are unknown, the training data for the LSTM are the first 2275 HI points. The parameters of the LSTM are configured as shown in Table 3. The number of neurons in the input, hidden, and output layers of the LSTM are 360.29 and 1, respectively. The internal parameters of the LSTM are optimized using the Adam optimizer, and the learning rate is set to 0.07.

Table 3. Hyperparameter configuration of LSTM.

Parameters	Value
The number of neurons in the input layer	360
The number of neurons in the hidden layer	29
The number of neurons in the output layer	1
Learning rate lr	0.07
Optimizer	Adam

To demonstrate the superiority of the HI extraction method proposed in this paper on RUL prediction, two state-of-the-art deep learning-based HI construction methods are used for comparison. One is the recurrent convolutional neural network (RCNN), which was proposed in the literature [33], and the other is the CNN proposed in the literature [25]. The RCNN is an end-to-end HI extraction framework consisting of a convolutional neural network with residual structure and an LSTM serially connected with the specific structural parameters described in the literature [33]. The CNN-based HI construction method proposed in the literature [25] is divided into two steps, first using two convolutional layers for feature extraction and later using a fully connected layer for HI construction. It is worth noting that both methods, as in this paper, use the original vibration signal as the input to the model without a manual feature extraction step. All three methods are trained using the same training data, and the RUL prediction ability of the extracted HI is verified on Bearing1_1 and Bearing1_3. For convenience, the HI extracted by the proposed MSCAE, RCNN and CNN are MSCAE-HI, CRNN-HI and CNN-HI.

The RUL prediction results of HI extracted by the three models are shown in Figures 12–14. From Figure 13, it can be found that RUL prediction using CRNN-HI requires artificially set thresholds, and the choice of thresholds directly affects the RUL prediction results. If the set threshold is large, it will cause the prediction to stop early, and conversely, it will cause the prediction value of RUL to be larger than the real one, or even the phenomenon that it cannot converge to the set threshold. In this paper, the failure threshold of CRNN-HI is set to 0.2. In Figure 13a, the HI of Bearing1_1 extracted using the CRNN method shows an increasing trend at the end of degradation, which deviates from the actual degradation threshold and can affect the prediction ability of the LSTM. In Figure 13b, there are a few outliers in the HI of Bearing1_3, and there are HI points less than the threshold value when the bearing first starts running. It can be found in Figure 14 that there is a significant divergence of CNN-HI at the late degradation stage of the bearing, resulting in the predicted HI value being more likely to be close to the failure threshold when using CNN-HI for RUL prediction, resulting in the predicted RUL often being smaller than the true RUL value. In Figure 12, the MSCAE-HI constructed in this paper has less fluctuation and a smoother degradation trend, which can effectively reflect the real degradation trend of the bearing. Using MSCAE-HI for RUL prediction, the predicted RUL results are closer to the true values, and the predicted HI is more consistent with the true degradation trend. Combining the above analysis, the MSCAE-HI method proposed in this paper is more suitable for the RUL prediction of bearings.

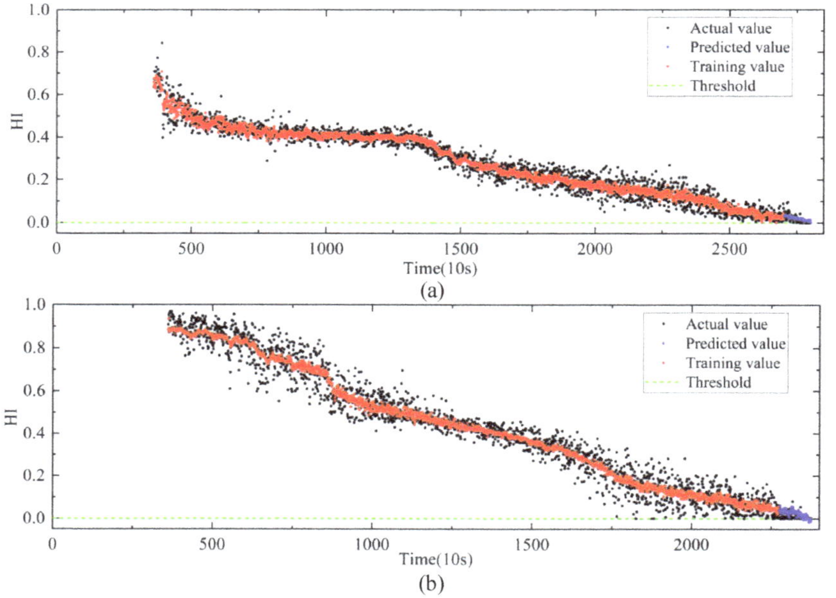

Figure 12. RUL prediction results of MSCAE-HI. (**a**) Bearing1_1. (**b**) Bearing1_3.

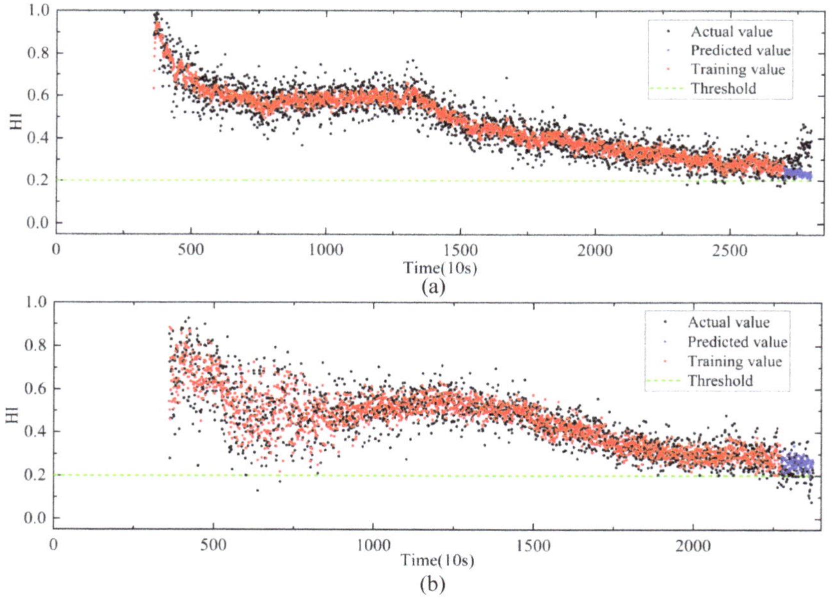

Figure 13. RUL prediction results of RCNN-HI. (**a**) Bearing1_1. (**b**) Bearing1_3.

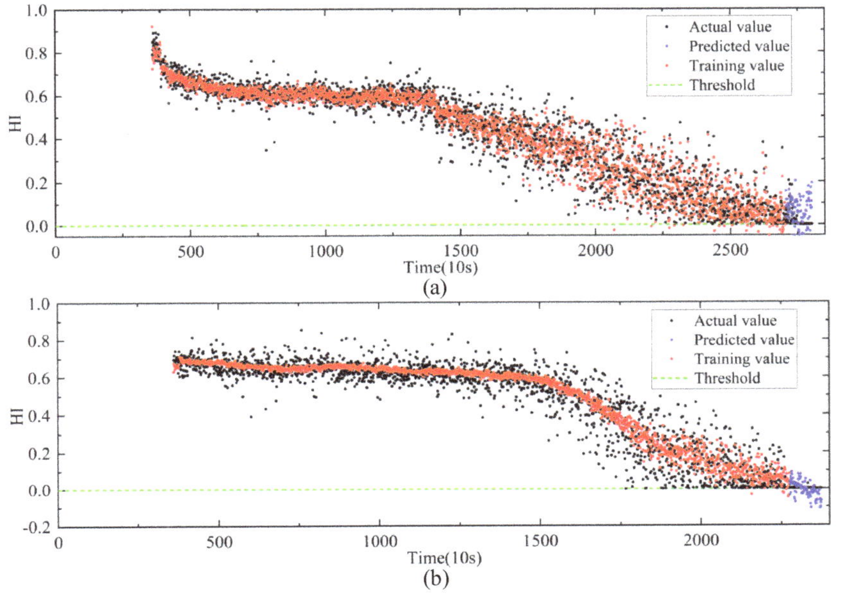

Figure 14. RUL prediction results of CNN-HI. (**a**) Bearing1_1. (**b**) Bearing1_3.

To further validate the superiority of MSCAE-HI, five metrics for evaluating the prediction accuracy of RUL were used to compare the prediction performance of the three methods. These five evaluation metrics are score scoring function, mean absolute error (MAE), normalized root mean square error (NRMSE), root mean square error (RMSE) and mean absolute percentage error (MAPE), which are defined in the literature [23]. Three methods were used to conduct five prediction experiments for Bearing1_1 and Bearing1_3, and the final evaluation results are shown in Table 4. In Table 4, the MAE, NRMSE, RMSE and MAPE of CNN-HI are the maximum among three methods. This phenomenon can be explained that the fluctuation of CNN-HI is very obvious at the late degradation stage, and the HI predicted by the LSTM is easily smaller than the degradation threshold, resulting in a large deviation of the predicted from the true RUL. RUL prediction using MSCAE-HI obtained the maximum score values and the minimum MAE, MRMSE, RMSE, and MAPE values on both Bearing1_1 and Bearing1_3, which once again confirmed the superiority of the HI indicator extraction method of MSCAE proposed in this paper.

Table 4. Results of RUL prediction evaluation for 3 HI.

		Score	
	MSCAE-HI	CRNN-HI	CNN-HI
Bearing1_1	**0.3315**	0.2334	0.1379
Bearing1_3	**0.4979**	0.1366	0.0712
		MAE	
	MSCAE-HI	CRNN-HI	CNN-HI
Bearing1_1	**152**	276	602
Bearing1_3	**214**	356	808
		NRMSE	
	MSCAE-HI	CRNN-HI	CNN-HI
Bearing1_1	**0.1516**	0.2601	1.5480
Bearing1_3	**0.2938**	0.3596	4.2882

Table 4. *Cont.*

	RMSE		
	MSCAE-HI	CRNN-HI	CNN-HI
Bearing1_1	**160.13**	296.58	616.13
Bearing1_3	**230.95**	378.31	823.33
	MAPE		
	MSCAE-HI	CRNN-HI	CNN-HI
Bearing1_1	**15.2**	27.6	60.2
Bearing1_3	**21.4**	35.6	80.8

5. Conclusions

In this paper, a novel framework for HI extraction, called MSCAE, is proposed. It can overcome the disadvantages of traditional methods that require manual extraction of time-frequency domain indicators as features and the need to set failure thresholds by experience in RUL prediction. It relies solely on the raw sensor vibration signal to extract HI and does not require additional determination of the failure threshold. MSCAE can use convolutional kernels of different sizes to effectively exploit the global and local information of vibration signals, enhancing the HI extraction capability. A quadratic function-based label is first constructed for the original vibration data, after which the model is trained using the training data, and the internal parameters are optimized using a compound loss function. Then, HI is extracted using the test-bearing data to verify the validity of MSCAE. Finally, the RUL prediction is performed using LSTM. The HI extraction capability of MSCAE is verified to be superior to that of CAE models using a single scale with the PHM2012 dataset. Furthermore, it is compared with two state-of-the-art HI construction methods, CRNN and CNN, to judge the prediction performance of RUL using five evaluation metrics. The comparison results confirm the superiority of the proposed MSCAE-extracted HI for RUL prediction. In this paper, HI extraction and RUL prediction for rolling bearings achieved excellent results; however, the generalization capability for mechanical components such as gears and engines needs further validation. The future direction is to apply MSCAE to HI extraction and RUL prediction of other mechanical components.

Author Contributions: Data curation, Q.Z.; Methodology, Z.Y.; Project administration, S.S., T.N. and Y.Z.; Resources, Q.Z.; Supervision, S.S., T.N. and Y.Z.; Validation, S.S.; Writing—original draft, Z.Y.; Writing—review & editing, Z.Y. and Q.Z. All authors have read and agreed to the published version of the manuscript.

Funding: This research was funded by Shaanxi Provincial Natural Science Foundation Project: 2022JQ-344 and Fourteenth Five-Year Plan Advance Research Project: 50902060401.

Data Availability Statement: Not applicable.

Conflicts of Interest: The authors declare no conflict of interest.

References

1. Ren, L.; Cui, J.; Sun, Y.; Cheng, X. Multi-bearing remaining useful life collaborative prediction: A deep learning approach. *J. Manuf. Syst.* **2017**, *43*, 248–256. [CrossRef]
2. Pan, Z.; Meng, Z.; Chen, Z.; Gao, W.; Shi, Y. A two-stage method based on extreme learning machine for predicting the remaining useful life of rolling-element bearings. *Mech. Syst. Signal Process.* **2020**, *144*, 106899. [CrossRef]
3. Gao, S.; Han, Q.; Zhou, N.; Pennacchi, P.; Chatterton, S.; Qing, T.; Zhang, J.; Chu, F. Experimental and theoretical approaches for determining cage motion dynamic characteristics of angular contact ball bearings considering whirling and overall skidding behaviors. *Mech. Syst. Signal Process.* **2022**, *168*, 108704. [CrossRef]
4. Ambrożkiewicz, B.; Syta, A.; Gassner, A.; Georgiadis, A.; Litak, G.; Meier, N. The influence of the radial internal clearance on the dynamic response of self-aligning ball bearings. *Mech. Syst. Signal Process.* **2022**, *171*, 108954. [CrossRef]
5. Zhang, Y.; Wang, J.; Du, H.; Yin, P. Error Evaluation of the Crown Profile of Logarithmic Generatrix Roller. *J. Phys. Conf. Ser.* **2021**, *1948*, 012065. [CrossRef]

6. Zhu, J.; Chen, N.; Peng, W. Estimation of Bearing Remaining Useful Life Based on Multiscale Convolutional Neural Network. *IEEE Trans. Ind. Electron.* **2019**, *66*, 3208–3216. [CrossRef]
7. Zhao, H.; Liu, H.; Jin, Y.; Dang, X.; Deng, W. Feature Extraction for Data-driven Remaining Useful Life Prediction of Rolling Bearings. *IEEE Trans. Instrum. Meas.* **2021**, *70*, 1–10. [CrossRef]
8. Deng, K.; Zhang, X.; Cheng, Y.; Zheng, Z.; Peng, J. A remaining useful life prediction method with long-short term feature processing for aircraft engines. *Appl. Soft Comput.* **2020**, *93*, 106344. [CrossRef]
9. Xiang, S.; Qin, Y.; Luo, J.; Pu, H.; Tang, B. Multicellular LSTM-based deep learning model for aero-engine remaining useful life prediction. *Reliab. Eng. Syst. Saf.* **2021**, *216*, 107927. [CrossRef]
10. Zeng, F.; Li, Y.; Jiang, Y.; Song, G. A deep attention residual neural network-based remaining useful life prediction of machinery. *Measurement* **2021**, *181*, 109642. [CrossRef]
11. Song, Y.; Gao, S.; Li, Y.; Jia, L.; Pang, F. Distributed Attention-Based Temporal Convolutional Network for Remaining Useful Life Prediction. *IEEE Internet Things J.* **2020**, *8*, 9594–9602. [CrossRef]
12. Chen, Z.; Wu, M.; Zhao, R.; Guretno, F.; Li, X. Machine Remaining Useful Life Prediction via an Attention Based Deep Learning Approach. *IEEE Trans. Ind. Electron.* **2020**, *68*, 2521–2531. [CrossRef]
13. Que, Z.; Jin, X.; Xu, Z. Remaining Useful Life Prediction for Bearings Based on a Gated Recurrent Unit. *IEEE Trans. Instrum. Meas.* **2021**, *70*, 1–11. [CrossRef]
14. Meng, Z.; Li, J.; Yin, N.; Pan, Z. Remaining useful life prediction of rolling bearing using fractal theory. *Measurement* **2020**, *156*, 107572. [CrossRef]
15. Wang, B.; Lei, Y.; Li, N.; Li, N. A Hybrid Prognostics Approach for Estimating Remaining Useful Life of Rolling Element Bearings. *IEEE Trans. Reliab.* **2018**, *69*, 1–12. [CrossRef]
16. Ren, L.; Sun, Y.; Cui, J.; Zhang, L. Bearing remaining useful life prediction based on deep autoencoder and deep neural networks. *J. Manuf. Syst.* **2018**, *48*, 71–77. [CrossRef]
17. Liu, L.; Song, X.; Chen, K.; Hou, B.; Ning, H. An enhanced encoder–decoder framework for bearing remaining useful life prediction. *Measurement* **2020**, *170*, 108753. [CrossRef]
18. Wang, B.; Lei, Y.; Li, N.; Wang, W. Multi-Scale Convolutional Attention Network for Predicting Remaining Useful Life of Machinery. *IEEE Trans. Ind. Electron.* **2020**, *68*, 7496–7504. [CrossRef]
19. Meng, M.; Zhu, M. Deep Convolution-based LSTM Network for Remaining Useful Life Prediction. *IEEE Trans. Ind. Inform.* **2020**, *17*, 1658–1667.
20. Cao, Y.; Ding, Y.; Jia, M.; Tian, R. A novel temporal convolutional network with residual self-attention mechanism for remaining useful life prediction of rolling bearings. *Reliab. Eng. Syst. Saf.* **2021**, *215*, 107813. [CrossRef]
21. Yao, D.; Li, B.; Liu, H.; Yang, J.; Jia, L. Remaining useful life prediction of roller bearings based on improved 1D-CNN and simple recurrent unit. *Measurement* **2021**, *175*, 109166. [CrossRef]
22. Hai, Q.; Lee, J.; Jing, L.; Gang, Y. Robust performance degradation assessment methods for enhanced rolling element bearing prognostics. *Adv. Eng. Inform.* **2003**, *17*, 127–140.
23. Qin, Y.; Chen, D.; Xiang, S.; Zhu, C. Gated Dual Attention Unit Neural Networks for Remaining Useful Life Prediction of Rolling Bearings. *IEEE Trans. Ind. Inform.* **2021**, *17*, 6438–6447. [CrossRef]
24. Chen, Y.; Peng, G.; Zhu, Z.; Li, S. A novel deep learning method based on attention mechanism for bearing remaining useful life prediction. *Appl. Soft Comput.* **2019**, *86*, 105919. [CrossRef]
25. Guo, L.; Lei, Y.; Li, N.; Yan, T.; Li, N. Machinery health indicator construction based on convolutional neural networks considering trend burr. *Neurocomputing* **2018**, *292*, 142–150. [CrossRef]
26. Li, N.; Lei, Y.; Lin, J.; Ding, S.X. An Improved Exponential Model for Predicting Remaining Useful Life of Rolling Element Bearings. *IEEE Trans. Ind. Electron.* **2015**, *62*, 7762–7773. [CrossRef]
27. Guo, L.; Li, N.; Jia, F.; Lei, Y.; Lin, J. A recurrent neural network based health indicator for remaining useful life prediction of bearings. *Neurocomputing* **2017**, *240*, 98–109. [CrossRef]
28. Chen, D.; Qin, Y.; Luo, J.; Xiang, S. Gated Adaptive Hierarchical Attention Unit Neural Networks for the Life Prediction of Servo Motors. *IEEE Trans. Ind. Electron.* **2022**, *69*, 9451–9461. [CrossRef]
29. Rumelhart, D.E.; McClelland, J.L. Learning Internal Representations by Error Propagation. In *Parallel Distributed Processing: Explorations in the Microstructure of Cognition: Foundations*; MIT Press: Cambridge, MA, USA, 1987; pp. 318–362.
30. Hochreiter, S.; Schmidhuber, J. Long Short-Term Memory. *Neural Comput.* **1997**, *9*, 1735–1780. [CrossRef]
31. Chen, D.; Qin, Y.; Wang, Y.; Zhou, J. Health indicator construction by quadratic function-based deep convolutional auto-encoder and its application into bearing RUL prediction. *ISA Trans.* **2021**, *114*, 44–56. [CrossRef]
32. Nectoux, P.; Gouriveau, R.; Medjaher, K.; Ramasso, E.; Varnier, C. PRONOSTIA: An experimental platform for bearings accelerated degradation tests. In Proceedings of the IEEE International Conference on Prognostics and Health Management, San Francisco, CA, USA, 17–20 June 2012.
33. Chen, L.; Xu, G.; Zhang, S.; Yan, W.; Wu, Q. Health indicator construction of machinery based on end-to-end trainable convolution recurrent neural networks. *J. Manuf. Syst.* **2020**, *54*, 1–11. [CrossRef]

applied sciences

MDPI

Article

Early Prediction of Remaining Useful Life for Rolling Bearings Based on Envelope Spectral Indicator and Bayesian Filter

Haobin Wen [1], Long Zhang [2] and Jyoti K. Sinha [1,*]

[1] Dynamics Lab, School of Engineering, The University of Manchester, Manchester M13 9PL, UK; haobin.wen@manchester.ac.uk

[2] The Department of Electrical and Electronic Engineering, The University of Manchester, Manchester M13 9PL, UK; long.zhang@manchester.ac.uk

* Correspondence: jyoti.sinha@manchester.ac.uk

Abstract: On top of the condition-based maintenance (CBM) practice for rotating machinery, the robust estimation of remaining useful life (RUL) for rolling-element bearings (REB) is of particular interest. The failure of a single bearing often results in secondary defects in the connected structure and catastrophic system failures. The prediction of RUL facilitates proactive maintenance planning to ensure system reliability and minimize financial loss due to unscheduled downtime. In this paper, to acquire early and reliable estimations of useful life, the RUL prediction of REBs is formulated into nonlinear degradation state estimation tackled by the combination of the envelope spectral indicator (ESI) and extended Kalman filter (EKF). By fusing the spectral energy of the bearing fault characteristic frequencies (FCFs) in the averaged envelope spectrum, the ESI is crafted to remove the interference from rotor-dynamics and reveal the bearing deterioration process. Once the fault is identified, the recursive Bayesian method based on EKF is utilized for estimating the bearing end-of-life time via the exponential state-space model. The distinctive advantage of the proposed approach lies in its ability to make an early prediction of RUL using a small number of ESI observations, offering an efficient practice for predictive health management at the early stage of bearing fault. The performance of the proposed method is validated using publicly available experimental bearing vibration data across three different operating conditions.

Keywords: predictive maintenance; remaining useful life; rolling-element bearings; degradation estimation; Bayesian methods; extended Kalman filter; nonlinear state estimation

1. Introduction

The reliability of equipment is crucial for smooth production processes in modern industry and particularly vital in smart manufacturing where mechanical systems are highly automated and integrated. Rolling-element bearings (REB) are critical structural components in a great variety of mechanical systems such as electric motors, gas turbines, aircraft engines, etc. Bearing faults are one of the root causes of secondary defects developed on connected structures [1,2], leading to potential safety hazards, unscheduled system downtime, significant maintenance cost and downtime loss. In the field of predictive maintenance, the estimation of remaining useful life (RUL) for bearings has attracted growing attention given its fundamental role in reducing systematic overhaul and preventing catastrophic failures through timely replacement.

The RUL of a piece of equipment is defined as the duration from the current time to the end of the service life [3]. The RUL of bearings can be regarded as the duration from the current inspection until functional failures occur. Different from the mean time to failure (MTTF) [4], which is a reliability measure representing the average time to replacement due to unrepairable failure, the bearing RUL is dynamically estimated (or a random variable) based on the current health condition and the predefined end-of-life

(EOL) threshold. Indeed, even under the same operating conditions, variations in bearing lifespan can be prominent due to manufacturing errors, mounting mistakes and differences in working conditions, as evidenced by a few public bearing lifetime experiments, e.g., the IMS data [5], the PROGNOSTIA data [6] and the XJTU-SY data [7]. Thereby, RUL prediction for individual bearings is significant for a system to operate at peak performance through the optimization of the maintenance schedule.

To date, researchers have made considerable efforts in the development of RUL prediction models for the predictive maintenance of REBs with promising performance in preliminary applications. In general, these models can be categorized into physics-based models and data-driven models.

The physics-based models rely on the carefully crafted full bearing life models based on the Hertz contact stress theory and the fatigue life limit, such as the Weibull fatigue life model, the Lundberg–Palmgren model and the Ioannides–Harris model, which associate the bearing fatigue life with the critical shear stress and the number of stress cycles to failure [8]. The 10-percent life or L10 life model widely used in the industry provides a reference for the basic rating life of different types of rolling bearings using the equivalent dynamic loads, as specified in the international standard ISO 281:2007 [9]. Specifically, researchers from the US National Aeronautics and Space Administration presented a technical tutorial outlining the theoretical variations and applications of various bearing life prediction models [10]. These models provide important references to determine bearing life and reliability, while the complexity of failure dynamics and individual differences may still undermine the practicability of the RUL estimations in a dynamical fashion. For example, as pointed out by Zaretsky, the use of inconsistent fatigue limit could lead to the over-prediction of bearing life [10].

On the other hand, data-driven models attempt to establish an RUL predictor using statistical (or) machine learning (ML) approaches based on historical condition monitoring (CM) measurements, such as temperature data, oil debris data, vibration data, acoustic emission data, etc. In ML regression, the bearing RUL is regarded as the response variable of a linear or nonlinear system based on the CM data input, where dominant features are recognized to efficiently map the input data to the target RUL. Examples of recent applications of ML models to bearing RUL predictions include support vector regression [11], random forest regression [12] and deep learning networks with sophisticated architectures [13]. ML models have been shown to facilitate a high level of prediction accuracy, while the performance may not be guaranteed without a substantial volume of full lifetime data, which is often difficult to obtain. Alternatively, Bayesian statistical approaches are another appealing branch within the realm of data-driven models given their superiority in integrating prior or domain information; they are free of reliance on massive quantities of historical data and also interpretable with uncertainty quantification. The RUL of a bearing can be viewed as a random variable depending on the current health conditions observed from the CM measurements. In this sense, the probability density of the RUL conditioned on historical observations becomes the main interest of the statistical model [3], where Bayesian inference is often adopted to obtain the posterior estimates of the RUL.

Recently, stochastic processes [14,15] and recursive Bayesian filters [16,17] have demonstrated promising performance in RUL estimation and offer a flexible framework for incorporating prior knowledge and updating beliefs based on new observations. The Kalman filter (KF) and its variants [18,19], the extended Kalman filter (EKF) [20] and unscented Kalman filter (UKF) [21], are efficient state estimation tools for dynamical systems given inaccurate models and noisy observations and have been frequently utilized in various industrial applications, such as the tracking of satellite trajectories [22] and the positioning of autonomous vehicles [23]. In the case of bearing condition monitoring, the latent degradation level is the target of state estimation behind the observable CM measurements, such as vibration acceleration signals, or the bearing health indicators (HI) obtained from certain feature engineering techniques. For example, Li et al. established the Wiener process-based exponential model for bearing degradation and applied the KF for removing random noise in the RUL prediction results [15]. According to [19], the KF is endowed with the capability of providing the optimal state estimation under the Gaussian noise assumption, which minimizes the mean square error between the true and the estimated state. Still, the effectiveness of KF can be limited due to the linear assumption of the system dynamics. The EKF generalizes the KF to handle nonlinear system dynamics by introducing local linearization based on Taylor's expansion. Considering the complexity of the bearing degradation process, Singleton et al. applied EKF to predict the RUL of bearings based on the entropy level of the time-frequency distribution and the root-mean-square (RMS) value of vibration signals [16]. Two distinct exponential state-space models are hypothesized and compared to represent the trajectories of the health indicators and estimate the true health state. The applications to the PROGNOSTIA data also reveal the efficiency of EKF for state estimation, where convergent prediction of RUL is achieved as the bearing approaches the end of its lifespan [16].

Despite notable progress, some challenges persist in the realm of bearing RUL prediction. As with the importance of feature extraction to robust fault diagnosis [24,25], feature extraction is one crucial step that affects the reliability of the RUL prediction. Conventional bearing health indicators based on statistical measures of vibration signals may encounter interference components arising from synchronous speed vibrations of the connecting structure, such as a faulty rotor with misalignment. These interfering components could greatly undermine the robustness of an RUL prediction. In addition, despite the flexibility of the Bayesian filters, their sensitivity to prior assumptions and initial states often lead to a large estimation error within the early predictions of RUL, which is less favoured in engineering practice and should be further improved.

To address the challenges in RUL prediction with limited historical data, the Bayesian filter is leveraged, combining the benefits of physics-based models and data-driven approaches by the integration of domain knowledge about bearing fault and the uncertainty estimation. In this study, the RUL prediction approach for rolling-element bearings is presented based on a combination of the envelope spectral indicator (ESI) and the extended Kalman filter. First, the ESI is proposed as the bearing degradation indicator such that it is solely related to the bearing faults. The ESI is derived from the spectral amplitude of the bearing fault characteristic frequencies (FCFs) in the averaged envelope spectrum (AES), which is computed via multiple envelope signal segments for the compression of spurious frequency peaks. By isolating the bearing defect components present in the vibration data, the ESI is more aligned with the actual degradation state of the bearing. Also, the ESI allows for the exclusion of unwanted factors such as rotor dynamics and faulty components linked to shafts and gears, which fosters a more realistic prediction of the bearing RUL. Secondly, the incipient fault in the bearing is detected based on the envelope analysis to facilitate the determination of the first detection time (FDT) of a fault, marking the start of the bearing degradation stage. Once the fault is detected, the EKF is leveraged to predict the bearing end-of-life time as early as possible given only a few ESI observations. By converting the RUL prediction to a state estimation problem, the simplified exponential state-space model is introduced to represent the acceleratingly incremental trajectory of the bearing degradation state, which is observed through the ESI. The EKF-based RUL prediction technique is evaluated on the public experimental bearing vibration data in [7]. The applications to the test bearings under various working conditions further verify the effectiveness of the proposed method in the early detection of bearing fault and the timely prediction of RUL. The main contributions of this study are summarized as follows:

- It determines the first detection time of fault for life prediction based on envelope spectrum inspection for the XJTU-SY dataset.
- The bearing degradation indicator, ESI, is utilized for more realistic RUL prediction.
- Bearing RUL is predicted at the early fault stage based on EKF state estimation.

The rest of the paper is organized as follows: Section 2 discusses the settings of the run-to-failure experiments on the publicly available bearing vibration data. Section 3 presents the overall methodology for bearing RUL prediction based on the combination of the envelope spectral indicator and the EKF, where detailed explanations on the development of the ESI and EKF are elaborated. The applications of the proposed methods and observations of the results are discussed in Section 4. Finally, the discussion and the concluding remarks are provided in Sections 5 and 6.

2. Experimental Settings

To develop data-driven models for the RUL prediction of rolling bearings, the publicly available run-to-failure dataset provided by the Institute of Design Science and Basic Component at the Xi'an Jiaotong University and the Changxing Sumyoung Technology Co., Ltd., the XJTU-SY dataset [7], is further utilized in this study. This dataset is chosen given its coverage of most bearing failure modes and multiple working conditions among the accelerated run-to-failure experiments.

The schematic diagram in Figure 1 shows the overall setup of the bearing rig in [7], where the test bearing is installed at the end of the shaft and loaded from the horizontal direction using the hydraulic loading system. From the healthy condition to the failure threshold specified as ten times the acceleration amplitude of the healthy stage, test bearings operating under three speed and loading conditions are available for analysis. For reference, the specifications of the analyzed data are provided in Table 1. Further details on the bearing data can be referred to in [7].

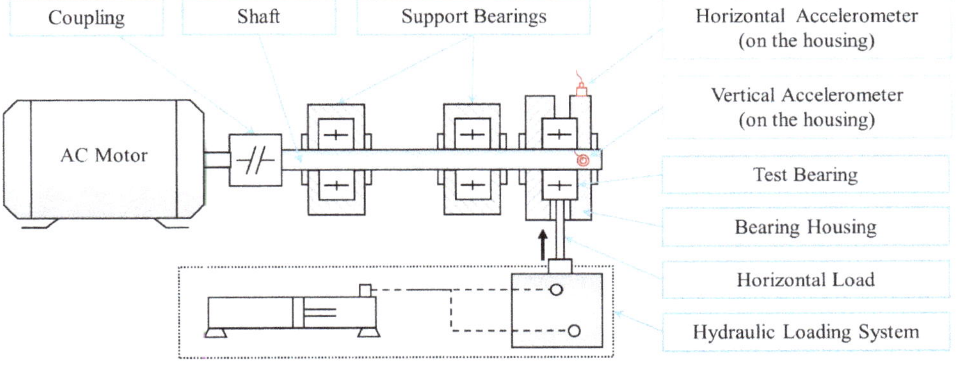

Figure 1. The schematic diagram of the experimental rig (reorganized from [7]).

Table 1. Specifications of the analyzed bearing data (reorganized from [7]).

Sampling Settings	Operating Condition (Speed/Load)	Bearing ID	Life Duration/min	Defect Components *
Rate: 25.6 kHz Length: 1.28 s/min	C1: 35Hz/12kN	1	161	Outer-Race
		2	158	Outer-Race
	C2: 37.5Hz/11kN	3	491	Inner-Race
		4	161	Outer-Race
	C3: 40Hz/10kN	5	371	Inner-Race
		6	1515	Inner-Race

* The fault type is revealed after the end of each experiment.

3. RUL Prediction Methodology

In this section, the overall methodology for the RUL prediction of bearings is presented. First, the envelope analysis with the averaging operation is introduced for the robust identification of bearing faults, based on which the envelope spectral indicator (ESI) can be formulated to reveal the bearing deterioration condition. Once a fault is detected in the bearing, the Bayesian filtering approach based on the extended Kalman filter is further utilized for RUL estimation at the early degradation stage.

3.1. Envelope Analysis for Bearing Fault Detection

3.1.1. Signal Pre-Processing: High-Pass Filtering

Signal processing and feature extraction play a crucial role in the reliability of data-driven RUL prediction. In vibration-based condition monitoring, the acceleration signals associated with the localized bearing faults tend to exhibit a cyclic pattern of high-frequency impulse responses occurring at resonance frequencies. These responses are induced by the series of quasi-periodic impacts generated by the moving components passing the fault [26], thereby dominating the high-frequency range. To mitigate the influence of low-frequency components, the high-pass filter is applied to the vibration acceleration signal, with the cut-off frequency at 1 kHz.

3.1.2. Averaged Envelope Spectrum

In general, the vibration acceleration signal of faulty bearings can be characterized as a series of impulsive components recurring at the fault characteristic frequencies (FCFs), wherein it can be modelled as the amplitude modulation on the resonance responses of bearing faults [26,27]. Given a real-valued time-domain acceleration signal $a(t)$, one way to extract the repetition components related to the bearing fault is to compute the envelope of its complex analytic signal, $a_c(t)$,

$$a_c(t) = a(t) + \widetilde{a}(t)j, \tag{1}$$

where $\widetilde{a}(t)$ denotes the Hilbert transform of $a(t)$ [27] and j is a complex imaginary unit. The envelope signal is defined as the magnitude of the complex analytic signal, which is given by

$$e(t) = \left|a(t) + \widetilde{a}(t)j\right| = \sqrt{a(t)^2 + \widetilde{a}(t)^2}. \tag{2}$$

After the high-pass filtering, rather than directly calculating the Fourier spectrum of the envelope signal, the operation of power spectral density (PSD) is performed based on multiple segments of the original envelope with a selected degree of overlapping. This process results in the averaged envelope spectrum (AES) which attempts to suppress the interference of spurious frequency peaks and random noise. As indicated by Sinha [27], due to the randomness in choosing the start point and length of the signal segments, by introducing a segmentation and averaging process, the averaged spectrum computed via PSD can be closer to the real frequency peaks within the signal. The averaged envelope spectrum $S_e(f_k)$ based on the PSD of the envelope signal $e(t)$ can be written as

$$S_e(f_k) = E[S(f_k)S^*(f_k)] = \frac{\sum_{i=1}^{N_s} S_i(f_k)S_i^*(f_k)}{N_s}, \tag{3}$$

where N_s is the total number of segments of the envelope signal $e(t)$, $S_i(f_k)$ denotes the Fourier transform of the i-th envelope segment at the frequency location $f_k(0 \leq f_k \leq f_s/2)$, and $S_i^*(f_k)$ is accordingly the complex conjugate of $S_i(f_k)$. $k = 0, 1, \ldots, L/2-1$ denotes the index of the frequency bin for the L-point Fourier transform for each segment and f_s the sampling frequency.

3.1.3. Bearing FCFs for Fault Identification

Based on the envelope analysis, the modulating components that carry the bearing defect information can be extracted from the acceleration signal. Also, the fault type can be efficiently recognized based on the identification of FCF within the averaged envelope spectrum. Specifically, the FCFs of different bearing components, respectively, the ball-pass frequency of inner-race fault (BPFI), the ball-pass frequency of outer-race fault (BPFO), the ball (or roller) spin frequency (BSF) and the fundamental train frequency (FTF, or the cage frequency) can be computed based on the bearing dimensions and the operating speed, as in [27]. As a reference, for the test bearings utilized in the XJTU-SY data [7], the bearing dimensions and their FCFs under different operating conditions are listed in Table 2.

Table 2. Specifications of test bearings and FCFs.

	Number of Balls	**Inner Diameter**	**Outer Diameter**	**Contact Angle**
Dimension	8	29.30 mm	39.80 mm	0°
FCF	**FTF (Hz)**	**BSF (Hz)**	**BPFO (Hz)**	**BPFI (Hz)**
Condition 1	13.49	72.33	107.91	172.09
Condition 2	14.45	77.50	115.62	184.38
Condition 3	15.42	82.66	123.32	196.68

The spectral patterns of the envelope spectrum can differ due to the varying formations of amplitude modulation in different bearing faults. Based on the presence of the FCFs in the envelope spectrum, different types of bearing faults can be identified. A schematic illustration of the envelope spectrum is shown in Figure 2. For example, the impact stroke by the inner-race fault will be modulated by the shaft speed due to the fault entering and exiting the load zone within one shaft revolution. Thus, the BPFI and its higher-order harmonics are accompanied by a series of side bands at the integer multiples of the shaft speed. For roller-element faults, the even harmonics of BSF dominates the envelope spectrum as the fault strikes both inner and outer races in one ball spin [26]; meanwhile, it is modulated by the cage speed (FTF). On the above basis, spectral inspection is conducted on the AES for robust detection and identification of early bearing faults.

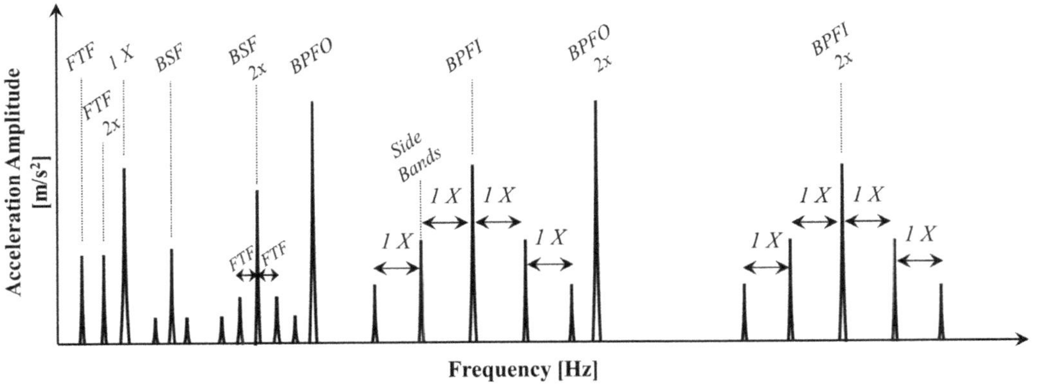

Figure 2. A typical example of the envelope spectrum with multiple bearing FCFs.

3.2. The Envelope Spectral Indicator

Accurate and reliable health indicators play a vital role in uncovering the health state of rolling bearings for reliable life prediction. Since creating a direct indicator specifically for assessing the bearing life may be challenging, commonly used HIs rely on physical or statistical quantities to evaluate the machine condition, such as the impulsiveness measure and the spectral energy of vibration signals. However, given the intricate nature of rotor-bearing systems, traditional HIs may contain extraneous components that are unrelated to bearing faults, which could compromise their reliability in RUL prediction. For example, synchronous speed vibration with rotor imbalance and misalignment due to manufacturing and assembly errors introduces extra components to the bearing vibration. Furthermore, as the bearing deteriorates, the vibration response can be more complex when compound defects develop on the connected structure.

In this context, in order to ensure that the chosen HI accurately represents the deteriorating condition of rolling bearings, the envelope spectral indicator (ESI) is developed based on the harmonics of the bearing FCFs in the average envelope spectrum. As a result, the rotor dynamics are eliminated, leaving only the components related to bearing faults. Consequently, the subsequent prediction of RUL may become more realistic since the presence of gear or rotor faults is not taken into account. Specifically, as a discrete time stochastic process, the ESI at time step t is given by

$$ESI(t) = \sum_{n=1}^{N} \left[S_e^{(t)}(nf_{FTF}) + S_e^{(t)}(nf_{BSF}) + S_e^{(t)}(nf_{BPFO}) + S_e^{(t)}(nf_{BPFI}) \right], \quad (4)$$

where n represents the order of the FCF harmonics and $S_e^{(t)}(f)$ denotes the acceleration amplitude of the average envelope spectrum of time t at the specific frequency location. Typically, the diagnostic informative components of the envelope spectrum concentrate within the low-frequency range. Also, it is observable (from Table 2) that for the analyzed test bearings, the third multiple of BSF is close to the second harmonic of BPFO. In this work, to ensure the reliability of the bearing health indicator, the FCF harmonics are utilized up to the third multiple, i.e., $N = 3$.

3.3. *First Detection Time, End-of-Life Time and RUL Prediction Time*

3.3.1. First Detection Time (FDT) of Fault

Based on the fault occurrence time, the bearing health condition during the whole lifespan can be divided into two stages: the healthy stage and the degradation stage. The change point of the bearing health state is the earliest time when the bearing fault is detected, which is defined as the first detection time (FDT) t_{FDT} and determined through identifying the FCFs based on the envelope analysis in Section 3.1.3.

3.3.2. End-of-Life (EOL) Time and EOL Threshold

As can be concluded from multiple endurance experiments on REBs [5–7], bearings operated under identical loading conditions could end up with different failure modes, either single or compound faults, and the individual lifespans of the same type of bearings could vary greatly due to multiple factors such as manufacturing and mounting errors. For this reason, the determination of the accurate EOL threshold for bearing health indicators is one of the key challenges in RUL prediction.

Assume that the spectral energy of the FCF components accumulates to the highest level when functional failures occur, i.e., the replacement should be made; the EOL time of the rolling bearing t_{EOL} can be selected as the time when the extracted ESI reaches the maximum,

$$t_{EOL} = \underset{t}{\text{argmax}} ESI(t), \quad (5)$$

Empirically, the EOL threshold for RUL prediction is calculated based on the average of the maximum ESI values computed from each set of test bearing data across the lifespan [7], as given by

$$\overline{\gamma} = \frac{1}{J} \sum_{j}^{J} ESI_j(t_{EOL}), \quad (6)$$

where $ESI_j(t_{EOL})$ denotes the ESI amplitude at the end-of-life time of the j-th bearing and J the total number of available test bearings.

3.3.3. RUL Prediction Time (PT)

As the RUL is a dynamic variable dependent on the current health condition, the prediction of RUL at the healthy stage may be untenable. On the other hand, the rate of fault deterioration is not known in advance and is often increasingly fast due to the accumulation of historical defects. To allow the optimal predictive maintenance planning, the inspection window between the FDT (t_{FDT}) and the PT of RUL (t_{PT}) is expected to be as short as possible to enable timely prediction of RUL at the early degradation stage.

On the above basis, a typical example of a bearing degradation process with respect to the machine operation time based on the health indicator is shown in Figure 3. The HI is assumed to start from a relatively stable plateau at the healthy stage and step up to enter the degradation stage once the first fault event is detected. Then, it further increases with the accumulation of faults and potential failures. As defined in Equation (6), the bearing is thought to suffer from functional failures when the HI first passes the average EOL threshold $\overline{\gamma}$ at time t_{EOL}. In the case of ESI, the threshold $\overline{\gamma}$ is determined based on the mean maximal magnitude of all available historical ESI trajectories, each of which extracts the spectral energy of the bearing FCF (defect) components in the averaged envelope spectrum. It is considered that the selected $\overline{\gamma}$ signifies the utmost extent of bearing deterioration, reaching a critical point where immediate replacement is required. Thereby, the threshold also facilitates the determination of the end-of-life time for the RUL prediction in practical applications.

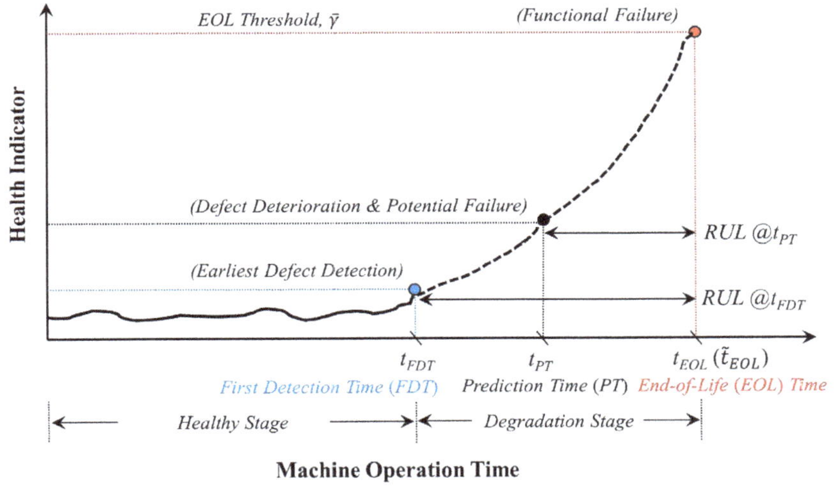

Figure 3. An example of bearing degradation process. Bearing health indicator vs. operation time. The bearing lifetime is typically divided into the healthy stage and the degradation stage.

3.4. Bayesian RUL Prediction Based on Extended Kalman Filter

3.4.1. Definition of Bearing RUL

Based on the illustration in Figure 3, the RUL of bearings is a time-dependent variable defined as the length from the current time t to the end-of-life time t_{EOL} jointly determined by the selected health indicator, ESI, and the EOL threshold $\overline{\gamma}$,

$$rul_t = t_{EOL} - t \; (with \; t_{EOL} \geq t). \tag{7}$$

In practical prediction of RUL, the estimated EOL time $\tilde{t}_{EOL}$ is extrapolated as the time when the health indicator reaches the EOL threshold. Hence, let x_t be the state of bearing degradation at time t; the estimated RUL at time t is derived as the interval from t to the estimated EOL time $\tilde{t}_{EOL}$, or the inferior limit of ΔT that drives the degradation state to exceed the threshold $\overline{\gamma}$, i.e.,

$$\widetilde{rul}_t = \tilde{t}_{EOL} - t = inf\{\Delta T : x_{t+\Delta T} \geq \overline{\gamma}\}. \tag{8}$$

3.4.2. Extended Kalman Filter

Kalman filtering is one of the Bayesian filtering approaches commonly used for state estimation, the problem of which consists of estimating the state vector x that describes the true behaviours of a dynamical system based on limited and noisy observations z which are often imperfect or even unknown due to the inaccuracy of sensor measurements and the restrictions of data acquisition [22]. The Kalman filter (KF) has been applied to many engineering applications such as target tracking [28], navigation [29] and system identification [30]. Recently, the KF has also been introduced to the field of equipment condition monitoring and predictive maintenance. For example, Batzel et al. proposed the RUL prediction model for aircraft power generators based on the Kalman filter by assuming the time-dependent state-space model to relate the estimated state with the RUL [31]. However, as the degradation of machine equipment often involves nonlinear dynamics, the performance of the KF may be restricted by the assumption of a linear system and Gaussian noise. The extended Kalman filter (EKF) is one of the generalizations of the Kalman filter to nonlinear state estimation and has been applied by Sepasi et al. to the health management of the Li-ion battery pack for the state of charge estimation [32]. In bearing health management, Singleton et al. used the EKF for estimating the RUL of rolling bearings based on the exponential state model assumed to govern the trends of the statistical indicators of the vibration signals [16]. In EKF, the state being estimated is assumed to follow a nonlinear discrete-time stochastic process [19],

$$\boldsymbol{x}_t = f(\boldsymbol{x}_{t-1}, \boldsymbol{u}_{t-1}, \boldsymbol{v}_{t-1}), \tag{9}$$

where $\boldsymbol{x}_t$ and $\boldsymbol{u}_t$ are the state vector and the input at the time step t, f is the nonlinear state transition function and $\boldsymbol{v}_t$ is the random process noise represented by the Gaussian process with zero mean and covariance matrix $\boldsymbol{Q}_t$. The sensor measurements can be modelled by

$$\boldsymbol{z}_t = h(\boldsymbol{x}_t, \boldsymbol{w}_t), \tag{10}$$

where $\boldsymbol{z}_t$ is the measurement vector at time step t, h is the nonlinear transition function of measurements relating $\boldsymbol{z}_t$ with the system state $\boldsymbol{x}_t$, and $\boldsymbol{w}_t$ denotes the measurement noise with zero mean and covariance matrix $\boldsymbol{R}_t$. Specifically, $\boldsymbol{v}_t$ and $\boldsymbol{w}_t$ are the uncorrelated processes of white noises defined as follows [33],

$$\boldsymbol{v}_t \sim N(\boldsymbol{0}, \boldsymbol{Q}_t), with\ E[\boldsymbol{v}_t] = \boldsymbol{0}, and\ E\left[\boldsymbol{v}_t\boldsymbol{v}_{t'}^T\right] = \boldsymbol{Q}_t\delta_t, \tag{11}$$

$$\boldsymbol{w}_t \sim N(\boldsymbol{0}, \boldsymbol{R}_t), with\ E[\boldsymbol{w}_t] = \boldsymbol{0}, and\ E\left[\boldsymbol{w}_t\boldsymbol{w}_{t'}^T\right] = \boldsymbol{R}_t\delta_t, \tag{12}$$

$$E\left[\boldsymbol{w}_t\boldsymbol{v}_{t'}^T\right] = \boldsymbol{0}, \tag{13}$$

where δ_t denotes the Kronecker delta function which is equal to 1 only if $t = t'$, and 0 otherwise.

Assuming a sufficient description of the system's nonlinearity via local linearization, the EKF approximates the nonlinear state model f and the measurement model h around the current estimate based on the first-order Taylor series expansion. Thus, the state vector and measurement vector are rewritten as

$$\boldsymbol{x}_t \approx \widetilde{\boldsymbol{x}}_{t|t-1} + \boldsymbol{F}_t\left(\boldsymbol{x}_{t-1} - \widetilde{\boldsymbol{x}}_{t-1|t-1}\right) + \boldsymbol{G}_t\boldsymbol{v}_{t-1}, \tag{14}$$

$$\boldsymbol{z}_t \approx \widetilde{\boldsymbol{z}}_{t|t-1} + \boldsymbol{H}_t\left(\boldsymbol{x}_t - \widetilde{\boldsymbol{x}}_{t|t-1}\right) + \boldsymbol{W}_t\boldsymbol{w}_t, \tag{15}$$

where $\widetilde{\boldsymbol{x}}_{t|t-1}$ and $\widetilde{\boldsymbol{z}}_{t|t-1}$ are, respectively, the noise-free a priori estimate of the state and measurement at time t [19],

$$\widetilde{\boldsymbol{x}}_{t|t-1} = f\left(\widetilde{\boldsymbol{x}}_{t-1|t-1}, \boldsymbol{u}_t, \boldsymbol{0}\right), \tag{16}$$

$$\tilde{z}_{t|t-1} = h\left(\tilde{x}_{t|t-1}, \mathbf{0}\right). \tag{17}$$

F_t, G_t, H_t, and W_t are the Jacobian matrices of the partial derivatives with respect to state vector x, noise v and w, respectively, defined as follows:

$$F_t = \nabla f_x = \left.\frac{\partial f(x,u,v)}{\partial x}\right|_{(\tilde{x}_{t|t}, u_t, 0)}, \tag{18}$$

$$G_t = \nabla f_v = \left.\frac{\partial f(x,u,v)}{\partial v}\right|_{(\tilde{x}_{t|t}, u_t, 0)}, \tag{19}$$

$$H_t = \nabla h_x = \left.\frac{\partial h(x,w)}{\partial x}\right|_{(\tilde{x}_{t|t-1}, 0)}, \tag{20}$$

$$W_t = \nabla h_w = \left.\frac{\partial h(x,w)}{\partial w}\right|_{(\tilde{x}_{t|t-1}, 0)}. \tag{21}$$

Based on the difference between the actual measurements z_t and the a priori prediction of measurement $\tilde{z}_t$, the prediction error of the measurement is obtained and scaled by the filter gain K_t before adding back to correct the a priori state estimation, which gives an a posteriori estimate of state.

$$\tilde{x}_{t|t} = \tilde{x}_{t|t-1} + K_t\left(z_t - \tilde{z}_{t|t-1}\right), \tag{22}$$

where the Kalman gain K_t serves to balance the state model belief and the measurement and is chosen based on the optimality criteria regarding the stochastic nature of the state process and the measurement dynamics [19].

3.4.3. State Process Model for Bearing Degradation Estimation

Due to the accelerating deterioration of the bearing as defects accumulate, the behaviour of the bearing health indicator is often approximated as an exponential function of time. For example, in the work of [16], $a_t e^{b_t \cdot t}$ is used for modelling the variance of vibration signals during the bearing lifespan and was shown to be more suitable than the form $a_t + b_t e^{c_t \cdot t}$ for RUL prediction. Note that the model parameters (i.e., a_t, b_t, and c_t) are time-variant and need to be updated dynamically in the EKF. To reduce the need for prior assumptions on the initial conditions, the bearing degradation model with the Morkovian property is formulated as

$$x_t = f(x_{t-1}) = e^{b_t \Delta t} x_{t-1}, \tag{23}$$

where x_t denotes the bearing degradation state indirectly observed via the proposed ESI. This can be regarded as a simplification of Singleton's model [16] by fixing the parameter a_t, leading to a multiplication relationship between two consecutive degradation levels x_{t-1} and x_t, as $e^{b_t \cdot \Delta t}$ is equivalent to the form $\frac{e^{b_t t}}{e^{b_{t-1}(t-1)}}$ with $\Delta t = 1$.

Assume that the ESI since the FDT follows an accelerated incremental trend before reaching the end-of-life threshold; the true bearing degradation level can be described by the state vector x_t, and the measurements z_t are the available historical observations of ESI at the prediction time. The state process model for EKF can be expressed as

$$x_t = f(x_{t-1}, v_t) = \begin{pmatrix} e^{b_{t-1}\Delta t} x_{t-1}(1) + v_t \\ b_{t-1} \end{pmatrix}, \tag{24}$$

where the interval between two adjacent time steps is one minute ($\Delta t = 1$) and the input u_t is omitted as there is no active control. Also, the measurement model for the ESI observations is given by

$$z_t = h(x_t, w_t) = x_t(1) + w_t. \tag{25}$$

The local linearization of the state model and the measurement model is achieved based on the Jacobian matrices given in Equation (18) to (21), which writes

$$F_t = \nabla f_x = \begin{pmatrix} e^{b_{t-1}\Delta t} & e^{b_{t-1}\Delta t} x_{t-1}(1) \\ 0 & 1 \end{pmatrix}, \tag{26}$$

$$G_t = \nabla f_v = \begin{pmatrix} 1 \\ 0 \end{pmatrix}, \tag{27}$$

$$H_t = \nabla h_x = \begin{pmatrix} 1 & 0 \end{pmatrix}, \tag{28}$$

$$W_t = \nabla h_w = 1. \tag{29}$$

3.4.4. EKF-Based RUL Prediction Algorithm

The EKF-based RUL estimation algorithm using the available ESI observations at the prediction time (t_{PT}) can be summarized as the following steps:

Step 1: Initialization. It is assumed that the initial state x_0 can be taken from the first few observations of the ESI. The covariance matrix of the state process error $\widetilde{P}_0$ is chosen empirically, as given by

$$\widetilde{x}_0 = \begin{pmatrix} x_0 \\ b_0 \end{pmatrix} = \begin{pmatrix} z_0 \\ \ln\left(\frac{ESI(t_{PT})}{ESI(t_{PT}-1)}\right) \end{pmatrix}, \tag{30}$$

$$\widetilde{P}_0 = \begin{pmatrix} 0.1 & 0 \\ 0 & 0.1 \end{pmatrix}, \tag{31}$$

where z_0 takes the ESI at the fault detection time, $ESI(t_{FDT})$. In this case, the covariance matrices of the state process noise and the measurement noise are scalar and set to $Q_t = \sigma_v^2$, $R_t = \sigma_w^2$.

Step 2: Prediction. Obtain the noise-free a priori estimate of the state vector $\widetilde{x}_{t|t-1}$, the measurement $\widetilde{z}_{t|t-1}$ and the covariance matrix of state error $\widetilde{P}_{t|t-1}$ based on the previous time step,

$$\widetilde{x}_{t|t-1} = f\left(\widetilde{x}_{t-1|t-1}, \mathbf{0}\right), \tag{32}$$

$$\widetilde{z}_{t|t-1} = h\left(\widetilde{x}_{t|t-1}, \mathbf{0}\right), \tag{33}$$

$$\widetilde{P}_{t|t-1} = F_{t-1} P_{t-1|t-1} F_{t-1} + G_{t-1} Q_{t-1} G_{t-1}^T, \tag{34}$$

Step 3: Update. Compute the filter gain K_t by

$$K_t = \widetilde{P}_{t|t-1} H_t^T \left(H_t \widetilde{P}_{t|t-1} H_t^T + W_t R_t W_t^T\right)^{-1}. \tag{35}$$

Subsequently, take in the observations of the bearing health indicator, ESI, as the actual measurement z_t to calculate the measurement prediction error $z_t - \widetilde{z}_{t|t-1}$, and update the prior state estimation to attain the a posteriori state estimation $\widetilde{x}_{t|t}$,

$$\widetilde{x}_{t|t} = \widetilde{x}_{t|t-1} + K_t\left(z_t - h\left(\widetilde{x}_{t|t-1}, \mathbf{0}\right)\right). \tag{36}$$

The a posteriori estimate for the state error covariance matrix $\widetilde{P}_{t|t}$ can be updated as given by

$$\widetilde{P}_{t|t} = (I - K_t H_t)\widetilde{P}_{t|t-1}. \tag{37}$$

For steps $t = 1, \ldots, (t_{PT} - t_{FDT-n})$ corresponding to time stamps $t_{FDT-n}, t_{FDT-n+1}, \ldots, t_{PT}$, the prediction step and the update step are recursively implemented until the last available observation of the ESI. n denotes the number of observations prior to t_{FDT} taken in the EKF for the stability of parameter estimation.

Step 4: RUL estimation. Finally, the RUL at the prediction time t_{PT}, $rul_{t_{PT}}$, can be extrapolated based on the end-of-life (EOL) threshold, $\overline{\gamma}$, according to the definition in Equations (7) and (8),

$$\widetilde{rul}_{t_{PT}} = \inf\left\{\Delta T : f\left(\widetilde{x}_{t_{PT}+\Delta T-1}\right) \geq \overline{\gamma}\right\} = \left(\ln \frac{\overline{\gamma}}{\widetilde{x}_{t_{PT}}(1)}\right) / b_{t_{PT}}. \quad (38)$$

Step 5: Uncertainty Estimation. Compute the 95% credible interval (CI) of the predicted RUL based on Section 3.4.5.

3.4.5. Uncertainty Estimation

Estimating the uncertainty of the predicted RUL plays an important role in the evaluation of prediction reliability. Based on the Gaussian assumption of the state process noise and the measurement noise, the confidence level of the RUL prediction can be calculated based on the a posteriori estimate of the covariance matrix for state process error, $\widetilde{\boldsymbol{P}}_{t|t}$, which is updated based on the filter gain as in (37). As the first diagonal element of $\widetilde{\boldsymbol{P}}_{t|t}$ is the variance of the estimated state at each time step, the 95% credible interval of state can be determined as follows

$$\widetilde{x}_{t,lower} = \widetilde{\boldsymbol{x}}_{t|t}(1) - 1.96\sqrt{\widetilde{\boldsymbol{P}}_{t|t}(1,1)}. \quad (39)$$

$$\widetilde{x}_{t,upper} = \widetilde{\boldsymbol{x}}_{t|t}(1) + 1.96\sqrt{\widetilde{\boldsymbol{P}}_{t|t}(1,1)}. \quad (40)$$

Substituting (39) and (40) to (38), the lower bound and the upper bound of the estimated RUL can be analogously extrapolated based on the end-of-life threshold.

For implementation reference, the pseudo-code of the EKF-based RUL estimation algorithm is shown in Table 3.

Table 3. The EKF-based RUL estimation algorithm.

Input: The ESI observations, $\boldsymbol{z}_{t_{FDT-n}:t_{PT}}$, EOL threshold, $\overline{\gamma}$, RUL prediction time t_{PT}, state process noise variance, σ_v^2, and the measurement noise variance, σ_w^2
Initialization: State vector $\widetilde{\boldsymbol{x}}_0$, covariance of state process error, $\widetilde{\boldsymbol{P}}_0$
For steps $t = 1, \ldots, (t_{PT} - t_{FDT-n})$, if $x_t < \overline{\gamma}$, implement the following: 1) Predict a prior estimate of state $\widetilde{\boldsymbol{x}}_{t\|t-1}$ by (32), measurement $\widetilde{\boldsymbol{z}}_{t\|t-1}$ by (33), and the covariance of state process error $\widetilde{\boldsymbol{P}}_{t\|t-1}$ by (34). 2) The extended Kalman filter gain $\boldsymbol{K}_t$ is computed by (35). 3) Obtain ESI observation at time step t, $\boldsymbol{z}_t$, to calculate measurement prediction error $\left(z_t - \widetilde{z}_{t\|t-1}\right)$, and update the posterior estimations of state $\widetilde{\boldsymbol{x}}_{t\|t}$ by (36) and covariance matrix $\widetilde{\boldsymbol{P}}_{t\|t}$ by (37). 4) $t = t + 1$, repeat from 1). Extrapolate the RUL at the prediction time $\widetilde{rul}_{t_{PT}}$ based on the state transition model via (38). Compute the 95% credible interval of the estimated RUL based on analogous extrapolation by (39) and (40).
Output: RUL estimates at the prediction time $\widetilde{rul}_{t_{PT}}$ with its credible interval.

On the above basis, the overall procedure for the proposed methodology for the RUL prediction of rolling bearings can be summarized in Figure 4.

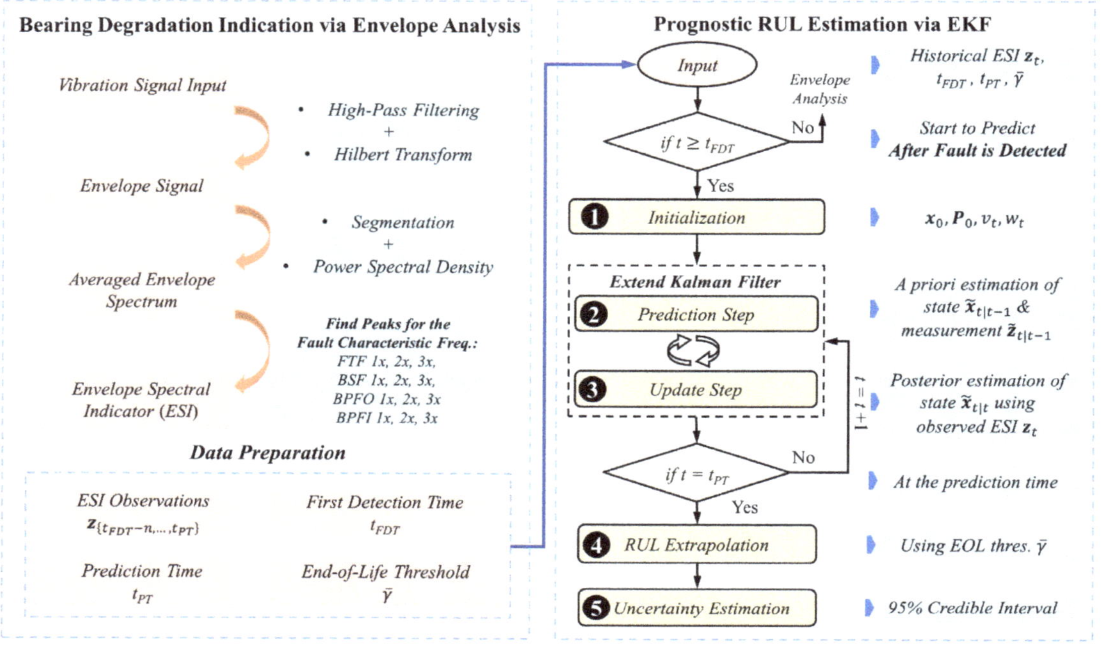

Figure 4. The overall procedure of the EKF-based RUL prediction method.

4. Results and Observations

4.1. Fault Detection Based on Averaged Envelope Spectrum

As the radial force of test bearing is imposed with a hydraulic loading from the horizontal direction, the raw vibration signals of the horizontal accelerometers are utilized for further analysis; they are shown in Figure 5.

In order to identify faults in each bearing at an early stage, the proposed averaged envelope spectrum is calculated according to Section 3.1.2 for each acceleration signal during the experiment. Subsequently, a spectral inspection is carried out on each envelope spectrum to pinpoint the bearing's fault characteristic frequencies. This process is performed for every signal measurement throughout the bearing's lifespan which helps in identifying the initiation of the bearing fault and facilitates the generation of a precise detection time t_{FDT} for the further prediction of RUL.

The results of the fault detection are summarized in Table 4, where t_{FDT} denotes the first detection time of faults and the bearing EOL time is further discussed in Section 4.2. For illustration, Figures 6–8 present the spectral changes during the initial detection of a bearing fault for Bearings 2, 4 and 6, marking the transition from the healthy state to the degradation stage.

As shown in Figure 6, the vibration signals of Bearing 2 at the 58th and the 59th minute are both dominated by the frequency component at the shaft speed. However, the peaks at the BPFO and its second multiple become more prominent in the latter spectrum (Figure 6b). Despite the weaker energy compared to the rotation frequency ('1X'), the BPFO components indicate the onset of the outer-race fault which occurs after the 59th minute during the run-to-failure experiment.

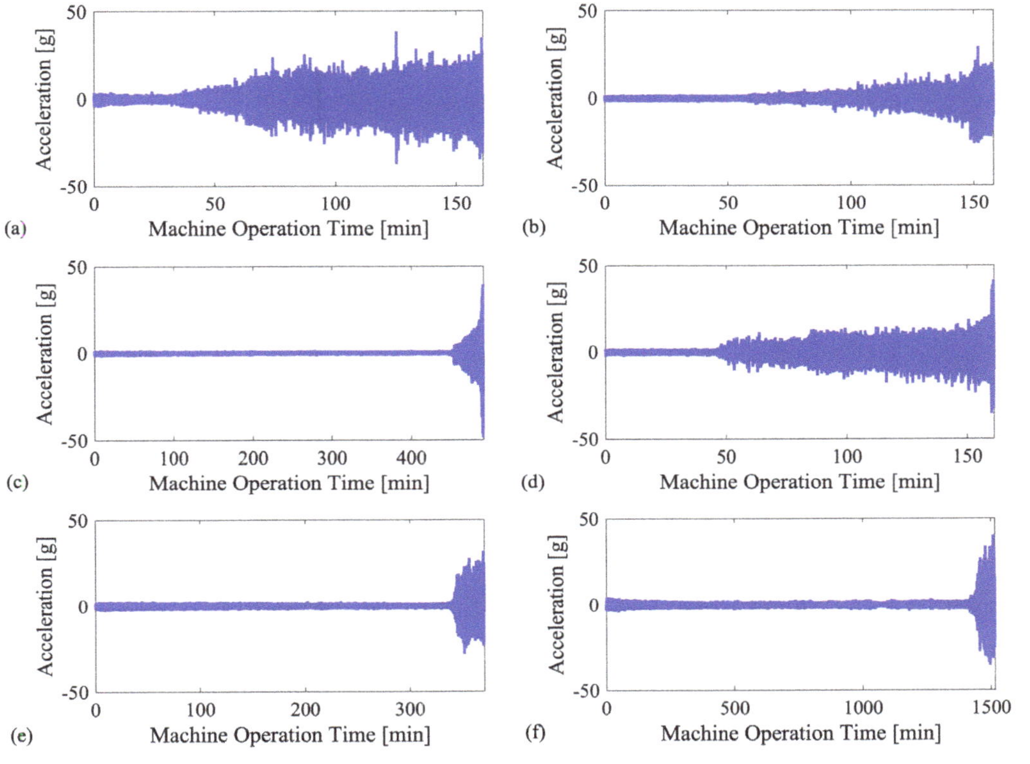

Figure 5. The horizontal acceleration signals of the analyzed test bearings: (**a–f**): Bearings 1 to 6.

Table 4. Results of bearing fault detection based on envelope spectrum inspection.

Operating Condition (Speed/Load)	Bearing ID	First Detection Time of Fault t_{FDT}	End-of-Life Time t_{EOL}	Fault Type
C1: 35Hz/12kN	1	35 min	126 min	Outer-Race Fault
	2	59 min	151 min	Outer-Race Fault
C2: 37.5Hz/11kN	3	446 min	490 min	Inner-Race Fault
	4	47 min	160 min	Outer-Race Fault
C3: 40Hz/10kN	5	327 min	353 min	Inner-Race Fault
	6	1418 min	1478 min	Inner-Race Fault

Similarly, the spectral variations at the fault initiation of Bearing 4 are presented in Figure 7. It is seen that the envelope spectrum of the 46th minute is dominated by the shaft rotation frequency (Figure 7a), whereas the frequency components at the BPFO and its harmonic intensify and dominate the envelope spectrum at the 47th minute. With further inspection of the full life-cycle data, it is verified that the outer-race fault has manifested in Bearing 4 since the 47th minute.

Likewise, Figure 8 shows the envelope spectra during the fault onset for Bearing 6. It is seen that the envelope spectrum in Figure 8a is governed by the shaft speed but also masked with random components. However, although dominated by the shaft rotation frequency, the envelope spectrum at the 1418th minute exhibits evident frequency components at the BPFI and its second multiple. Furthermore, the presence of sidebands at the synchronous speed ('1X') suggests a strong modulation of the BPFI by the shaft speed, which indicates the occurrence of the inner-race fault in Bearing 6 since the 1418th minute.

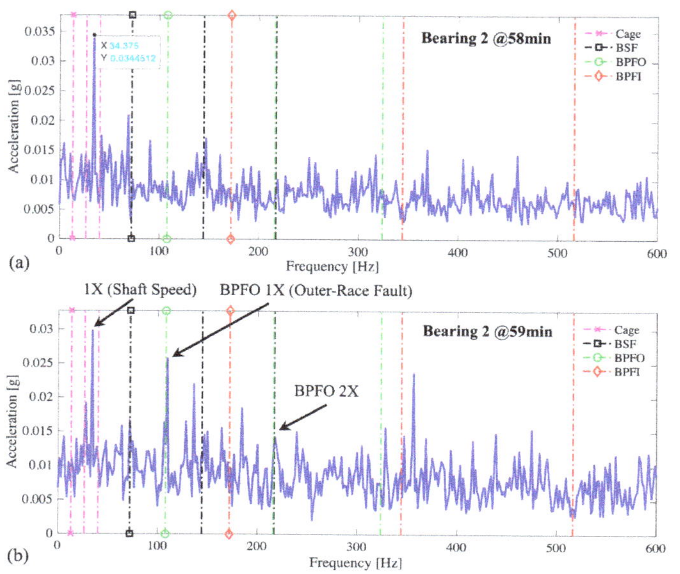

Figure 6. The averaged envelope spectra indicating the fault initiation of Bearing 2: (**a**) The envelope spectrum at the 58th minute. (**b**) The envelope spectrum at the 59th minute, where the outer-race fault is identified.

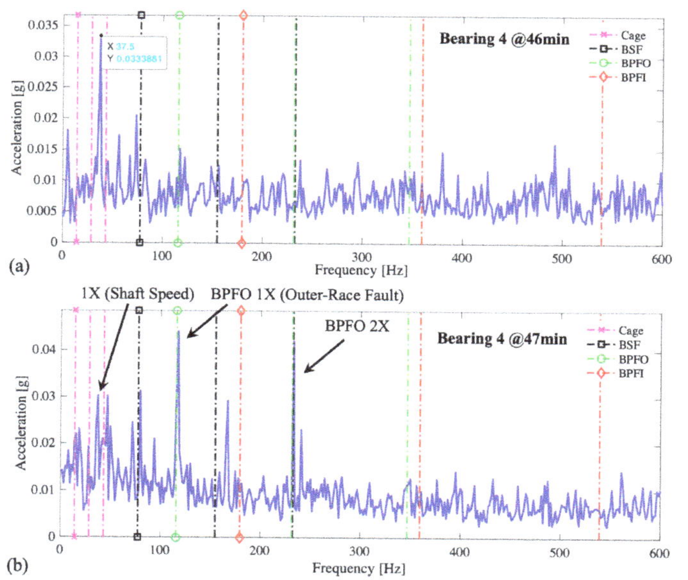

Figure 7. The averaged envelope spectra indicating the fault initiation of Bearing 4: (**a**) The envelope spectrum at the 46th minute. (**b**) The envelope spectrum at the 47th minute when the outer-race fault is identified.

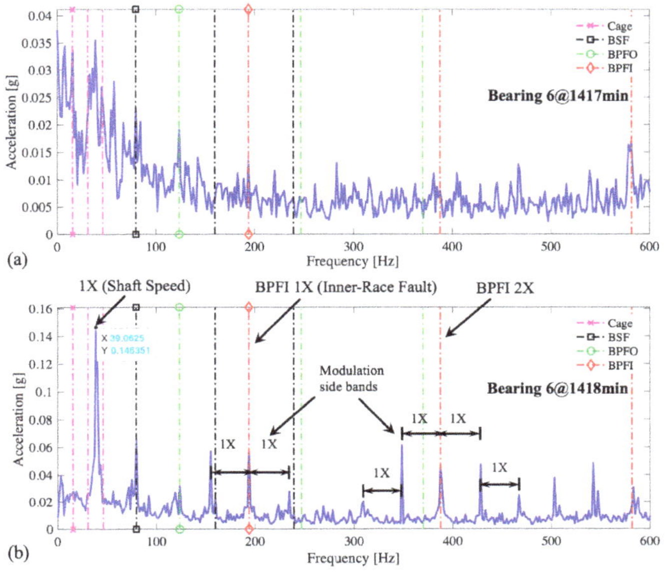

Figure 8. The averaged envelope spectra indicating the fault initiation of Bearing 6: (**a**) The envelope spectrum at the 1417th minute. (**b**) The envelope spectrum at the 1418th minute when the inner-race fault is identified.

4.2. Envelope Spectral Indicators

Based on the results of the averaged envelope spectrum, the bearing fault characteristic frequencies can be identified, the spectrum amplitudes of which are combined to compute the proposed ESI for measuring the bearing degradation level. Accordingly, the ESIs of the test bearings are shown in Figure 9. For each test bearing, the first detection time of faults, t_{FDT}, identified based on the inspection of the averaged envelope spectrum, is marked by the black dotted line.

It is seen that each of the ESI plots maintains a stable platform at the healthy stage while exhibiting an overall ascending trajectory from the time when the bearing fault initiates, indicative of an accelerating trend of bearing degradation. However, it is noteworthy that these ESIs are not strictly monotonic and display substantial fluctuations as they approach the end of machine operation. Except Bearings 3 and 4, the ESI trajectories of other test bearings show noticeable declines at the end of the experiments. This could be explained by the theory of 'healing phenomena' during the bearing deterioration [5], where the defect surfaces caused by wear or fatigue may partially restore the original smoothness due to several potential factors such as the redistribution of lubrication or the removal of debris, leading to a small performance improvement in friction and vibration. However, it should be noted that this phenomenon does not guarantee the prevention of subsequent degradation.

As a representation of the functional failures of rolling bearings, the end-of-life time is formulated as the time when the extracted ESI reaches the maximum, indicating the peak energy of the vibration components at the FCFs, as defined in Equation (4). Respectively, the end-of-life time t_{EOL} for each bearing is listed in Table 4 and marked with a black dashed line in Figure 9.

Based on the computation of the ESI for all the available test bearings, the average level of the ESI at t_{EOL} is utilized as the failure threshold for further prediction of RUL, which is $\overline{\gamma} = 3.24$.

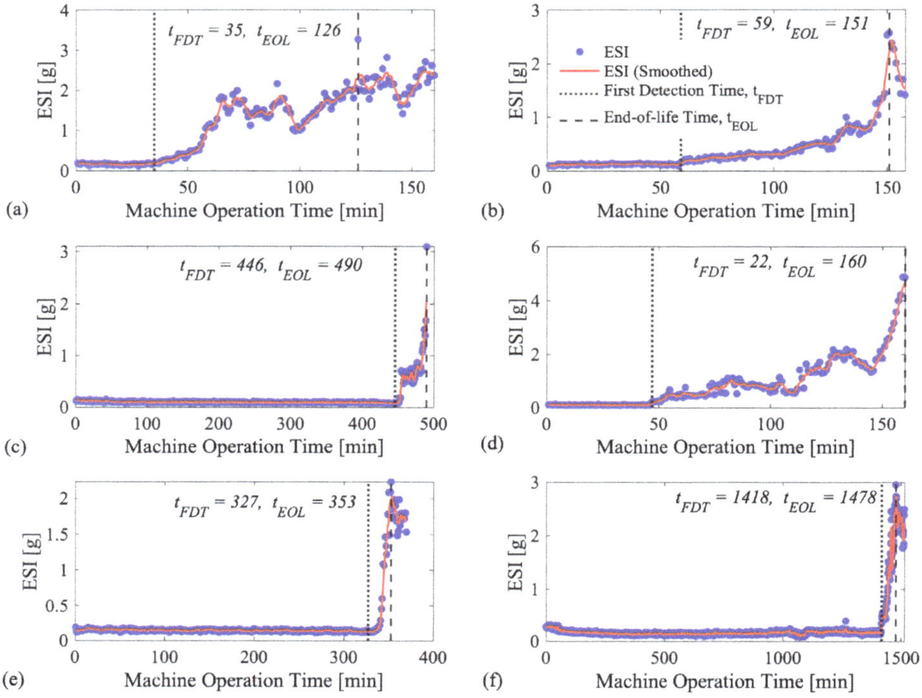

Figure 9. The envelope spectral indicator (ESI) of bearing degradation state: (**a–f**): Bearings 1 to 6. (Blue dotted scatter plot: the extracted ESI. Red solid line: the smoothed ESI using 4-point moving average).

4.3. RUL Prediction of Test Bearings under Condition C1

Once the early fault is detected, the EKF is applied to estimate the latent degradation level of bearings for RUL prediction. As defined in Section 3.4.3, the degradation state of the test bearing is assumed to follow an accelerating increasing trend before reaching the average end-of-life threshold $\overline{\gamma}$.

In Figure 10a, the degradation-stage ESI of Bearing 1 is presented as a grey solid line. The first detection time of fault t_{FDT} is the 35th minute as marked by the black dotted line. Observations reveal that the ESI after t_{FDT} exhibits a gradual upward trend with more fluctuation at the later time. The maximum value of ESI corresponds to the end-of-life time of Bearing 1 ($t_{EOL} = 126$) as represented by the red dash-dot line, indicative of the need for replacement. Considering the uncertainty in the initial states, 10 additional data points preceding t_{FDT} are incorporated to update the state estimations in the EKF. Henceforth, the ESIs starting from the 25th minute, specifically marked with red circles, are utilized for the EKF-based RUL prediction model.

After initializing the model according to Section 3.4.4 and setting the noise levels empirically as $v_t = 0.1$, $w_t = 0.05$, the EKF algorithm is implemented for degradation state estimation until the prediction time t_{PT}, at which the estimated failure time $\tilde{t}_{EOL}$ is predicted based on the state extrapolation and the average EOL threshold $\overline{\gamma}$. Figure 10a–d show the extrapolation results of the bearing degradation state at four prediction times: the 40th, 45th, 50th and 55th minutes, which are, respectively 5, 10, 15 and 20 minutes after the fault detection. It can be observed that the estimated EOL time, $\tilde{t}_{EOL}$, in each prediction is closely in the vicinity of the baseline, t_{EOL}.

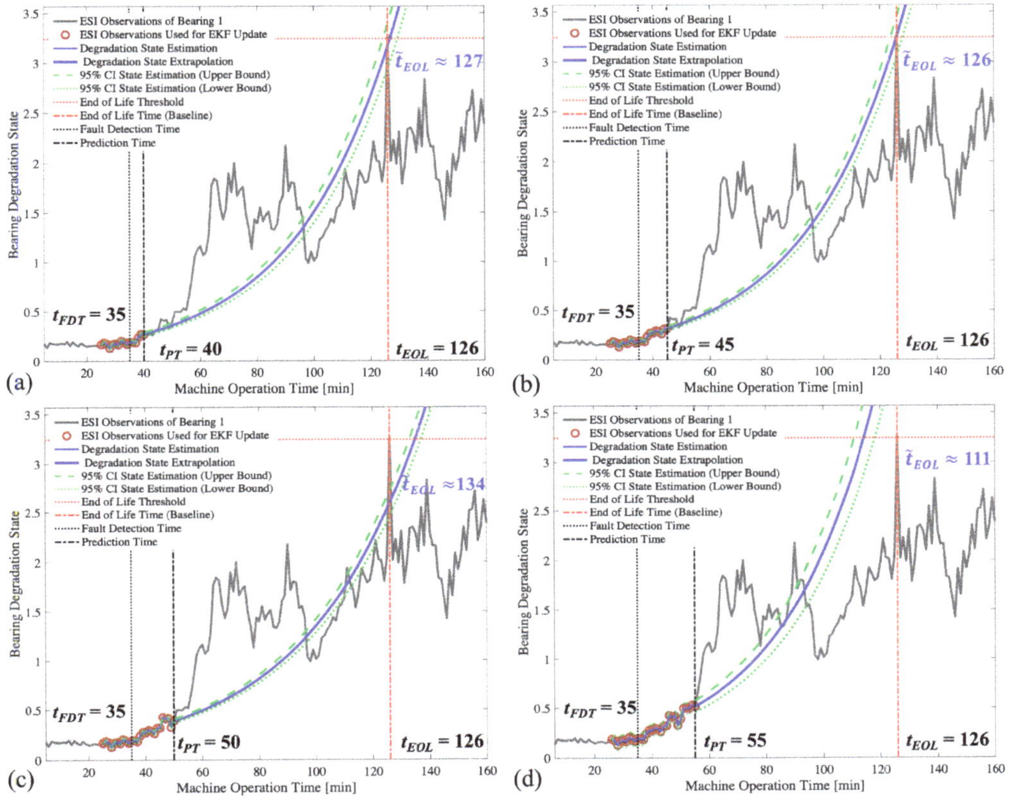

Figure 10. Bearing 1 degradation state prediction based on the EKF: (**a**) State prediction at the 40th minute. The corresponding estimated RUL is = 87 $\tilde{rul}_{40} = \tilde{t}_{EOL} - t_{PT} = 87$ minutes. (**b**) State prediction at the 45th minute, with $\tilde{rul}_{45} = 81$ minutes. (**c**) State prediction at the 50th minute, with $\tilde{rul}_{50} = 84$ minutes. (**d**) State prediction at the 55th minute, with $\tilde{rul}_{55} = 56$ minutes.

As for Bearing 2, Figure 11 demonstrates that the ESI of the bearing follows an accelerating increasing pattern, reaching the maximum at the 151st minute. An outer-race fault is detected at the 59th minute based on the spectral inspection results of envelope analysis (see Section 4.1). Similarly, the ESI data from the 49th minute (marked with red circles) are employed for state estimation in the EKF algorithm. Using the noise levels at $v_t = 0.1$, $w_t = 0.05$, the estimated EOL time at various predicting times can be obtained.

Figure 11a–d present the results of four t_{PT} at the 66th, 69th, 71st and 74th minutes. It is observed that despite the early prediction at the 66th minute in Figure 11a, the prediction of the EOL time $\tilde{t}_{EOL}$ in Figure 11b is exactly the same as the baseline EOL time, and that the extrapolations of $\tilde{t}_{EOL}$ at the later times are also accurate, as in Figure 11c,d.

To evaluate the performance of the proposed method for bearing RUL prediction, the state estimation results with uncertainty estimation for Bearings 1 and 2 are summarized in Table 5. The result shows that the established degradation model well expresses the initial upward trend in the ESI during the early stage of bearing degradation using one dynamic model parameter, b_t. It also demonstrates that the EKF algorithm is capable of estimating the RUL of bearings with a few ESI data points since the first detection of faults.

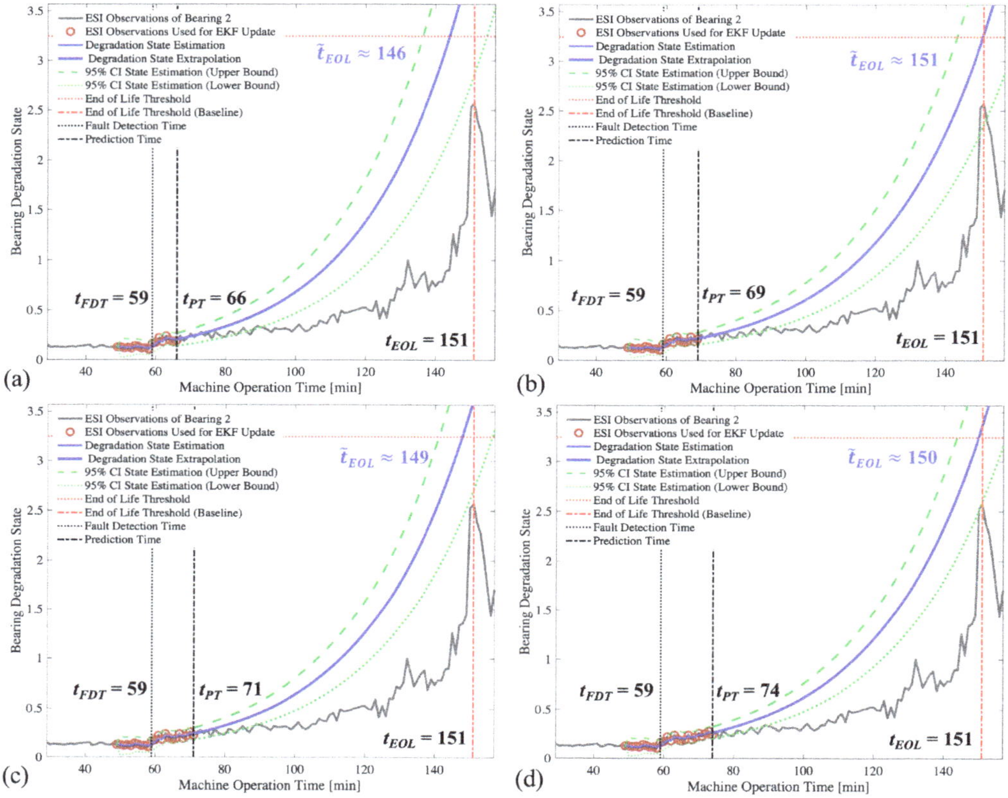

Figure 11. Bearing 2 degradation state prediction based on the EKF: (**a**) State prediction at the 66th minute. The corresponding estimated RUL is $\widetilde{rul}_{66} = \tilde{t}_{EOL} - t_{PT} = 80$ minutes. (**b**) State prediction at the 69th minute, with $\widetilde{rul}_{69} = 82$ minutes. (**c**) State prediction at the 71st minute, with $\widetilde{rul}_{71} = 78$ minutes. (**d**) State prediction at the 74th minute, with $\widetilde{rul}_{74} = 76$ minutes.

Table 5. RUL Prediction Results of Bearing 1 and Bearing 2 (Condition 1).

Bearing ID	**Baseline FDT *** t_{FDT}**/minute**	**Baseline EOLT *** t_{EOL}**/minute**	**RUL PT *** t_{PT}**/minute**	**Estimated End-of-Life Time** $\tilde{t}_{EOL}$**/minute**	**Baseline RUL /minute**	**Estimated RUL (Residual) /minute**	**RUL 95% CI /minute**
1	35	126	40	127	86	87 (+1)	[84,89]
			45	126	81	**81 (0)**	[79,83]
			50	134	76	84 (+8)	[83,87]
			55	111	71	56 (−15)	[56,63]
2	59	151	66	146	85	80 (−5)	[70,88]
			69	151	82	**82 (0)**	[74,91]
			71	149	80	78 (−2)	[70,85]
			74	150	77	76 (−1)	[69,83]

* Baseline FDT: first detection time of fault. Baseline EOLT: bearing end-of-life time of ground truth. PT: RUL prediction time.

4.4. RUL Prediction of the Test Bearings under Condition C2

To validate the prediction performance of the proposed method, the bearings under Condition 2 are analyzed, in which the shaft speed is 37.5 Hz with the loading at 11 kN. The EKF model is further applied to the ESI of Bearings 3 and 4 for RUL prediction at the early stage of bearing degradation.

According to the envelope analysis results in Table 4, the presence of the inner-race fault is identified in Bearing 3 after 446 minutes. Then, the ESI data from the 436th minute to the RUL prediction time t_{PT} are employed for the degradation state estimation using the proposed EKF models. The state process noise and the measurement noise are, respectively, set as $v_t = 0.02$, $v_t = 0.05$. After recursive filtering using the ESI observations (marked with a red circle in Figure 12), the estimated EOL time $\widetilde{t}_{EOL}$ at the prediction time can be extrapolated based on the updated state at t_{PT} according to (38). Four predictions within 20 minutes after the fault is detected (t_{FDT}) are shown in Figure 12.

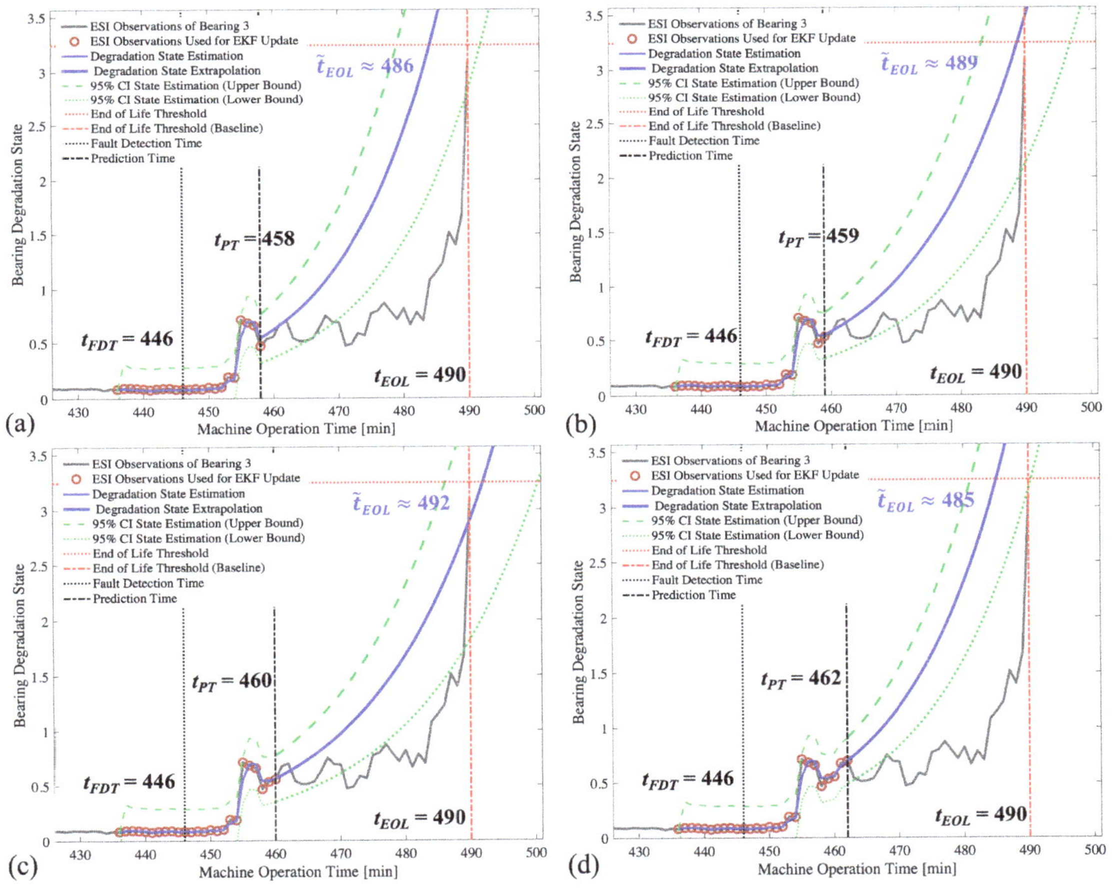

Figure 12. Bearing 3 degradation state prediction based on the EKF: (**a**) State prediction at the 458th minute. The corresponding estimated RUL is $\widetilde{rul}_{458} = \widetilde{t}_{EOL} - t_{PT} = 28$ minutes. (**b**) State prediction at the 459th minute, with $\widetilde{rul}_{459} = 30$ minutes. (**c**) State prediction at the 460th minute, with $\widetilde{rul}_{460} = 32$ minutes. (**d**) State prediction at the 462nd minute, with $\widetilde{rul}_{462} = 23$ minutes.

In the prediction result, an evident surge in the ESI of Bearing 3 is observed following the initiation of a fault. This surge results in the prediction of an early failure time at $\tilde{t}_{EOL} = 486$, which is extrapolated from the state estimation at $t_{PT} = 458$, as in Figure 12a. Meanwhile, the further predictions of the EOL time at $t_{PT} = 459$ and $t_{PT} = 460$ (Figure 12b,c) are well projected to the vicinity of the baseline EOL time at $t_{EOL} = 490$. Overall, the 95% credible intervals of the four predictions provided in the green dashed lines consistently cover the baseline end-of-life time.

Analogously, Figure 13 displays the state estimation results by applying the EKF algorithm for the ESI of Bearing 4. The noise levels for the state process and the observations are, respectively, set at $v_t = 0.05$, $w_t = 0.08$. Notably, the state extrapolation at different prediction times effectively represents the overall pattern of the ESI (in the grey solid line). The predicted end-of-life times at $t_{PT} = 67$ minute and $t_{PT} = 68$ minute are estimated, respectively, as $\tilde{t}_{EOL} = 159$ minute and $\tilde{t}_{EOL} = 158$ minute, which closely correspond to the baseline end-of-life time at $t_{EOL} = 160$ minute.

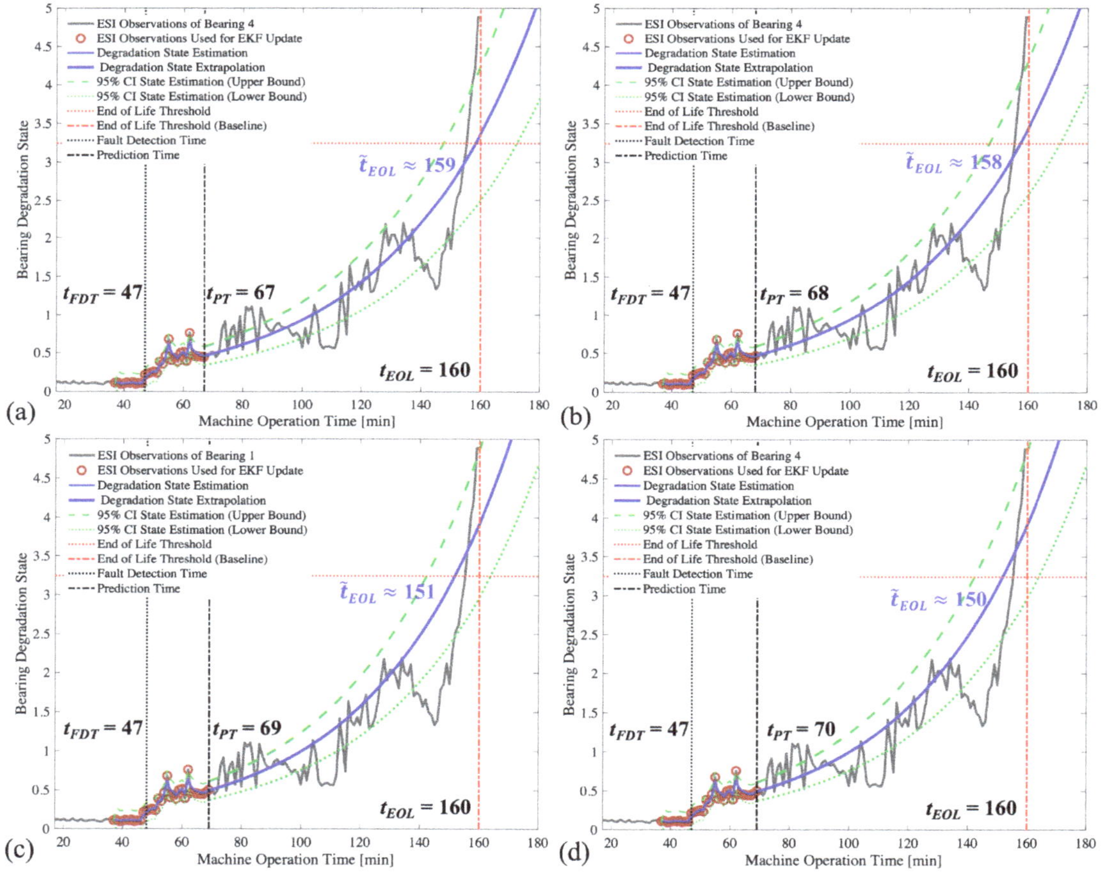

Figure 13. Bearing 4 degradation state prediction based on the EKF: (**a**) State prediction at the 67th minute. The corresponding estimated RUL is $\tilde{rul}_{67} = \tilde{t}_{EOL} - t_{PT} = 92$ minutes. (**b**) State prediction at the 68th minute, with $\tilde{rul}_{68} = 90$ minutes. (**c**) State prediction at the 69th minute, with $\tilde{rul}_{69} = 82$ minutes. (**d**) State prediction at the 70th minute, with $\tilde{rul}_{70} = 80$ minutes.

For the assessment of the prediction accuracy, the state estimation results and the corresponding RUL are provided in Table 6, where the mean estimated RUL, the residuals between the estimated RUL and the baseline RUL, and the 95% credible intervals are summarized. It is seen that the overall prediction of RUL is fairly accurate and tends to be conservative given the negative residuals of the RUL estimate. Also, the absolute errors of the predicted RUL are no more than 5 minutes for Bearing 3, where the predicting times of the RUL are within 20 minutes after the fault detection. In addition, for Bearing 4, the absolute errors of the RUL prediction are less than 10 minutes with the predicting time of the RUL at around 20 minutes after the fault is detected.

Table 6. RUL Prediction Results of Bearing 3 and Bearing 4 (Condition 2).

Bearing ID	Baseline FDT * t_{FDT} /minute	Baseline EOLT * t_{EOL} /minute	RUL PT * t_{PT} /minute	Estimated End-of-Life Time $\tilde{t}_{EOL}$ /minute	Baseline RUL /minute	Estimated RUL (Residual) /minute	RUL 95% CI /minute
3	446	490	458	486	32	28 (−4)	[21,34]
			459	489	31	**30 (−1)**	[24,38]
			460	492	30	32 (+2)	[26,41]
			462	485	28	23 (−5)	[19,28]
4	47	160	67	159	93	**92 (−1)**	[80,105]
			68	158	92	90 (−2)	[78,102]
			69	151	91	82 (−9)	[73,95]
			70	150	90	80 (−10)	[71,91]

* Baseline FDT: first detection time of fault. Baseline EOLT: bearing end-of-life time of ground truth. PT: RUL prediction time.

4.5. Further Validation on the Test Bearings under Condition C3

Next, further validation is introduced to the test bearings operating under condition 3, where the shaft speed is 40 Hz with 10 kN hydraulic loading.

For Bearing 5, the ESI exhibits a steady plateau before the inner-race fault is found at the 327th minute, after which it surges rapidly to the maximum value at the 353rd minute. This indicates that Bearing 5 undergoes a rapid degradation stage that lasts for merely 26 minutes. Based on the EKF algorithm outlined in Section 3.4.4, the bearing end-of-life time can be estimated using the ESI observations from the 327th minute to the various prediction times, t_{PT}. With the noise levels of the state process and the measurement set at $v_t = 0.1$ and $w_t = 0.1$, the results of the state estimation and further extrapolation based on the EKF are shown in Figure 14.

As for Bearing 6, after the detection of the inner-race fault at the 1418th minute (t_{FDT} = 1418), the ESI observations since the 1407th minute (as marked with red circles in Figure 15) are used for degradation state estimation in the EKF. It is seen that the state extrapolation captures the overall increasing trend of the ESI and provides a fairly precise estimate of end-of-life time, $\tilde{t}_{EOL}$.

For performance evaluation, the RUL at the prediction time of Bearings 5 and 6 are summarized in Table 7. For Bearing 5, it can be concluded that the absolute residuals of the predicted RUL are less than 5 minutes. As the degradation state represented by the ESI steeply rises to approach the end-of-life threshold, a prompt replacement is suggested for Bearing 5 to avoid critical failures.

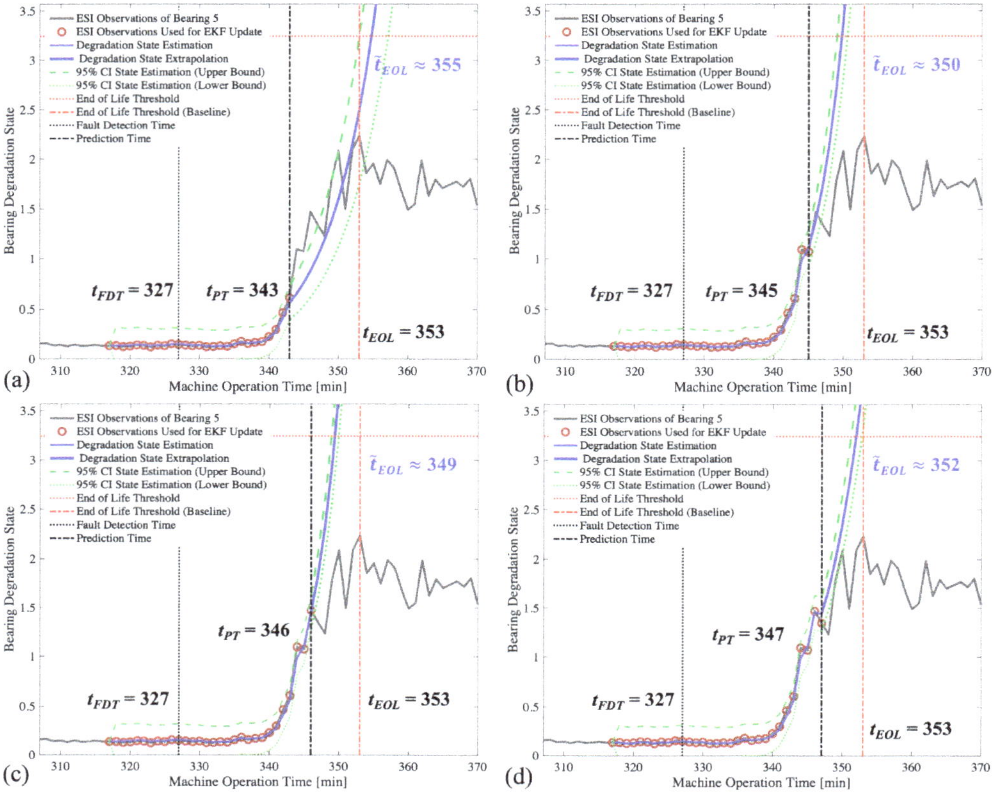

Figure 14. Bearing 5 degradation state prediction based on the EKF: (**a**) State prediction at the 343rd minute. The corresponding estimated RUL is $\widetilde{rul}_{343} = \tilde{t}_{EOL} - t_{PT} = 12$ minutes. (**b**) State prediction at the 345th minute, with $\widetilde{rul}_{345} = 5$ minutes. (**c**) State prediction at the 346th minute, with $\widetilde{rul}_{346} = 3$ minutes. (**d**) State prediction at the 347th minute, with $\widetilde{rul}_{347} = 5$ minutes.

Table 7. RUL Prediction Results of Bearing 5 and Bearing 6 (Condition 3).

Bearing ID	Baseline FDT * t_{FDT} /minute	Baseline EOLT * t_{EOL} /minute	RUL PT * t_{PT} /minute	Estimated End-of-Life Time $\tilde{t}_{EOL}$	Baseline RUL /minute	Estimated RUL (Residual) /minute	RUL 95% CI /minute
5	327	353	343	355	10	12 (+2)	[10,14]
			345	350	8	5 (−3)	[4,6]
			346	349	7	3 (−4)	[2,4]
			347	352	6	**5 (−1)**	[4,6]
6	1418	1478	1428	1472	50	44 (−6)	[38,50]
			1429	1476	49	47 (−2)	[41,53]
			1430	1477	48	47 (−1)	[41,54]
			1441	1478	37	**37 (0)**	[33,41]

* Baseline FDT: first detection time of fault. Baseline EOLT: bearing end-of-life time of ground truth. PT: RUL prediction time.

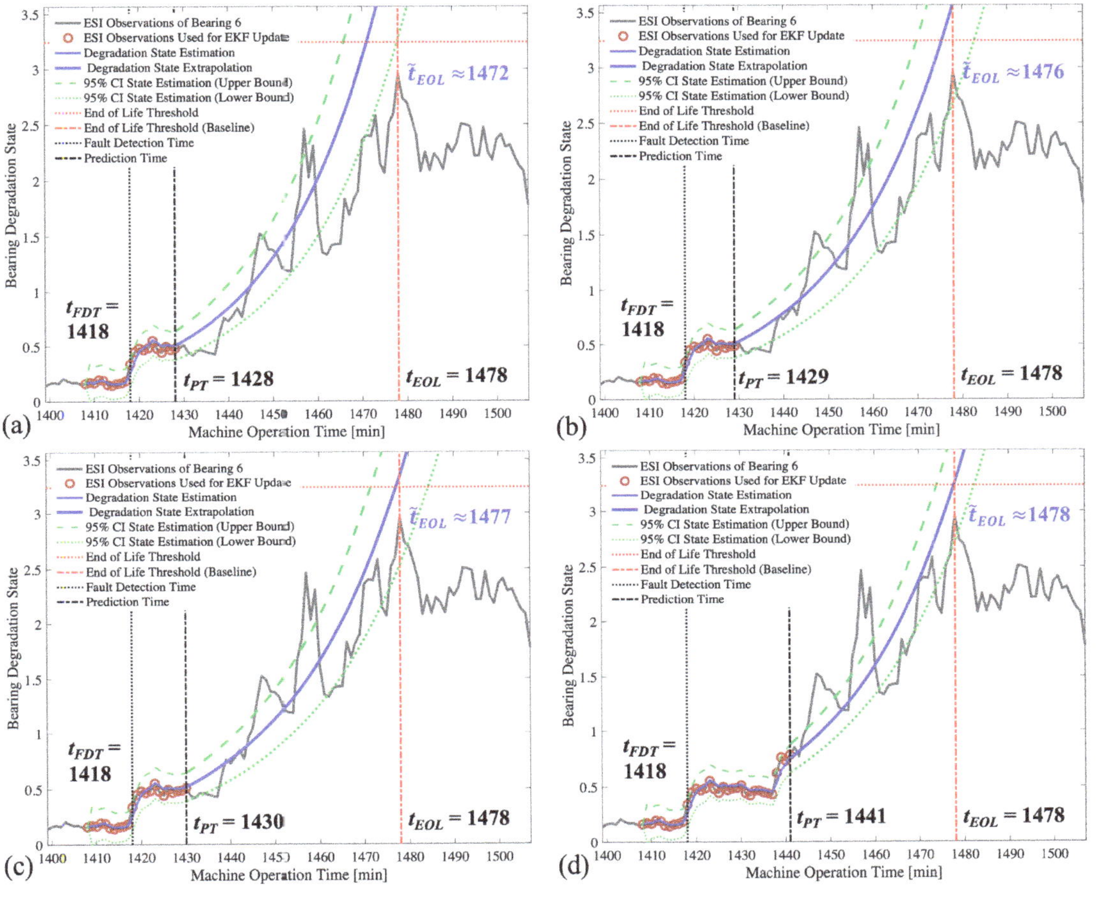

Figure 15. Bearing 6 degradation state prediction based on the EKF: (**a**) State prediction at the 1428th minute. The corresponding estimated RUL is $\widetilde{rul}_{1428} = \tilde{t}_{EOL} - t_{PT} = 44$ minutes. (**b**) State prediction at the 1429th minute, with $\widetilde{rul}_{1429} = 47$ minutes. (**c**) State prediction at the 1430th minute, with $\widetilde{rul}_{1430} = 47$ minutes. (**d**) State prediction at the 1441st minute, with $\widetilde{rul}_{1441} = 37$ minutes.

As for Bearing 6, the estimated RULs at the prediction times $t_{PT} = 1428$ and $t_{PT} = 1429$ are close to but earlier than their baselines. Note that the prediction times are merely about 10 minutes away from the detection of fault. Additionally, the RUL estimations at $t_{PT} = 1430$ and $t_{PT} = 1441$ are more precise since their residuals have reduced to less than one minute.

5. Discussion

As mentioned in Section 1, the pure physics-based models for RUL prediction are beneficial when a clear understanding of the failure mechanism is available, while the deterioration dynamics of the bearing is not yet fully unveiled and often in practice, not all the prior physics information (e.g., defect dimension) is available. Therefore, the prediction accuracy of the pure physics-based methods may be limited in practical applications. In general, the pure data-driven approaches and statistical approaches are not strictly separated. Although pure statistics methods (e.g., survival analysis) may require a large amount of historical data that provide consistent statistical meaningful results, such as the data requirement in pure data-driven approaches (e.g., neural networks) to ensure generalization performance, both have this advantage: no detailed knowledge of the failure mechanism is needed.

In this study, the recursive Bayesian filter is applied combining the advantages of both physics-based and data-driven approaches. The envelope spectral indicator is crafted relating to bearing fault characteristics and is incorporated to the EKF for sequential estimation of the bearing degradation state with uncertainty estimation.

The experimental results on the test bearings under various operating conditions have demonstrated promising performance of the ESI in representing the bearing degradation process and the EKF in RUL prediction. In the above cases, the EKF-based RUL predictions at the discrete time steps shortly after the fault detection exhibit commendable accuracy and precision. The prediction results suggest that the EKF could effectively integrate the bearing ESI data and update the state estimation, which underscores the capability of the proposed model to capture short-term trends and variations in the ESI during bearing deterioration, providing valuable insights into the impending failure and the RUL of bearings.

On the other hand, it is also important to acknowledge that the prediction based on more ESI observations may exhibit a more intricate pattern. In the cases of continuous RUL prediction, it is justifiable to claim that there is a convergence of prediction accuracy towards the end of the bearing life because of the health indicators approaching the pre-specified end-of-life threshold. Yet, the significance of the late-stage RUL predictions might be diminished due to their proximity to the bearing failure, thereby limiting their practical value for maintenance decision making. In contrast, the significance of the presented method is highlighted for the ability to provide meaningful RUL predictions during the early phase of the bearing degradation.

Regarding the computation cost of the proposed methods, the proposed averaged envelope spectrum can be efficiently computed based on the Hilbert transform and the fast Fourier transform. As for the EKF, the state process model for the bearing degradation has been further optimized and only one-dimensional measurement is leveraged. Also, the study targets the early phase RUL prediction by using only a few observations of the health indicators at the fault detection phase, which results in low computation requirements.

In summary, the effectiveness of the EKF-based RUL prediction methodology has been demonstrated on the experimental bearing vibration data of various operating conditions. The bearing RUL can be predicted shortly after the fault detection based on the ESI at the early phase of bearing degradation. The findings collectively highlight the promising potentials of the proposed method combining the ESI and the EKF for the early prediction of RUL, which contributes to the practical implementation of advanced condition monitoring and predictive maintenance methodologies.

6. Conclusions

The presented work proposes the envelope spectral indicator combining the vibration energy of the bearing fault characteristic frequencies to closely represent the bearing degradation state, and utilizes the extended Kalman filter algorithm for the degradation state estimation given only a few ESI observations about the fault detection time. The proposed ESI is formulated based on the bearing FCF components in the averaged envelope spectrum, which is enhanced against the random noise through the PSD of the envelope signal segments. The ESI is beneficial in representing the latent deterioration condition of rolling-element bearings as the constituting components of the ESI are solely related to the bearing defects. This indicates that the ESI is free of the interference from the synchronous speed component and the fault components arising from the coupled mechanical structures such as rotors and gear pairs. Along with the computation of ESI, the bearing fault can be efficiently identified through the spectral inspection of the averaged envelope spectrum.

On the above basis, the early RUL prediction approach for REBs is achieved based on the combination of ESI and EKF. Once the fault is detected in the bearing, only a small number of the ESI observations from the early degradation process are employed for degradation state estimation and end-of-life time prediction. The predictive performance is validated on the public experimental vibration data measured from a bearing rig. The thorough results and observations presented have demonstrated the efficacy of the proposed method for predicting the RUL of rolling bearings even at the early stage of degradation. Therefore, by enabling the early detection and prediction of RUL, the proposed method may foster more efficient and cost-effective maintenance practices and is identified as having great potential to optimize predictive maintenance strategies for engineering practice.

Author Contributions: Conceptualization, H.W., L.Z. and J.K.S.; methodology, software, writing—original draft preparation, H.W.; resources, supervision, project administration, L.Z. and J.K.S.; writing—review and editing, H.W., L.Z. and J.K.S. All authors have read and agreed to the published version of the manuscript.

Funding: This research received no external funding.

Institutional Review Board Statement: Not applicable.

Informed Consent Statement: Not applicable.

Data Availability Statement: The XJTU-SY bearing dataset studied in this paper is available through link https://biaowang.tech/xjtu-sy-bearing-datasets/, accessed on 27 October 2023.

Acknowledgments: Haobin Wen acknowledges the financial support jointly provided by the University of Manchester and the China Scholarship Council (CSC).

Conflicts of Interest: The authors declare no conflicts of interest.

References

1. Dupuis, R. Application of Oil Debris Monitoring for Wind Turbine Gearbox Prognostics and Health Management. *Annu. Conf. PHM Soc.* **2010**, *2*, 1–15.
2. Liang, L.; Ding, X.; Wen, H.; Liu, F. Impulsive Components Separation Using Minimum-Determinant KL-Divergence NMF of Bi-Variable Map for Bearing Diagnosis. *Mech. Syst. Signal Process.* **2022**, *175*, 109129. [CrossRef]
3. Si, X.S.; Wang, W.; Hu, C.H.; Zhou, D.H. Remaining Useful Life Estimation—A Review on the Statistical Data Driven Approaches. *Eur. J. Oper. Res.* **2011**, *213*, 1–14. [CrossRef]
4. Lienig, J.; Bruemmer, H. Reliability Analysis. In *Fundamentals of Electronic Systems Design*; Springer International Publishing: Cham, Switzerland, 2017; pp. 45–73.
5. Qiu, H.; Lee, J.; Lin, J.; Yu, G. Wavelet Filter-Based Weak Signature Detection Method and Its Application on Rolling Element Bearing Prognostics. *J. Sound Vib.* **2006**, *289*, 1066–1090. [CrossRef]
6. Nectoux, P.; Gouriveau, R.; Medjaher, K.; Ramasso, E.; Morello, B.; Zerhouni, N.; Varnier, C. PRONOSTIA: An Experimental Platform for Bearings Accelerated Degradation Tests. In Proceedings of the IEEE International Conference on Prognostics and Health Management, PHM'12, Denver, CO, USA, 18–21 June 2012.
7. Wang, B.; Lei, Y.; Li, N.; Li, N. A Hybrid Prognostics Approach for Estimating Remaining Useful Life of Rolling Element Bearings. *IEEE Trans. Reliab.* **2020**, *69*, 401–412. [CrossRef]
8. Poplawski, J.V.; Peters, S.M.; Zaretsky, E.V. Effect of Roller Profile on Cylindrical Roller Bearing Life Prediction—Part I: Comparison of Bearing Life Theories. *Tribol. Trans.* **2008**, *44*, 339–350. [CrossRef]
9. International Organization for Standardization ISO 281:2007 Rolling Bearings: Dynamic Load Ratings and Rating Life. Available online: https://www.iso.org/standard/38102.html (accessed on 11 July 2023).
10. Zaretsky, E.V. *Rolling Bearing Life Prediction, Theory, and Application*; NASA Tech. Reports 2013, No. NASA/TP-2013-215305; NASA: Washington, DC, USA, 2013.
11. Loutas, T.H.; Roulias, D.; Georgoulas, G. Remaining Useful Life Estimation in Rolling Bearings Utilizing Data-Driven Probabilistic E-Support Vectors Regression. *IEEE Trans. Reliab.* **2013**, *62*, 821–832. [CrossRef]
12. Alfarizi, M.G.; Tajiani, B.; Vatn, J.; Yin, S. Optimized Random Forest Model for Remaining Useful Life Prediction of Experimental Bearings. *IEEE Trans. Ind. Inform.* **2023**, *19*, 7771–7779. [CrossRef]
13. Mao, W.; He, J.; Zuo, M.J. Predicting Remaining Useful Life of Rolling Bearings Based on Deep Feature Representation and Transfer Learning. *IEEE Trans. Instrum. Meas.* **2020**, *69*, 1594–1608. [CrossRef]
14. Wang, H.; Ma, X.; Zhao, Y. An Improved Wiener Process Model with Adaptive Drift and Diffusion for Online Remaining Useful Life Prediction. *Mech. Syst. Signal Process.* **2019**, *127*, 370–387. [CrossRef]
15. Li, Y.; Huang, X.; Ding, P.; Zhao, C. Wiener-Based Remaining Useful Life Prediction of Rolling Bearings Using Improved Kalman Filtering and Adaptive Modification. *Meas. J. Int. Meas. Confed.* **2021**, *182*, 109706. [CrossRef]
16. Singleton, R.K.; Strangas, E.G.; Aviyente, S. Extended Kalman Filtering for Remaining-Useful-Life Estimation of Bearings. *IEEE Trans. Ind. Electron.* **2015**, *62*, 1781–1790. [CrossRef]

17. Mosallam, A.; Medjaher, K.; Zerhouni, N. Data-Driven Prognostic Method Based on Bayesian Approaches for Direct Remaining Useful Life Prediction. *J. Intell. Manuf.* **2016**, *27*, 1037–1048. [CrossRef]
18. Särkkä, S. *Bayesian Filtering and Smoothing*; Cambridge University Press: Cambridge, UK, 2013; ISBN 9781107030657.
19. Andrade-Cetto, J. The Kalman Filter. *Tech. Rep. Inst. Robot. i Inform. Ind.* 2002. CSIC-UPC (IRI-DT-02-01). Available online: https://www.iri.upc.edu/publications/show/545 (accessed on 1 May 2023).
20. Haykin, S. *Kalman Filtering and Neural Networks*; Wiley: Hoboken, NJ, USA, 2001; Volume 5, ISBN 9780471369981.
21. Wan, E.A.; Van Der Merwe, R. The Unscented Kalman Filter for Nonlinear Estimation. In Proceedings of the IEEE 2000 Adaptive Systems for Signal Processing, Communications, and Control Symposium (Cat. No.00EX373), Lake Louise, AB, Canada, 4 October 2000; pp. 153–158. [CrossRef]
22. Coelho, M.d.F.; Bousson, K.; Ahmed, K. An Improved Extended Kalman Filter for Radar Tracking of Satellite Trajectories. *Designs* **2021**, *5*, 54. [CrossRef]
23. Singh, M.; Lakra, S.; Das, S.; Mishra, S.K.; Sahoo, A.K.; Acharya, B. Extended Kalman Filter-Based Position Estimation in Autonomous Vehicle Applications. *Lect. Notes Electr. Eng.* **2023**, *887*, 427–440. [CrossRef]
24. Liang, L.; Wen, H.; Liu, F.; Li, G.; Li, M. Feature Extraction of Impulse Faults for Vibration Signals Based on Sparse Non-Negative Tensor Factorization. *Appl. Sci.* **2019**, *9*, 3642. [CrossRef]
25. Wen, H.; Zhang, L.; Sinha, J.K. Adaptive Band Extraction Based on Low Rank Approximated Nonnegative Tucker Decomposition for Anti-Friction Bearing Faults Diagnosis Using Measured Vibration Data. *Machines* **2022**, *8*, 694. [CrossRef]
26. Randall, R.B.; Antoni, J. Rolling Element Bearing Diagnostics—A Tutorial. *Mech. Syst. Signal Process.* **2011**, *25*, 485–520. [CrossRef]
27. Sinha, J.K. *Vibration Analysis, Instruments, and Signal Processing*; CRC Press: Boca Raton, FL, USA, 2014; ISBN 9781482231458.
28. Dai, Y.; Yu, S.; Yan, Y.; Yu, X. An EKF-Based Fast Tube MPC Scheme for Moving Target Tracking of a Redundant Underwater Vehicle-Manipulator System. *IEEE/ASME Trans. Mechatron.* **2019**, *24*, 2803–2814. [CrossRef]
29. Feng, D.; Wang, C.; He, C.; Zhuang, Y.; Xia, X.-G. Kalman-Filter-Based Integration of IMU and UWB for High-Accuracy Indoor Positioning and Navigation. *IEEE Internet Things J.* **2020**, *7*, 3133–3146. [CrossRef]
30. Lei, Y.; Xia, D.; Erazo, K.; Nagarajaiah, S. A Novel Unscented Kalman Filter for Recursive State-Input-System Identification of Nonlinear Systems. *Mech. Syst. Signal Process.* **2019**, *127*, 120–135. [CrossRef]
31. Batzel, T.D.; Swanson, D.C. Prognostic Health Management of Aircraft Power Generators. *IEEE Trans. Aerosp. Electron. Syst.* **2009**, *45*, 473–483. [CrossRef]
32. Sepasi, S.; Ghorbani, R.; Liaw, B.Y. Improved Extended Kalman Filter for State of Charge Estimation of Battery Pack. *J. Power Sources* **2014**, *255*, 368–376. [CrossRef]
33. Simon, D. *Optimal State Estimation: Kalman, H∞, and Nonlinear Approaches*; Wiley: Hoboken, NJ, USA, 2006; ISBN 9780471708582. [CrossRef]

Disclaimer/Publisher's Note: The statements, opinions and data contained in all publications are solely those of the individual author(s) and contributor(s) and not of MDPI and/or the editor(s). MDPI and/or the editor(s) disclaim responsibility for any injury to people or property resulting from any ideas, methods, instructions or products referred to in the content.

applied sciences

MDPI

Article

Predicting Failure Probability in Industry 4.0 Production Systems: A Workload-Based Prognostic Model for Maintenance Planning

Giuseppe Converso †,‡, Mosè Gallo †,‡, Teresa Murino †,‡ and Silvestro Vespoli *,‡

Dipartimento di Ingegneria Chimica, dei Materiali e della Produzione Industriale, Università degli Studi di Napoli Federico II, 80125 Napoli, Italy; giuseppe.converso@unina.it (G.C.); mose.gallo@unina.it (M.G.); teresa.murino@unina.it (T.M.)

* Correspondence: silvestro.vespoli@unina.it

† Current address: Dipartimento di Ingegneria Chimica, dei Materiali e della Produzione Industriale (DICMAPI), Piazzale Tecchio 80, 80125 Napoli, Italy.

‡ These authors contributed equally to this work.

Abstract: Maintenance of equipment is a crucial issue in almost all industrial sectors as it impacts the quality, safety, and productivity of any manufacturing system. Additionally, frequent production rescheduling due to unplanned and unintended interruptions can be very time consuming, especially in the case of centrally controlled systems. Therefore, the ability to estimate the likelihood that a monitored machine will successfully complete a predefined workload, taking into account both historical data from the machine's sensors and the impending workload, may be essential in supporting a new approach to scheduling activities in an Industry 4.0 production system. This study proposes a novel approach for integrating machine workload information into a well-established PHM algorithm for Industry 4.0, with the aim of improving maintenance strategies in the manufacturing process. The proposed approach utilises a logistic regression model to assess the health condition of equipment and a neural network computational model to estimate its failure probability according to the scheduled workloads. Results from a prototypal case study showed that this approach leads to an improvement in the prediction of the likelihood of completing a scheduled job, resulting in improved autonomy of CPSs in accepting or declining scheduled jobs based on their forecasted health state, and a reduction in maintenance costs while maximising the utilisation of production resources. In conclusion, this study is beneficial for the present research community as it extends the traditional condition-based maintenance diagnostic approach by introducing prognostic capabilities at the plant shop floor, fully leveraging the key enabling technologies of Industry 4.0.

Keywords: Industry 4.0; prognostics; cyber-physical system; smart maintenance; machine workload

1. Introduction

Today, we live in a world that could only be described as science fiction. With the rise of the Internet of Things (IoT), everything has become smarter and connected in a way that resembles the industrial revolution of the western world 35 years ago [1]. The modern industry must adapt and maintain maximum efficiency while facing new challenges such as high reliability, low environmental risk, and human safety. This has led to the emergence of the digital economy, also known as Industry 4.0, which focusses on the logic that governs the entire manufacturing process rather than just technological advancements in manufacturing processes. The introduction of cyber-physical systems (CPS) has made each production resource (e.g., machines, equipment, and operators) a part of the "cyber production system" connected to the IoT. However, this new production scenario also presents challenges such as the need to build and standardise CPSs and the implementation of new communication standards among CPSs [2].

On the other hand, this new industrial paradigm also offers numerous opportunities. A network of CPSs can help achieve goals such as product customisation, a reduction in production lead times, a more efficient control of manufacturing processes, and an increased product quality and reliability. In this CPS configuration, which aligns with the Industry 4.0 paradigm, machines and production equipment are connected at a higher level of awareness, allowing them to become aware of their process capabilities. One area of focus in auxiliary processes is maintenance activity, which "combines various methods, tools, and techniques to reduce maintenance costs while increasing equipment reliability, availability, and safety" [3]. In fact, maintenance stops, whether expected or unexpected, account for a significant portion of operating expenses in various industry sectors, with nearly 30% of maintenance costs being attributed to inefficient policies (e.g., unscheduled downtime due to sudden failures) [4]. Industry 4.0 technologies enable the development of new maintenance strategies, such as "smart maintenance".

Since 1950, various maintenance logics have been developed, including time-based maintenance (TBM). TBM involves regularly scheduled preventive maintenance in order to slow down the deterioration process of components. The TBM maintenance strategy assumes that the system's failure distribution is known either experimentally or statistically [5]. However, the real evolution of maintenance has occurred in the past three decades with the emergence of condition-based maintenance (CBM), which is a "decision-making strategy that enables real-time diagnosis of failure and prognosis of future equipment health" [6]. By integrating sensors into operating machines, the system's health state can be quantified in a proactive manner. It is desirable for the CBM policy to evolve from a diagnostic approach based on fault detection to a prognostic approach based on degradation and fault prediction. The "prognostic process" allows for the prediction and anticipation of machine failure, resulting in numerous benefits in the fields of safety, economics, and resource management [3].

Prognostics and health management (PHM) "refers specifically to the phase involved with predicting future behaviour, including remaining useful life (RUL), in terms of current operating state and the scheduling of required maintenance actions to maintain system health" [7]. There have been many algorithms developed for predicting the health state of a machine within the context of PHM [6,8,9]. However, to the best of our knowledge, only a limited number of approaches are currently available for estimating the probability that a monitored machine will successfully complete a predefined workload by integrating data about its historical behaviour and current health state. While the number of these algorithms has significantly increased in recent years, most of them are only suitable for machines operating under constant workload conditions. In manufacturing, it is important for a prognostic tool to consider both sensor data and information about previous and future workloads [10].

The objective of this study is to explore the potential benefits and challenges of Industry 4.0 and its impact on maintenance strategies, specifically focusing on the integration of prognostics and health management (PHM) into the manufacturing process. In particular, this work aims to: (i) propose the integration of machine workload information, previously not considered in other studies, into a well-established PHM algorithm; (ii) formulate a methodology for integrating machine workload information into any given PHM algorithm in order to explore the potential benefits and challenges of Industry 4.0 on maintenance strategies and to contribute to the development of more efficient and effective maintenance strategies in the industry.

The relevance of this work to practitioners lies in the potential for improved decision-making in maintenance strategies for Industry 4.0 production systems. By incorporating machine workload information into a well-established prognostic model, practitioners can benefit from more accurate forecasting of failure probability, leading to more efficient utilisation of production resources and lower maintenance costs. Additionally, the proposed approach can be applied to various monitored machines and equipment without any specific constraints, making it a versatile solution for practitioners in various industrial contexts. Overall, this research aims to provide practitioners with a practical and effective solution for improving the autonomy of CPSs in accepting or declining scheduled jobs according to their forecasted health state, ultimately leading to improved production efficiency and cost savings.

The proposed prognostic approach consists of four different phases. The first stage involves acquiring signals from the sensors of the monitored machine. In the second step, a degradation assessment model is developed using the logistic regression tool. In the third phase, a feed-forward neural network (FNN) is used to build the prognostic model. Finally, in the final step, a set of auxiliary artificial networks (ANNs) are added to the model to integrate workload information into the data. This model, due to the versatility of the analytical tools employed, can be applied to various monitored machines and equipment without any specific constraints or hypotheses on their behaviour or application context.

The remainder of this paper is organised as follows. In Section 2, the related literature on prognostics and the digital industry is discussed. Section 3 presents the context and formal description of the prognostic model, while in Section 4, a case study is considered and all computational results are presented. Finally, Section 5 concludes the work and discusses further research.

2. Literature Review

2.1. Literature Review Strategy

In order to thoroughly investigate the potential benefits and challenges of integrating machine workload information into a well-established PHM algorithm for Industry 4.0 and its impact on maintenance strategies in the manufacturing process, a comprehensive literature review was conducted. The literature review strategy included a search of academic databases, such as Scopus, Web of Sciences (WOS), ScienceDirect, and IEEE Xplore, using relevant keywords such as "Industry 4.0", "maintenance strategies", "prognostics and health management", "machine workload information", and "manufacturing". Additionally, a manual search of relevant journals and conference proceedings related to the fields was conducted to ensure a broad and diverse range of literature was included in the review.

The literature reviewed covers a range of topics including the definition and key components of Industry 4.0, the integration of CPSs and IoT in the manufacturing process, the development of new maintenance strategies within an Industry 4.0 context, and the integration of machine workload information into PHM algorithms. The literature reviewed also includes various case studies and examples of the implementation of these concepts in various industry sectors. The literature review aimed to provide a clear understanding of the current state of research in this field and identify gaps in the literature that this study aims to address.

2.2. Industry 4.0, Cyber-Physical System Architecture, and the First Integration with Prognostics

The term "Industrie 4.0" first appeared in 2011 at the Hannover Fair when the German government used it to designate a strategic approach to manufacturing based on a significant degree of manufacturing automation. Later, the English term "Industry 4.0" was used to identify the ongoing revolution involving manufacturing contexts and which is based upon a stricter interaction between manufacturing robots, humans, and machines themselves [11–15]. While in the scientific literature the essence and the very definition of this new production paradigm is still debated, we focus here on the established key components of Industry 4.0. Particularly, four main cornerstones epitomise I4.0; namely, (i) cyber-physical systems (CPS) (Figure 1), (ii) the Internet of Things (IoT), (iii) the Internet of Services (IoS) and (iv) smart factories [16].

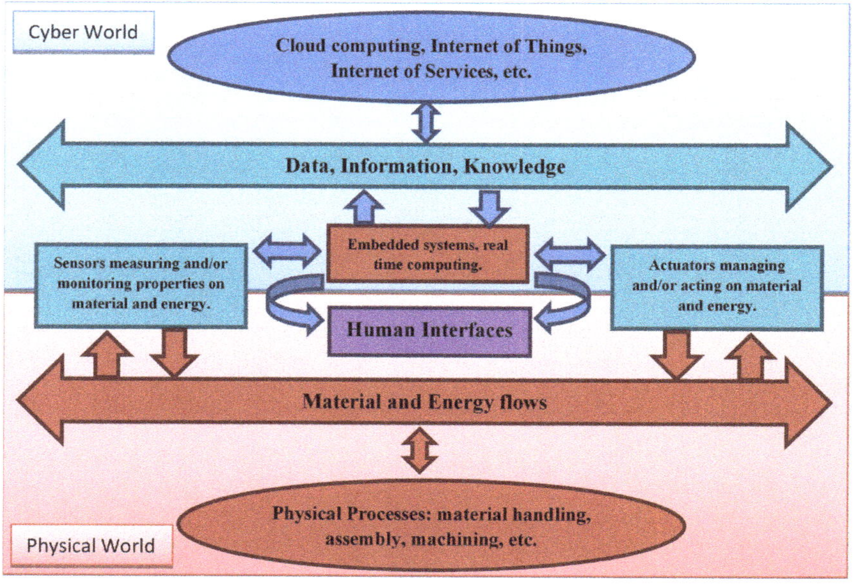

Figure 1. Cyber-physical system architectures.

If the CPS represents the "mechatronic hardware" that allows, with its varying capabilities, the fusion of the physical and virtual world [17,18], the IoT and IoS turn out to be the strictly connected pivotal elements whose integration into the manufacturing system permits the fourth industrial revolution to take place. In fact, the "Internet of Things" makes it possible to connect "things", according to a new architecture in which each CPS may cooperate through addressing schemas [19]. The "Internet of Services", on the other hand, represents the layer through which the various entities in the I4.0 ecosystem interact and share their services over the internet. These services could be provided and combined into value-added services by different suppliers, allowing a new way of value chain activities [20].

A functional integration of the above-mentioned components effectively synthesises the smart factory concept, defined "as a factory that context-aware assists people and machines in executions of their tasks" [21]. Therefore, smart factories have a modular infrastructure, where CPSs monitor physical processes, make decentralised decisions, and cooperate with other CPSs, humans, or other value chain contributors over the IoT and IoS. In a smart factory, not only is there a significant change in interactions among resources, but also in how production systems are operated and decisions are made. In this context, the large amount of data collected by the IoT–IoS architecture pave the way for a higher level of production system control [22].

From a smart factory perspective, Lee et al. in [23] introduce the unified 5C-level architecture as a guideline for the implementation of CPSs in I4.0 manufacturing systems. This work is to be considered as the first attempt to integrate the CPS concept into the current industrial environment, representing, as a matter of fact, the starting point of our research. In [23], the various architectures of the new manufacturing paradigm are analysed (i.e., smart connection, data-to-information conversion, cyber, cognition, and configuration levels), focusing on the cyber level. At this level, as recognised by the Authors, an algorithm is necessary to track changes in the state of a particular machine or equipment, hence deriving additional knowledge from its historical data. To this aim, the same Authors in [24,25] introduce the "Time Machine" approach consisting of three parallel sections: snapshot collection, similarity identification, and synthesis optimised future steps.

Furthermore, Lee et al. in [24,26,27] proposed an effective self-aware and self-maintaining machine system, showing the first example of a prognostics tool applied to a diesel engine, with the utilisation of several algorithms from the Watchdog Agent® toolbox. However, in the proposed tool, the Authors did not consider the information about the past, present, and future workload. In essence, their tool is focused on data-driven monitoring to identify any unusual patterns. In that regard, some recent studies try to integrate the past and future workload information into a failure probability estimation. In particular, Vespoli et al. [28] introduced a machine learning model approach through the use of an artificial neural network (ANN) for the degradation assessment of battery monitoring equipment. This methodology consisted of a unique processing level, and showed the enhanced failure probability estimation obtained from the integration of the workload information in a prognostic tool. Adu-Amankwa et al. [29] investigated small–medium enterprise CNC machine shops' readiness to adopt and sustain predictive maintenance within an I4.0 context and scrutinised the suitability of predictive maintenance practices in the considered context. Negri et al. [30] instead proposed an architecture to integrate a digital twin-based health prediction into the scheduling process of an industrial plant. It minimised the indirect revenue loss caused by a failure in the production schedule.

In the context of I4.0 and CPS implementation, the objective of this paper is to extend the work of the above-cited studies by integrating the historical sensor data of the monitored equipment with information deriving from previous and future workloads. The outcome is represented by a prognostic tool which gives a probability value representing the likelihood that the equipment will fail while performing the scheduled workload. As a result of the objective this research, in the following we briefly provide the readers with basic concepts on prognostics and health management (PHM) and the relevant literature.

2.3. *Prognostics and Health Management (PHM)*

Within the scientific literature, most researchers refer to prognostics as "an engineering discipline focused on predicting the time at which a system or a component will no longer perform its intended function" [3,7]. As far as prognostic models are concerned, Peng et al. [6] proposed a complete review of all the known methodologies of prognostics. They identified four different categories of prognostic approach: physical models, knowledge-based models, data-driven models, and hybrid models. The quality of the prognostic predictions, including steady-state, transient performance, unanticipated conditions, loads, and operational regimes, was the main advantage of a model-based approach [6], while its limitations were high implementation costs, the difficulty of building a good physical model, and the impossibility to adopt a specific model for other systems. The knowledge-based approach relies upon stochastic probabilistic reliability models of equipment/components, or upon experts' judgments. In their review, Peng et al. [6] suggested two main methodologies to this aim; namely, expert system and fuzzy logic. Data-driven approaches are appropriate when the understanding of the system's functioning principles is not comprehensive or when the system is complex and the development of an accurate physical model is too expensive [31,32].

Bektas et al. in [33] explained that during the planning and testing phases, the data-driven model faces the challenges posed by the complexity of real-world systems. In general, these include dealing with incomplete knowledge of future load conditions, inaccurate estimation of the current state of health, poor evolution models, sensor noise, and varying operating conditions [33]. The data-driven methodology is based on statistical and learning techniques, in which the collected data under nominal and degraded conditions are combined and, thanks to a generic machine-learning algorithm, it is possible to generate appropriate prognostic models. In this sense, two data-driven methodologies are identifiable: statistical and artificial approaches. In the literature, there are many application examples of the latter approach. The most used and experimented is the artificial neural network (ANN) and its variants (e.g., feed-forward neural network (FFNN), polynomial neural network (PNN), dynamic wavelet neural network (DWNN), self-organizing feature map (SOM), and multilayer perception neural network (MPNN)). An ANN is a generic algorithm that tries to mimic the human brain structure. It consists of simple processing elements (comparable to biological neurons, that comprises a node and weight) connected to a complex layer structure, which enables the model to approximate a complex nonlinear function with multi-inputs and a single or multi-output [34].

The use of intelligent approaches, notably ANN-based algorithms, are attractive in the areas of prognostics for their potential to reduce the complexity of a problem and to support the ability of time series prediction. The performance of existing data-driven methods leveraging deep learning models, such as the recurrent neural network (RNN), heavily depends on the quality and size of the available training data used to optimise their model parameters. However, gathering large amounts of run-to-failure training data can be very costly and time-consuming [32].

In the literature, generic ANNs are often used for prognostic purposes. Rivas et al. in [35], for example, developed a model based on recurrent neural networks to identify possible asset malfunctions. In particular, the proposed framework analyses a dataset of historical sensor data in order to perform predictive maintenance on a set of engines. Zhang and Ganesan [36] theorised an intelligent diagnostic in which a self-organised neural network was deployed for performing a multivariable trending of faults, applied to the RUL estimation of bearing systems. In fault diagnostics, the most used models of artificial networks are the feed-forward neural network (FFNN) configurations, in which information moves in only one direction [37–40]. In the literature, the use of FFNNs is proposed for gears and bearings [41–45], engines [46], and other types of mechanical components. Recently, the literature has also investigated hybrid approaches which combine data-driven prognostics and model-based prognostics, hence benefiting from both their advantages (precision and applicability) [47].

3. Problem Formulation and Proposed Model

3.1. Problem Definition, Hypothesis, and General Purpose of the Model

The interconnection opportunities introduced by Industry 4.0 show all their potentiality when continuous information exchange between CPSs is in place. The literature review declares that the most important feature of the fourth revolution is represented by the possibility of autonomous scheduling and (potentially) rescheduling in CPSs within a short period, to avoid, for instance, an unscheduled stop or a safety accident. In order to make this prospect concrete, it is essential to have in-depth knowledge of the CPS itself, with a prognostic tool able to predict the likelihood that an industrial asset may complete a predefined job.

Within the literature, the remaining useful life (RUL) is a well-established reference parameter in equipment prognostics. However, the Industry 4.0 capabilities agree to avoid the use of this parameter, uncertain of its nature, favouring the estimation of a failure probability value. This value, strictly related to the health state of the equipment, may allow a more efficient balancing between production and maintenance activities. Every component or equipment usually produces several kinds of signals when working conditions change over time (e.g., a higher temperature, different vibrations, or significant acoustic emissions). The problem is that there are many different equipment types, and, in a real vision, every equipment owns its history and its behaviour, due to previous unbalanced workload and uncertain material properties.

Therefore, the design of a prognostic algorithm, able to consider a different kind of workload reassuming it in a single value, is desirable. This value should reflect the failure probability of the future workload to be scheduled on the machine, resulting in a sort of equipment degradation indicator. This physical feature and, most of all, methodology to be calculated must be universally applicable on each CPS and not dependent on its specific technological processing, both for general reasons of plant maintenance management and certain reasons of communication policy between CPSs. From a theoretical perspective, it would be possible to conceive a particularised prognostics algorithm for each set of machines. However, this paper aims to derive a data-driven model that could be easily adapted to whatever machine and operational scenario, only by changing the involved training dataset.

The proposed self-detected condition-based maintenance (SdCBM) policy needs four main requirements to be developed [Figure 2]:

- Cyber-physical system signal acquisition;
- Degradation modelling of the monitored machine/equipment;
- Prognostics generation model;
- Failure probabilities normalisation as a function of the next scheduled job.

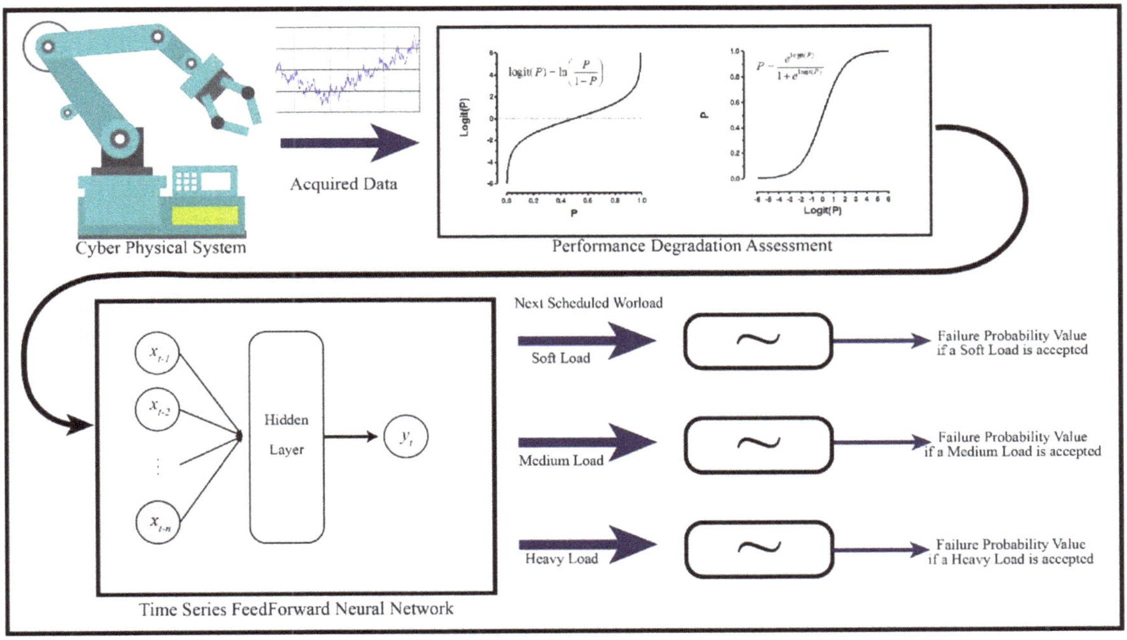

Figure 2. Self-detected condition-based maintenance framework.

In agreement with the previously discussed requirements for a widely applicable model that can be concretely implementable in a production plant, the main and most significant effort of this paper consists of achieving a model both able to respect all the emerged hypotheses and constraints, to be composed of tools largely used and computationally consolidated, and to provide a feature strictly related to the RUL, but innovatively defined by an appropriate combination of the selected computational tools.

3.2. The Proposed Model: CPS Signal Acquisition

From an operational perspective, the design and implementation of the CPS represent the most relevant concern of the proposed study, as they represent the connection between the physical world (the monitored machine/equipment) and the cyber world. As discussed in Section 2, Lee et al. [23] proposed a unified 5C-level architecture as a guideline for the implementation of CPSs in Industry 4.0 manufacturing systems (smart connector level). In this sense, the signal acquisition is part of the CPS architecture itself. Regarding the behaviour and structure of the monitored machine/equipment, it is essential to choose carefully:

- The typology and features of the sensors to be used (e.g., accuracy and response time);
- Their positioning within the system;
- A standardised communication protocol for data transfer, such as MTConnect [27].

By analysing the history of the failures of a generic system together with its design features, it is possible to derive some recurrent problems and critical components. It is worth noting that the analyst may also adopt more sophisticated top-down approaches, such as Failure Modes, Effects and Criticality Analysis. (FMECA, MIL-P-1629, 1949), to decompose the system into its constituent elements in order to identify system's critical failures. The state of critical components must be monitored and analysed. Even if our approach can be applied to every machine/equipment, the initial definition, deployment, and setup of the CPS need to be conducted with the help of maintenance technicians. Their initial contribution, in fact, together with the design team's support, turn out to be crucial for a successful implementation at least at the current stage of development. However, in the coming years, the achievements of big data analysis and the availability of data from different kinds of machines will permit an automatic or "supervised" choice of the sensors themselves.

3.3. The Proposed Model: Degradation Modelling of the Monitored Machine/Equipment

As discussed in the Literature Review section, the degradation assessment of industrial machines and equipment for maintenance purposes is often accomplished by means of logistic regression. For instance, Yan et al. [48] applied logistic regression for the degradation assessment of the motor motion of an elevator door with the use of an ARMA-based model for the estimation of the remaining useful life. Caesarendra et al. [49] instead proposed the use of logistic regression to estimate the failure degradation of bearings. While based on a run-to-failure data set, their research considered the result as target vectors of failure probabilities, exploiting an RVM (relevance vector machine) algorithm as an intelligent system for predicting the future failure probabilities. As a consequence of the previous studies and the flexibility of logistic regression, we adopted this regression model for the degradation assessment in the proposed algorithm. Logistic regression considers a binary "outcome"; that is, a binary dependent variable whose values are usually coded as zero for the negative outcome (no failure event) and as one for the positive outcome (failure event).

Equation (1) gives the probability of this event:

$$Prob(event) = P(\vec{x}) = \frac{e^{(\alpha+\beta_1 x_1+\beta_2 x_2+\ldots+\beta_n x_n)}}{1+e^{(\alpha+\beta_1 x_1+\beta_2 x_2+\ldots+\beta_n x_n)}} \quad (1)$$

where $\vec{x} = (x_1 + x_2 + \ldots + x_3)$ is the input vector of independent variables reproducing the system's state, while α, β_1, β_2, ..., β_n represent the regression coefficients. The estimation of these coefficients is one of the critical activities of model fine-tuning (as the regression coefficients synthesise the physical knowledge and behaviour of the monitored machine/equipment). If an extensive database of historical data is available, these coefficients can be derived with the maximum likelihood estimation method by finding the estimator values that maximise the probability of making the input data given as parameters [48,50].

If historical data are not available, it is possible to acquire only the independent vector variables, without the knowledge of conditions linked to the regular behaviour of the monitored machine/equipment. This knowledge is not sufficient for performing a regression analysis. However, in this case, it is possible to take advantage of expert maintenance technicians' expertise by running different tests on the monitored machine/equipment and setting in advance the corresponding performance level of the monitored condition. Another, and more intuitive, formulation of the logistic regression is shown in Equation (2). It relates to the effect of the dependent variable, not to the probability of the event, but the "odd" of the event (the relationship between the probability of an event and the probability of a non-event):

$$ln\left[\frac{Prob(\vec{x})}{1-Prob(\vec{x})}\right] = \alpha + \beta_1 x_1 + \beta_2 x_2 + \ldots + \beta_n x_n. \quad (2)$$

In Equation (2), the first term is the logarithm of the odds ratio, known as "logit", i.e., the term "Logistic Regression". In this form, there is an apparent linear association between the "logit" and the dependent variables, in contrast to the previous formulation (Equation (1)). In this formulation, logistic regression returns the probability of a specific event as a value in the range 0–1. In the proposed model, it is necessary to build a degradation model, which summarises in a single dependent variable the failure probability of the machine. It is worth pinpointing that, at this stage, we are still in the diagnostic phase. The next step is the prediction of dynamic failure probabilities for the future periods, starting from the performance degradation model.

3.4. The Proposed Model: Prognostics Generation Model

In the prognostic literature, different data-driven techniques are available to predict future failure probabilities. However, the most used methods in this context are relevance vector machine (RVM), Gaussian process regression (GPR), and neural networks (NNs). The flexibility offered by neural networks in the feed-forward configuration is crucial for the entire project's practical feasibility [51]. In this paper, we developed the prognostic model through feed-forward neural networks. However, the general logic and the framework as a whole is still valid in the case of RVM and GPR. Therefore, in the prognostic stage, we used a feed-forward neural network in a time series prediction configuration. This type of neural network application is known as a multi-step prediction in which the network employs the previously predicted value to forecast the future values iteratively. In this way, it is possible to predict a series of values depending on the required number of time steps. The first value predicted from the network (y_t) is a function of the last n recorded values at preceding time steps (x_{t-1}):

$$y_t = f(x_{t-n}, x_{t-n+1}, x_{t-n+2}, \ldots, x_{t-1}) \quad (3)$$

In Equation (3), each recorded value represents an input associated with a different node of the NN. In predicting the next value, y_{t+1}, the last $n-1$ recorded values and the last predicted value are used:

$$y_{t+1} = f(x_{t-n+1}, x_{t-n+2}, \ldots, x_{t-1}, y_t) \quad (4)$$

Then, it is possible to predict all the future desired h values iteratively:

$$y_{t+h} = f(x_{t-n+h}, x_{t-n+h+1}, \ldots, x_{t-1}, y_t, y_{t+1}, \ldots, y_{t+h-1}) \quad (5)$$

In this context, it is essential to define and control each aspect of its development and implementation. As a consequence, each network, which is linked to certain equipment or machine with unique features, presents a different neural structure (specifically, the number of nodes and the activation function of each different node). However, thanks to the uniformity of the value that logistic regression provides, it is possible to standardise the input data activity for each neural network. In our approach, the architecture of the neural network consists of an output layer composed of only one output (the predicted value) and an input layer consisting of the first n recorded values. According to Tan et al. [52], it is advisable to introduce nonlinearity into the network by using some nonlinear activation functions (i.e., the *tanh* or *logit* function) as a combination of functions from input to output layer.

Hence, it is important to find the dimension of the network (i.e., the number of nodes), the weight of nodes, and define each activation function. In particular, the number of active hidden neurons in the network is a critical factor. If the network has a small number of hidden nodes, the performance tends to be ineffective. Meanwhile, if the network has too many hidden nodes, the risk of data over-fitting arises and impedes the desired generalisation. In the literature, it is usual to define the dimension and the activation functions of a neural network with a trial and error strategy. When data acquisition takes place, a set of data is extracted and used for estimating the dimension of the neural network. Starting with the most straightforward network, composed of only one hidden and one output node, an iterative training of these networks (varying the number of nodes and the activation functions) is made with the previously collected data.

For each trial, root mean square deviation (RMSD), which represents a good measure of accuracy to compare forecasting errors of different models, is estimated until the minimum of this value is found. RMSD represents the standard deviation of the difference between the predicted and target value, and it can be calculated as follows:

$$RMSD = \sqrt{\frac{\sum_{i=1}^{n}(y_t - x_t)^2}{n}} \tag{6}$$

where n is the number of time series values, y_t is the predicted value, and x_t is the real monitored value. Once the dimension of the neural network is determined, the same dataset used for this estimation can be used for an optimal adjustment of each synaptic weight between neurons. It is worth mentioning here that these activities are to be personalised for each machine/equipment, as each has its own failure history and "standard" working conditions.

3.5. The Proposed Model: Failure Probability Normalisation as a Function of the Next Scheduled Job

The introduced model can be successfully applied in industrial settings to forecast the future failure probabilities of monitored machines/equipment, considering their performance history and without any information about their future workloads. Thus, the approach presented at this point is capable of functionally extending the prognostic algorithms already available in the literature. Indeed, while the previous steps of the introduced approach are easily replaceable with alternative algorithms and methodologies, we would like to concentrate here on further improving the quality of the predictions by providing these prognostic algorithms with a new piece of information, i.e., the workload, both completed and to be completed. For instance, while forecasting the failure probabilities of a machine whose recent workload has been considerably slight, the prevision error produced by the model could be relevant, especially in the case of value underestimation (e.g., due to a sudden increase in the mechanical load of next scheduled jobs).

For improved accuracy, it would be better to categorise machines into different categories according to the type of loads they are subject to. In the presence of stationary workloads, which may determine a slight or proportional change to a machine's working conditions, the model does not require any amendment (the workload influence turns out to be negligible in this scenario). Under variable workloads, however, the consideration of previous and future workload values for the monitored machine/equipment could be relevant for accuracy enhancement. Potential better performance of the proposed prognostic approach would benefit the algorithm for machine/equipment job scheduling as well. It may consider, as a further variable, the failure probabilities of workloads associated with the various jobs to be scheduled. This may lead to a more efficient utilisation of machines while reducing their RUL under a controlled risk of failure. Hence, we must tackle the new issue of finding the relationship between the failure probabilities, as derived from the logistic regression, and the workload values corresponding to the future scheduled jobs of the monitored machine/equipment.

From an operational point of view, the scheduled job is only an indication of the operations the machine/equipment has to perform in the immediate future. Therefore, identifying and classifying all the possible jobs from a workload perspective turns out to be an important activity. An expert technician, with specialised knowledge of the considered machine/equipment, may complete this activity in the initial stage of project development. However, this same task could be carried out with the help of a pattern recognition neural network. As an example, in the case of three different workload values (soft, medium, and heavy load) for the monitored machine/equipment, it would be easy for the technician to find the relation between the various jobs and their expected workload (e.g., mechanical load) on the machine (Table 1).

Table 1. Technician completed example of mechanical load of the monitored machine.

JOB	Expected Mechanical Load
Job 1	Soft Load
Job 2	Heavy Load
Job 3	Soft Load
Job 4	Medium Load
Job 5	Heavy Load

Although different classifications of machine workload values could be considered, once the relationship mentioned above is established, it will be necessary to introduce this information into the prognostic model. Among the several different technical opportunities, we chose neural networks as the tool to accomplish the task of identifying the link between the future expected workload values and one chosen as a reference. Taking the soft workload value as reference (Table 1), two feed-forward neural networks would be necessary: the first one to find the relationship between the medium workload and the reference workload and the second one to find the other relationship between the heavy workload and the reference workload. The number of hidden neurons and the combination of activation functions for these new neural networks can be found with a trial and error approach similar to that cited in the previous subsection. These networks are structured with one input node and one output node, with a variable number of hidden neurons to be determined. The input node receives as input the value of prognostic failure probability under the medium or heavy workload values, while the output one provides the failure probability that a soft workload value would have produced under the same degradation conditions.

The difficulty of this activity is the identification of valid failure probability values for training these auxiliary networks. In addition, it should be noted that the effect of the workload on machine degradation might differ in different degradation stages. Hence, it is advisable to conduct some pre-tests on the monitored machines/equipment at the various degradation stages. As an example, by running three exemplars of a specific monitored machine/equipment, each one with a different workload value, it is possible to collect failure probability values under different workload values and within the various degradation stages. Although this approach may appear burdensome and expensive, it is worth noting that this data collection may occur during the regular operation of machines/equipment, as well as in future periods, allowing, in this way, a future update of the model itself.

Hence, during its operation, the model knows the current scheduled job for the monitored machine/equipment and, if this job has an expected soft workload, it directly uses the failure probabilities provided from logistic regression to forecast the future output value. Otherwise, if the current job has an expected medium or heavy workload, the output value is then normalised according to the reference value with the help of the two auxiliary neural networks introduced above. This value is then used in the time series feed-forward neural network for forecasting the future failure probability. The output value of the prognostic model thus obtained is referred to as the reference workload (i.e., soft load) and it is necessary, therefore, to adjust this value to the expected workload of the next job.

To this extent, two other auxiliary neural networks, trained to find the relationship between the reference workload value and the expected workload of the next scheduled job, may perform this adjustment. The final output of the model represents the forecasted failure probability values of the monitored machine/equipment. It effectively reproduces the continuous degradation process of the machine/equipment with the knowledge of the future scheduled workload. With this enhanced accuracy, the proposed tool may be integrated into the production schedule activity of Industry 4.0 manufacturing plants (or even into CPSs themselves in an M2M setting), leading to more balanced scheduling of production resources.

4. Logistic Cyber-Physical Systems: A Prototypal Case Study

In the following, we present a case study implementation on a prototypal system. The innovative context of the proposed approach, the substantial lack of algorithms, and the subsequent unavailability of test data (necessary to compare different design solutions), represented a great difficulty of this research. This case study comes from a project experience where a prototypal logistic CPS was built. The objective of this prototype was to demonstrate the potential of Industry 4.0 concerning the material handling problem of a typical job–shop production system. The considered prototype consists of four processing stations, each connected to an automatic conveyor belt by means of a system designed to withdraw/release baskets that are conceived to handle loads of various magnitude. When the scheduled basket is approaching the designed processing station, the system automatically activates the withdraw/release system. From a technical point of view, we used NXT Bricks as actuators groups for the motors of the withdraw/release systems and STM32F4 NUCLEO by ST Microelectronics for the data collection from the various sensors.

The behaviour of our prototype showed failures and malfunctioning due to several different causes, including wear of mechanical parts, insufficient motor thrust, insufficient power supplied by the battery pack of the NXT Brick, and system movements that were out of nominal tolerances. In particular, the power level delivered from the Li-ion accumulator of the brick and the insufficient thrust of the motor (due to the high temperature of its windings) turned out to be the leading causes of the withdraw/release system failure. With reference to these failure causes, the project team identified the physical variables to monitor for an accurate estimation of the failure probability: (i) battery temperature (gauged by a thermocouple); (ii) battery voltage (measured with ADC voltmeter); (iii) motor windings temperature (gauged by a thermocouple); and (iv) rest position and speed of the withdraw/release system (controlled by an encoder). We collected all this data from sensors by means of the above mentioned STM32F4 NUCLEO, which allowed us to transfer in real-time all the data to a PC through a virtual serial connection through USB.

Thus, for the degradation assessment model phase, we took advantage of the assistance of technical experts in electric motors, who specified the parameters to be monitored and included in the logistic regression model previously described. Having defined the parameters to be monitored, nine experimental cycles at the fixed operating parameters were performed until component failure was observed. Thus, the combined data aided the inferential process (based on the maximum likelihood estimation in Section 3) of parameter estimation for the proposed logistic regression model. The considered model is shown in Equation (7):

$$Prob(failure) = Prob(\vec{x}) = \alpha + \beta_1 x_1 + \beta_2 x_2 + \beta_3 x_3 + \beta_4 x_4 + \beta_5 x_5 \tag{7}$$

where x_1 is the battery temperature, x_2 is the battery voltage, x_3 is the motor winding temperature, and x_4 and x_5 are, respectively, the rest position and the speed of the withdraw/release system (Figure 3).

As expected, the estimated regression coefficients showed that the battery voltage (β_2) and the motor winding temperature (β_3) coefficients have the highest values (meaning that these parameters are the most important causes of failure). After the validation of the degradation model, the experiments continued with six other runs, in which the information about each workload being allocated by the production scheduler to the processing station at every time period was available. It should be noted that, in the considered prototype, the workload was represented by the weight of the handled material. However, during the experiments, the scheduler assigned the various jobs to the various stations only to minimise the lead time, disregarding the failure probability value estimated by the CPS.

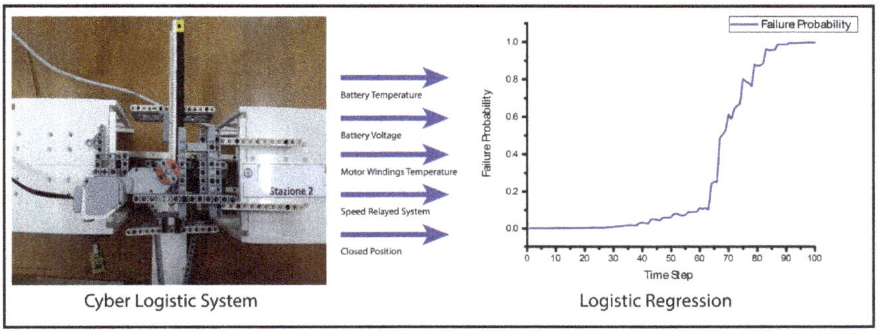

Figure 3. The generated degradation assessment model for our prototype.

In particular, data from six different runs in this configuration were collected until the material handling system failed to complete the assigned tasks (Table 2). After each run, the battery was completely charged and the motor winding was replaced with a regenerated one. Regarding data collection, the following data were collected at a frequency of 4 Hz:

- The battery temperature;
- The battery voltage;
- The motor winding temperature;
- The rest position of the releasing system;
- The speed of the releasing system.

From an accuracy point of view, data were collected with the in-built ADC of the above-mentioned NUCLEO STM32F401 with a resolution of 10 bit, and typical signal noise was removed from the registered data with the use of a low-pass filter. This phase is crucial to obtain smoother signals and hence allows the next phase's algorithm to focus more qualitatively on trends rather than on isolated signal spikes. The total dataset for the next phase was over 17 thousand lines in length. It should be noted that this dataset was divided into three parts; the first part (70%) was used for training, the second part (20%) was used for validation, and the third part (10%) was used for testing. This subdivision is conventional for an artificial neural network algorithm. The training datasets are used to train the neural network, the validation datasets are used to determine the value of the retrofitted gradient, and the test datasets are used to estimate the RMSD value.

Table 2. Recorded failures of our prototype.

Run Number	Jobs Completed before Failure
Run #1	72
Run #2	81
Run #3	58
Run #4	85
Run #5	69
Run #6	75

After collecting data, it was necessary to determine the ANN configuration (i.e., the number of neurons and the combination of activation functions) used for the prognostics generation model. Combining both sets of data (those collected for the degradation assessment model and those described previously), we started the training phase, in a trial and error fashion, with an ANN composed of one input and one hidden neuron, combining various activation functions (sigmoid, logit, tanh, etc.) and estimating the root mean square deviation (RMSD) for each trial. The best performance, in terms of the RMSD (e.g., a value of 10–4), was obtained in the considered case study with a neural network composed of ten input neurons and eleven hidden neurons all with a sigmoid activation function (Figure 4). It should be noted that this configuration allowed for a consistent number of hidden neurons, thereby avoiding the typical ANN over-fitting problem. The MSE of this latter training phase of the artificial neural network is reported in Figure 5. As the neural network under consideration operates in a regressive mode and not as a classifier, while precision, accuracy, and recall could be evaluated, they are not as relevant as the MSE and RMSD values. Regarding the estimation of synaptic weight, a conventional backpropagation algorithm was used during the training phase.

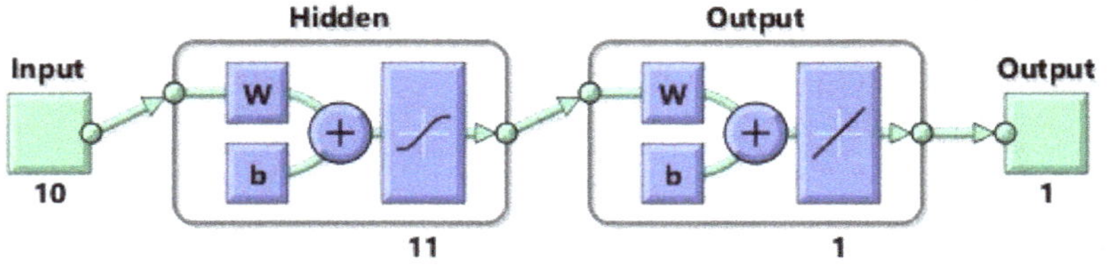

Figure 4. Resulting artificial neural network architecture.

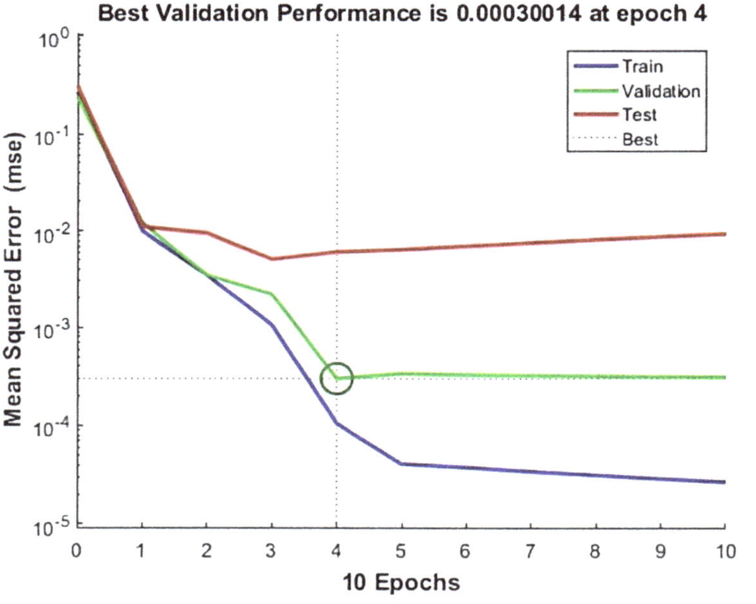

Figure 5. Performance plot of the artificial neural network training phase.

The resulting ANN was used to predict the future failure probability of the withdraw/release system neglecting the current and future workload. The next phase of the proposed approach is the workload normalisation, where new neural networks should be trained with the failure probability values obtained during the normalisation process, hence taking into consideration previous workload information. To this aim, it is necessary to categorise the possible workload into predefined categories. For the considered case, three different categories were identified: a soft load for baskets less than 300 g, a medium load for baskets between 300 and 600 g, and a heavy load for baskets greater than 600 g. As there are three distinct workloads in the case under consideration, we assume that the workload with the lowest load, or "soft load", is the workload against which the others are normalised. Hence, two neural networks for normalisation purposes were implemented (each with a single input, three hidden nodes, and a single output). Then, two other additional neural networks (with the same structure) were constructed with the purpose of reconstructing the output for a total of four auxiliary neural networks. The training data for these neural networks were derived from the nine pre-test experiments conducted for the degradation assessment generation phase on three different prototypes of the same type, which operate under constant workloads. In fact, in these cases, the prototypes were used to collect data on operation under constant soft, medium, and heavy loads, respectively.

As a result, after constructing all necessary neural networks, we proceeded to test the proposed approach in an untrained scenario. The purpose of this study was to prove the benefits of including the workload information in the prognostic algorithm. To this aim, two distinct approaches were tested: the first, a traditional approach, in which the approach has been applied up to step 4, i.e., without incorporating workload information; and a second, the proposed approach, in which workload information was fully integrated. As it is possible to see from the results shown in Table 3, both approaches performed well; however, with the normalisation process of the input and output, the algorithm forecasted the failure probability of the future state with a better performance by an order of magnitude. Figures 6–9 show the failure probabilities of the given tests with a different range of prediction while, as said, Table 3 reports the RMSD values of the final estimation calculated according to Equation (6). These results highlight the improved performance of the predictions in the presence of workload normalisation.

Table 3. RMSD of the k-step final estimation of failure probabilities.

K-Step Predicted Failure	RMSD Values without Normalisation	RMSD Values with Normalisation
T + 1	5.098×10^{-1}	1.533×10^{-2}
T + 2	6.903×10^{-1}	2.658×10^{-2}
T + 3	8.412×10^{-1}	3.281×10^{-2}
T + 4	9.246×10^{-1}	4.307×10^{-2}
T + 5	1.153×10^{0}	6.138×10^{-2}

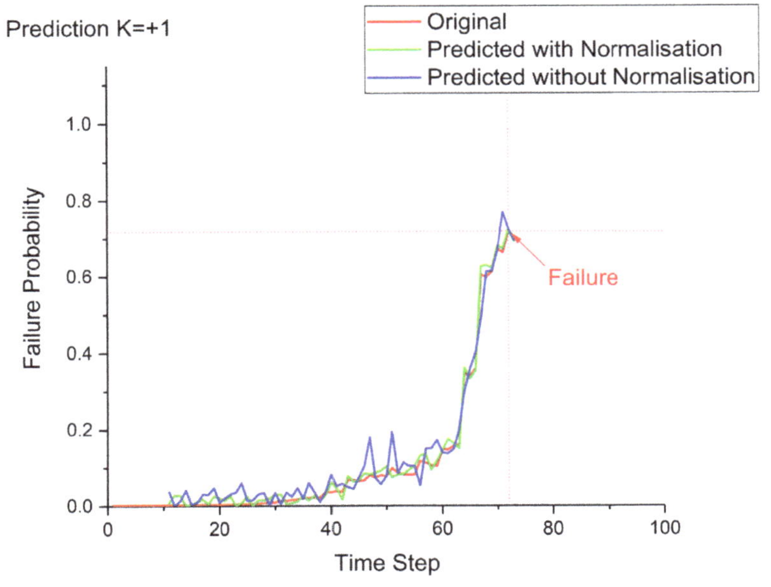

Figure 6. Failure prediction at T + 1.

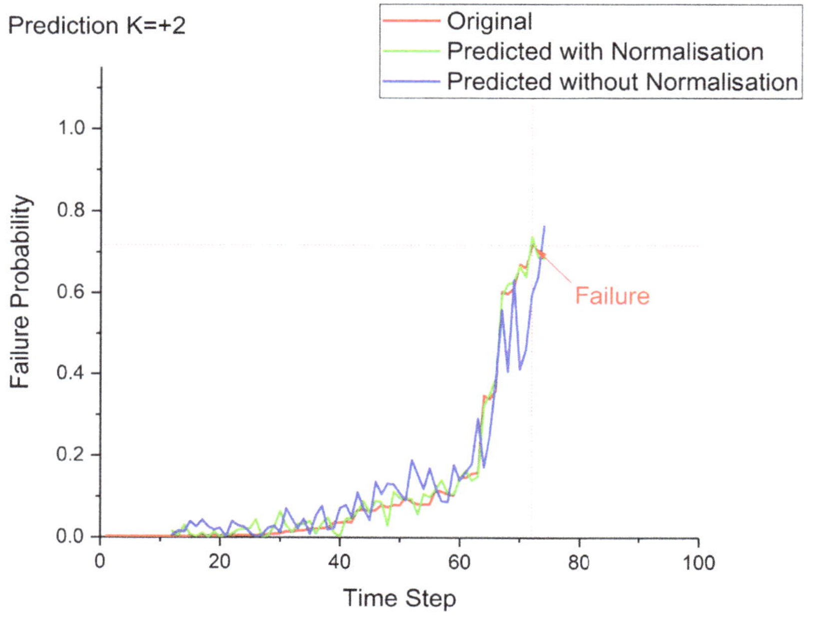

Figure 7. Failure prediction at T + 2.

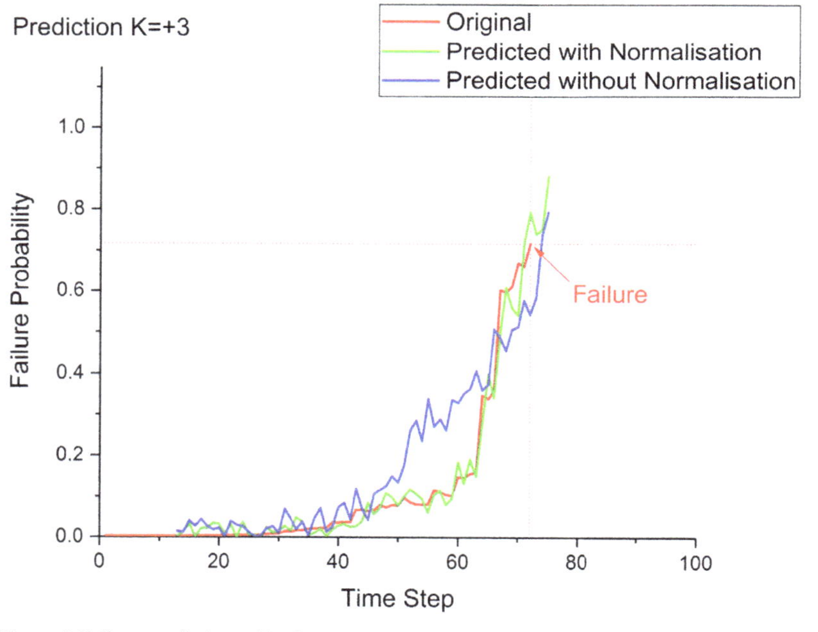

Figure 8. Failure prediction at T + 3.

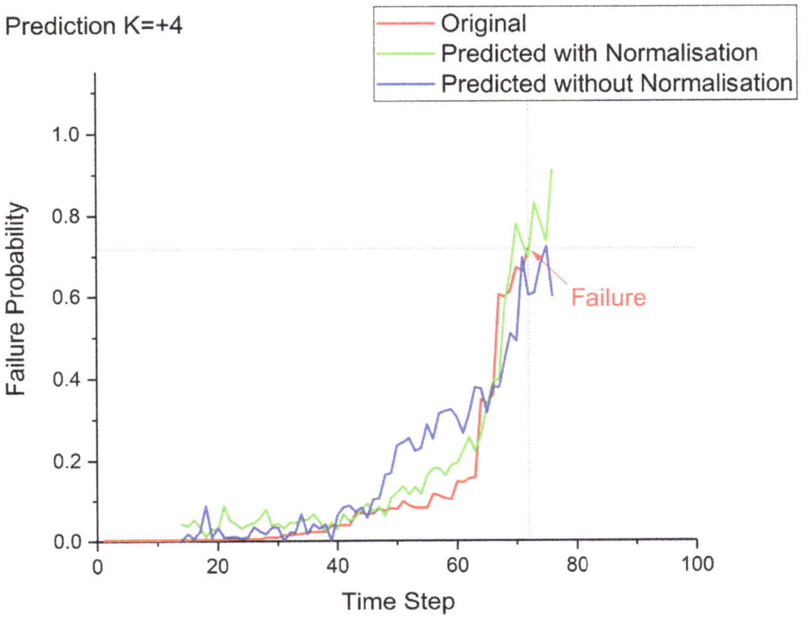

Figure 9. Failure prediction at T + 4.

This approach showed that one can use the most traditional prognostic techniques (logistic regression and neural networks) in emerging possible prognostic techniques. However, based on the promising results obtained and demonstrated, it is clear that whenever possible, information on past and future workload is valuable information that is frequently already available for possible use but has not been utilised for this objective. As a result, this paper showed that the use of such information must be considered in future prognostic applications tasked with predicting the state of health of machines subjected to varying magnitudes of loads.

5. Conclusions

In conclusion, this study aimed to explore the potential benefits and challenges of integrating machine workload information into a well-established PHM algorithm for Industry 4.0 and its impact on maintenance strategies in the manufacturing process. A comprehensive literature review revealed that while there is significant interest in this problem and its potential solutions, there is a lack of specific PHM algorithms geared towards practical implementation in the manufacturing domain due to the required computing power.

To address this gap, a novel approach that integrates machine workload information into a well-established PHM algorithm to estimate the failure probability of machines or equipment was proposed. This approach was tested on a real-world manufacturing system and the results showed that the integration of machine workload information into the PHM algorithm notably improved the accuracy of failure probability forecasting. In particular, for the considered case study, the integration of planned workload information into a classical PHM algorithm resulted in an increased accuracy estimation of the failure probability of 15%.

The results of this study suggest that this approach has the potential to improve the autonomy of CPSs in accepting or declining scheduled jobs according to their forecasted health state and to lower maintenance costs, while maximising the utilisation of production resources by providing CPSs with a more accurate awareness of their functioning capabilities. However, the proposed approach also presents challenges, such as the integration of this approach into a hierarchical manufacturing system and the potential conflicts that may arise from the communication and negotiation among CPSs within the plant. Therefore, future research should investigate these challenges and investigate the potential improvements that can be achieved by integrating machine workload information into PHM algorithms in different industrial contexts. Additionally, future research should focus on how to implement and deploy the proposed approach in the manufacturing domain.

Author Contributions: Conceptualization, G.C. and S.V.; Methodology, S.V.; Software, S.V.; Formal analysis, M.G.; Investigation, S.V.; Writing—original draft, G.C. and S.V.; Writing—review & editing, M.G.; Supervision, T.M. All authors have read and agreed to the published version of the manuscript.

Funding: This research received no external funding.

Institutional Review Board Statement: Not applicable.

Informed Consent Statement: Not applicable.

Data Availability Statement: Not applicable.

Conflicts of Interest: The authors declare no conflict of interest.

References

1. Giusto, D.; Iera, A.; Morabito, G.; Atzori, L. *The Internet of Things: 20th Tyrrhenian Workshop on Digital Communications*; Springer: New York, NY, USA, 2010.
2. Zuehlke, D. SmartFactory—Towards a factory-of-things. *Annu. Rev. Control.* **2010**, *34*, 129–138. [CrossRef]
3. Dragomir, O.; Gouriveau, R.; Dragomir, F.; Minca, E.; Zerhouni, N. Review of prognostic problem in condition-based maintenance. In *2009 European Control Conference (ECC)*; IEEE: Budapest, Hungary, 2009; pp. 1585–1592. [CrossRef]
4. Mobley, R. Predictive Maintenance. In *Plant Engineer's Handbook*; Elsevier: Burlington, MA, USA, 2002; pp. 867–888. [CrossRef]
5. Gertsbakh, I.B. *Gertsbakh: Models of Preventive Maintenance*; Elsevier: Amsterdam, The Netherlands, 1977.
6. Peng, Y.; Dong, M.; Zuo, M. Current status of machine prognostics in condition-based maintenance: A review. *Int. J. Adv. Manuf. Technol.* **2010**, *50*, 297–313. [CrossRef]
7. Vachtsevanos, G.; Lewis, F.; Roemer, M.; Hess, A.; Wu, B. *Intelligent Fault Diagnosis and Prognosis for Engineering Systems*; Wiley: Hoboken, NJ, USA, 2006. [CrossRef]
8. Achouch, M.; Dimitrova, M.; Ziane, K.; Sattarpanah Karganroudi, S.; Dhouib, R.; Ibrahim, H.; Adda, M. On Predictive Maintenance in Industry 4.0: Overview, Models, and Challenges. *Appl. Sci.* **2022**, *12*, 8081. [CrossRef]
9. Aliyu, R.; Mokhtar, A.A.; Hussin, H. Prognostic Health Management of Pumps Using Artificial Intelligence in the Oil and Gas Sector: A Review. *Appl. Sci.* **2022**, *12*, 11691. [CrossRef]
10. Ruschel, E.; Santos, E.; Loures, E. Industrial maintenance decision-making: A systematic literature review. *J. Manuf. Syst.* **2017**, *45*, 180–194. [CrossRef]
11. Peron, M.; Fragapane, G.; Sgarbossa, F.; Kay, M. Digital Facility Layout Planning. *Sustainability* **2020**, *12*, 3349. [CrossRef]
12. Ceruti, A.; Marzocca, P.; Liverani, A.; Bil, C. Maintenance in aeronautics in an Industry 4.0 context: The role of Augmented Reality and Additive Manufacturing. *J. Comput. Des. Eng.* **2019**, *6*, 516–526. [CrossRef]
13. Jamwal, A.; Agrawal, R.; Sharma, M.; Giallanza, A. Industry 4.0 Technologies for Manufacturing Sustainability: A Systematic Review and Future Research Directions. *Appl. Sci.* **2021**, *11*, 5725. [CrossRef]
14. Miśkiewicz, R.; Wolniak, R. Practical Application of the Industry 4.0 Concept in a Steel Company. *Sustainability* **2020**, *12*, 5776. [CrossRef]
15. Angelopoulos, A.; Michailidis, E.T.; Nomikos, N.; Trakadas, P.; Hatziefremidis, A.; Voliotis, S.; Zahariadis, T. Tackling Faults in the Industry 4.0 Era—A Survey of Machine-Learning Solutions and Key Aspects. *Sensors* **2019**, *20*, 109. [CrossRef]
16. Ruiz-Sarmiento, J.R.; Monroy, J.; Moreno, F.A.; Galindo, C.; Bonelo, J.M.; Gonzalez-Jimenez, J. A predictive model for the maintenance of industrial machinery in the context of industry 4.0. *Eng. Appl. Artif. Intell.* **2020**, *87*, 103289. [CrossRef]
17. Kagermann, H.; Reinhart, J. Strategic Enterprise Management (SEM). In *Fortschritte im Rechnungswesen*; Gabler Verlag: Wiesbaden, Germany, 2013; pp. 329–354. [CrossRef]
18. Bauer, W.; Horváth, P. Industrie 4.0—Volkswirtschaftliches Potenzial für Deutschland. *Controlling* **2014**, *27*, 515–517. [CrossRef]
19. Sittón-Candanedo, I.; Alonso, R.; Rodríguez-González, S.; García Coria, J.; De La Prieta, F. Edge Computing Architectures in Industry 4.0: A General Survey and Comparison. In *Advances in Intelligent Systems and Computing*; Springer International Publishing: Berlin/Heidelberg, Germany, 2019; Volume 950, pp. 121–131. [CrossRef]
20. Buxmann, P.; Hess, T.; Ruggaber, R. Internet of Services. *Business Inf. Syst. Eng.* **2009**, *1*, 341–342. [CrossRef]
21. Lucke, D.; Constantinescu, C.; Westkämper, E. Smart Factory—A Step towards the Next Generation of Manufacturing. In *Manufacturing Systems and Technologies for the New Frontier*; Springer: London, UK, 2008; pp. 115–118. [CrossRef]
22. Weichert, D.; Link, P.; Stoll, A.; Rüping, S.; Ihlenfeldt, S.; Wrobel, S. A review of machine learning for the optimization of production processes. *Int. J. Adv. Manuf. Technol.* **2019**, *104*, 1889–1902. [CrossRef]

23. Lee, J.; Bagheri, B.; Kao, H.A. A Cyber-Physical Systems architecture for Industry 4.0-based manufacturing systems. *Manuf. Lett.* **2015**, *3*, 18–23. [CrossRef]
24. Lee, J.; Ni, J.; Djurdjanovic, D.; Qiu, H.; Liao, H. Intelligent prognostics tools and e-maintenance. *Comput. Ind.* **2006**, *57*, 476–489. [CrossRef]
25. Lee, J.; Ardakani, H.; Yang, S.; Bagheri, B. Industrial Big Data Analytics and Cyber-physical Systems for Future Maintenance; Service Innovation. *Procedia CIRP* **2015**, *38*, 3–7. [CrossRef]
26. Lee, J.; Kao, H.A.; Yang, S. Service Innovation and Smart Analytics for Industry 4.0 and Big Data Environment. *Procedia CIRP* **2014**, *16*, 3–8. [CrossRef]
27. Athulan, V.; Will, S.; Armando, F.; David, D.; Paul, W. *Improving Machine Tool Interoperability Using Standardized Interface Protocols: MT Connect. Laboratory for Manufacturing and Sustainability*; ASME Publisher: New York, NY, USA, 2008.
28. Vespoli, S.; Guizzi, G.; Converso, G.; Popolo, V.; Tedesco, A. An electrical DC Motor Equivalent Circuit testbed for the battery Prognostic Health and Management. In *2019 II Workshop on Metrology for Industry 4.0 and IoT (MetroInd4.0;IoT)*; IEEE: Brescia, Italy, 2019; Volume 1, pp. 186–196. [CrossRef]
29. Adu-Amankwa, K.; Attia, A.; Janardhanan, M.; Patel, I. A predictive maintenance cost model for CNC SMEs in the era of Industry 4.0. *Int. J. Adv. Manuf. Technol.* **2019**, *104*, 3567–3587. [CrossRef]
30. Negri, E.; Ardakani, H.; Cattaneo, L.; Singh, J.; Macchi, M.; Lee, J. A Digital Twin-based scheduling framework including Equipment Health Index and Genetic Algorithms. *IFAC-PapersOnLine* **2019**, *52*, 43–48. [CrossRef]
31. Dvorzak, M.; Magnien, J.; Kleb, U.; Kraker, E.; Mücke, M. Bayesian Hierarchical Modelling for Uncertainty Quantification in Operational Thermal Resistance of LED Systems. *Appl. Sci.* **2022**, *12*, 63. [CrossRef]
32. Sadoughi, M.; Lu, H.; Hu, C. A Deep Learning Approach for Failure Prognostics of Rolling Element Bearings. In Proceedings of the 2019 IEEE International Conference on Prognostics and Health Management (ICPHM), San Francisco, CA, USA, 17–20 June 2019; pp. 1–7. [CrossRef]
33. Bektas, O.; Jones, J.; Sankararaman, S.; Roychoudhury, I.; Goebel, K. A neural network filtering approach for similarity-based remaining useful life estimation. *Int. J. Adv. Manuf. Technol.* **2019**, *101*, 87–103. [CrossRef]
34. Wen, F.; Yang, X.; Gong, X.; Lai, K. Multi-Scale Volatility Feature Analysis and Prediction of Gold Price. *Int. J. Inf. Technol. Decis. Mak.* **2017**, *16*, 205–223. [CrossRef]
35. Rivas, A.; Fraile, J.; Chamoso, P.; González-Briones, A.; Sittón, I.; Corchado, J. A Predictive Maintenance Model Using Recurrent Neural Networks. In *Advances in Intelligent Systems and Computing*; Springer International Publishing: Berlin/Heidelberg, Germany, 2019; Volume 950, pp. 261–270. [CrossRef]
36. Zhang, S.; Ganesan, R. Multivariable Trend Analysis Using Neural Networks for Intelligent Diagnostics of Rotating Machinery. *J. Eng. Gas Turbines Power* **1997**, *119*, 378–384. [CrossRef]
37. Jakkamputi, L.; Devaraj, S.; Marikkannan, S.; Gnanasekaran, S.; Ramasamy, S.; Rakkiyannan, J.; Xu, Y. Experimental and Computational Vibration Analysis for Diagnosing the Defects in High Performance Composite Structures Using Machine Learning Approach. *Appl. Sci.* **2022**, *12*, 12100. [CrossRef]
38. Roemer, M.; Hong, C.; Hesler, S. Machine Health Monitoring and Life Management Using Finite-Element-Based Neural Networks. *J. Eng. Gas Turbines Power* **1996**, *118*, 830–835. [CrossRef]
39. Fan, Y.; Li, C. Diagnostic rule extraction from trained feedforward neural networks. *Mech. Syst. Signal Process.* **2002**, *16*, 1073–1081. [CrossRef]
40. Battineni, G.; Sagaro, G.G.; Nalini, C.; Amenta, F.; Tayebati, S.K. Comparative Machine-Learning Approach: A Follow-Up Study on Type 2 Diabetes Predictions by Cross-Validation Methods. *Machines* **2019**, *7*, 74. [CrossRef]
41. Tao, L.; Yang, C.; Cheng, Y.; Lu, C.; Ragulskis, M. Machine component health prognostics with only truncated histories using geometrical metric approach. *Mech. Syst. Signal Process.* **2018**, *113*, 168–179. [CrossRef]
42. Larson, E.; Wipf, D.; Parker, B. Gear and bearing diagnostics using neural network-based amplitude and phase demodulation. In Proceedings of the 51st Meeting of the Society for Machinery Failure Prevention Technology, New York, NY, USA, 19–22 April 1997; pp. 511–521.
43. Li, B.; Chow, M.Y.; Tipsuwan, Y.; Hung, J. Neural-network-based motor rolling bearing fault diagnosis. *IEEE Trans. Ind. Electron.* **2000**, *47*, 1060–1069. [CrossRef]
44. Gebraeel, N.; Lawley, M. A Neural Network Degradation Model for Computing and Updating Residual Life Distributions. *IEEE Trans. Autom. Sci. Eng.* **2008**, *5*, 154–163. [CrossRef]
45. Bhavsar, K.; Vakharia, V.; Chaudhari, R.; Vora, J.; Pimenov, D.Y.; Giasin, K. A Comparative Study to Predict Bearing Degradation Using Discrete Wavelet Transform (DWT), Tabular Generative Adversarial Networks (TGAN) and Machine Learning Models. *Machines* **2022**, *10*, 176. [CrossRef]
46. Dong, D.; Hopfield, J.; Unnikrishnan, K. Neural networks for engine fault diagnostics. In Proceedings of the Neural Networks for Signal Processing VII. Proceedings of the 1997 IEEE Signal Processing Society Workshop, Amelia Island, FL, USA, 24–26 September 1997; Volume 7, pp. 636–644. [CrossRef]
47. Medjaher, K.; Zerhouni, N. Hybrid prognostic method applied to mechatronic systems. *Int. J. Adv. Manuf. Technol.* **2013**, *69*, 823–834. [CrossRef]
48. Yan, J.; Koç, M.; Lee, J. A prognostic algorithm for machine performance assessment and its application. *Prod. Plan. Control* **2004**, *15*, 796–801. [CrossRef]

49. Caesarendra, W.; Widodo, A.; Yang, B.S. Application of relevance vector machine and logistic regression for machine degradation assessment. *Mech. Syst. Signal Process.* **2010**, *24*, 1161–1171. [CrossRef]
50. Czepiel, S.A. Maximum Likelihood Estimation of Logistic Regression Models: Theory and Implementation. Available online: https://czep.net/stat/mlelr.pdf (accessed on 21 December 2023).
51. Zhao, J.; Xu, Z.X.; Zuo, D.P.; Wang, X.M. Temporal variations of reference evapotranspiration and its sensitivity to meteorological factors in Heihe River Basin, China. *Water Sci. Eng.* **2015**, *8*, 1–8. [CrossRef]
52. Tan, Y.; Van Cauwenberghe, A. Neural-network-based d-step-ahead predictors for nonlinear systems with time delay. *Eng. Appl. Artif. Intell.* **1999**, *12*, 21–35. [CrossRef]

Disclaimer/Publisher's Note: The statements, opinions and data contained in all publications are solely those of the individual author(s) and contributor(s) and not of MDPI and/or the editor(s). MDPI and/or the editor(s) disclaim responsibility for any injury to people or property resulting from any ideas, methods, instructions or products referred to in the content.

applied sciences

MDPI

Article

Intra-Domain Transfer Learning for Fault Diagnosis with Small Samples

Liangwei Zhang [1], Junyan Zhang [1], Yeping Peng [2] and Jing Lin [3,*]

1 Department of Industrial Engineering, Dongguan University of Technology, Dongguan 523808, China; liangwei.zhang@dgut.edu.cn (L.Z.); chanjiang0702@gmail.com (J.Z.)
2 Guangdong Key Laboratory of Electromagnetic Control and Intelligent Robots, College of Mechatronics and Control Engineering, Shenzhen University, Shenzhen 518060, China; yeping.peng@szu.edu.cn
3 Division of Operation and Maintenance, Luleå University of Technology, 97187 Luleå, Sweden
* Correspondence: janet.lin@ltu.se

Abstract: The concept of deep transfer learning has spawned broad research into fault diagnosis with small samples. A considerable covariate shift between the source and target domains, however, could result in negative transfer and lower fault diagnosis task accuracy. To alleviate the adverse impacts of negative transfer, this research proposes an intra-domain transfer learning strategy that makes use of knowledge from a data-abundant source domain that is akin to the target domain. Concretely, a pre-trained model in the source domain is built via a vanilla transfer from an off-the-shelf inter-domain deep neural network. The model is then transferred to the target domain using shallow-layer freezing and finetuning with those small samples. In a case study involving rotating machinery, where we tested the proposed strategy, we saw improved performance in both training efficiency and prediction accuracy. To demystify the learned neural network, we propose a heat map visualization method using a channel-wise average over the final convolutional layer and up-sampling with interpolation. The findings revealed that the most active neurons coincide with the corresponding fault characteristics.

Keywords: fault diagnosis; transfer learning; time-frequency spectrum; small samples; heat map

1. Introduction

Fault diagnosis is a crucial aspect of system safety and dependability [1], as it can pinpoint fault locations, identify fault types, and even gauge the severity of certain faults [2]. Existing approaches to fault diagnosis can generally be categorized as either model-based or data-driven [3]. The former approach necessitates a deep understanding of the subject matter since it builds a mathematical model of the subject of interest based on its first principles. The latter employs machine learning, statistics, signal processing, and other techniques to perform pattern recognition on monitoring data. Data-driven approaches are becoming more popular since they have a lower barrier to entry [4,5].

Data-driven approaches typically rely on a large amount of training samples to learn patterns that can be generalized to testing instances [6,7]. However, samples of faulty types are insufficient in many real-world applications, leading to an imbalanced classification problem [8]. Oversampling and down-sampling techniques may be used to create a more balanced dataset [9], but this introduces a bias, wherein more samples are selected from one class than another. Another workaround is data augmentation by learning the underlying generating mechanisms of scarce classes using generative models, such as a Variational Auto-Encoder (VAE) [10] or a Generative Adversarial Network (GAN) [11,12]. Though some success has been reported, it is a paradox to learn from data that are rare to synthesize more data.

Deep transfer learning has been adopted to address the issue of small samples in fault diagnosis, with the hope of reutilizing the knowledge learned from a source task in a target task [13]. Importantly, the number of samples required for training in the target task can be greatly reduced. In one study, an enhanced deep auto-encoder model was proposed to transfer the knowledge learned from a data-abundant source domain to a data-scarce target domain for the purpose of fault diagnosis [14]. Elsewhere, deep transfer learning was applied to transfer knowledge among various operating modes of rotating machinery, including rotational speed [15] and working load [16,17].

However, negative transfer may occur if source and target domains have significant differences; i.e., if domain-specific samples have large discrepancies in their probability distributions [18]. In other words, bringing source domain knowledge into the target task does not help, but rather hinders performance. The effect of negative transfer has been noted in several transfer-learning-enabled fault diagnosis applications [15,18,19], inspiring researchers to examine the causes of negative transfer. A review can be found in [20,21].

Although the factors contributing to negative transfer are multifaceted, the primary cause is rooted in a covariate shift between source and target domains [21,22]. Therefore, domain adaptation measures have been proposed to prevent negative transfer or mitigate its adverse impact, including source data filtering using an adversarial approach [22], new architectural design [23–25], the use of transitive transfer strategy [26], etc. Although many negative transfer countermeasures have been developed, transferring a pretrained model directly from a distant inter-domain to the target domain still poses problems [27,28]. More specifically, there is a knowledge gap on how to mitigate negative transfer in transfer learning with small samples in fault diagnosis tasks. This gap constitutes the major motivation of this study.

Inspired by the work in [26], where intermediate domains were implanted to bridge the gap between distant source and target domains, we propose an intra-domain learning strategy to explicitly utilize the knowledge of a distribution-alike source domain for the purpose of fault diagnosis with small samples. Concretely, we conduct a vanilla transfer from an off-the-shelf inter-domain deep learning model to a data-abundant source domain. The model is then transferred to the target task via shallow-layer freezing and finetuning. The intra-domain transfer learning strategy is the primary novelty of this research. We validate the proposed strategy in a case study with varying levels of small sample ratios and observe its efficacy over alternatives in terms of convergence speed and diagnostic accuracy. This study empirically proves the soundness and merit of transferring among distribution-alike domains in the field of fault diagnosis. We argue cautions against a direct transfer among inter-domains should be issued to reduce the risk of committing negative transfer.

To have an intuitive understanding of the learned model, we produce heat maps to visualize the learned high-level features of the input data. Unlike traditional Class Activation Mapping methods, our novel method uses a channel-wise average over the final convolutional layer and up-sampling with interpolation. Heat map visualization verifies the proposed strategy can extract fault-related characteristics.

The contributions of this research are twofold: (1) we propose an intra-domain transfer learning strategy and a fault diagnosis model for rotating equipment under the scenario of small samples; (2) we propose a novel heat map visualization method to demystify the transferred deep learning model.

The remaining sections are organized as follows. Section 2 explains the intra-domain transfer learning strategy and introduces a rotating equipment fault diagnosis model based on vibration data. Section 3 validates the proposed strategy and model in a case study of a gearbox dataset. Section 4 demonstrates the heat map method and provides a visual portrayal of the features extracted by the transferred deep learning model. Section 5 concludes the work.

2. Methodology

This section introduces the intra-domain transfer learning strategy and elucidates a fault diagnosis model using the proposed strategy.

2.1. Intra-Domain Transfer Learning Strategy

Transfer learning refers to the application of experiences gained in solving one task to a different but similar task. Experiences learned from the source domain are used as prior knowledge to solve target problems. Put formally, given a source domain D_S and a target domain D_T, transfer learning reuses the information embedded in D_S for better problem-solving in D_T. Through knowledge sharing, the requirements for supervisory instances in D_T can be greatly reduced, thus making it possible to learn even when samples are small.

The above statement holds true in general when the source and target domains are closely related. However, if there is a significant covariate shift between D_S and D_T, transfer learning might inversely hurt the target performance, i.e., negative transfer [22,26]; see Figure 1. In other words, inter-domain transfer may take the learning in the wrong direction in the target domain. To mitigate the impact of negative transfer, researchers have proposed many workarounds, including instance weighting [29], feature matching [30], and transitive transfer [31]. Inspired by the Distant Domain Transfer Learning (DDTL) approach in [26] where domain discrepancies are substantial, we propose an intra-domain transfer learning strategy.

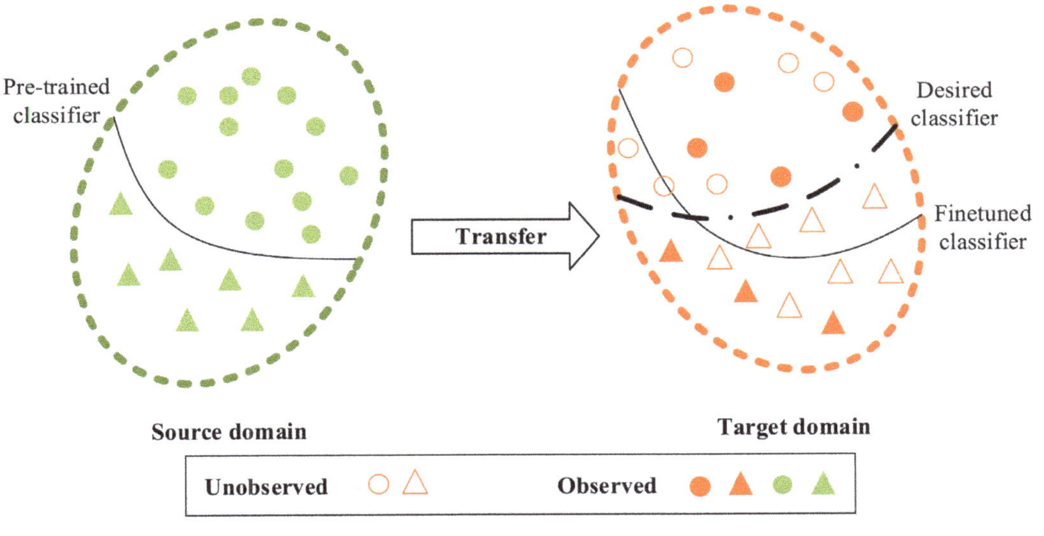

Figure 1. Illustration of negative transfer. The two oval dashed lines represent the distribution shape in the source and target domains. Transfer learning facilitates the training of a finetuned classifier in the target domain even with small samples, but distribution discrepancy may cause negative transfer, leading to poor generalization in target tasks.

The intra-domain transfer learning strategy assumes the source domain is different from but akin to the target domain. Even so, negative transfer can still occur. To preclude this, we propose a two-stage transfer scheme: first, we construct a pre-trained model in the source domain via a vanilla transfer from an off-the-shelf inter-domain deep neural network; then, we transfer the pretrained model to the target domain via shallow-layer freezing and finetuning. Figure 2 depicts the two-stage intra-domain transfer learning strategy.

In our problem setting, the source domain is data-abundant, and this prevents negative transfer, even though we use an inter-domain vanilla transfer. We also use the first-stage transfer to speed up model training in the source domain. In the second stage, the shallow-layer freezing is intended to ensure knowledge transfer between akin domains, while finetuning uses the limited number of samples in the target domain to further train the model for better performance in target tasks.

In addition to requiring relative proximity between source and target domains, the intra-domain transfer learning strategy requires the source domain to be data-abundant. Note that in the field of fault diagnosis, annotated faulty samples are less expensive to acquire in the laboratory than in real-world applications. Therefore, a deluge of labelled samples in the source domain can be generated via fault injection experiments or simply first-principal simulation. This constitutes another motivation of this research.

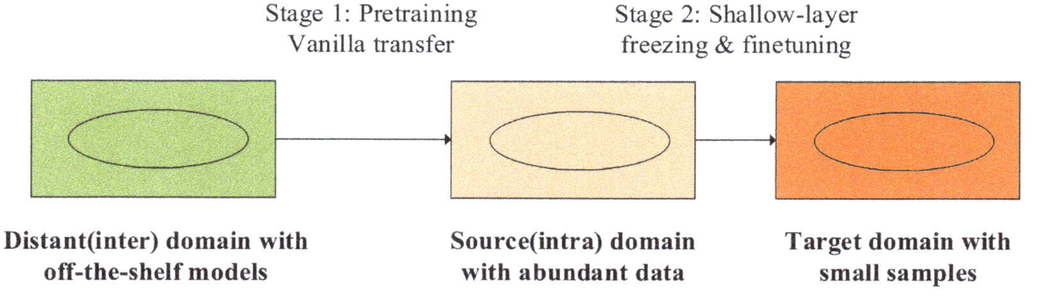

Figure 2. Illustration of the intra-domain transfer learning strategy.

2.2. Fault Diagnosis Model with Intra-Domain Transfer

Armed with the above intra-domain transfer strategy, in this section, we propose a fault diagnosis model for rotating machinery. Although many types of data (infrared images, temperature sequences, etc.) can be measured, vibration signals are still most prevalent in fault diagnosis applications. Without loss of generality, we assume both the source and the target domain use high frequency non-periodic vibration signals to diagnose faults.

Given their enormous capacity to represent knowledge, deep neural networks have been widely adopted as the carrier of knowledge in transfer learning applications. Specifically, convolutional neural networks (CNN) were early choices in the fields of computer vision and industrial fault diagnosis. Hereinafter, we use CNN as the main architecture for fault diagnosis modelling. Again, without loss of generality, we select ResNet-50 as our off-the-shelf inter-domain model since it is a benchmarking model [32]. Notably, the complexity (depth) of the selected model can vary depending on the problem to be solved, especially the complexity of the data input.

Since ResNet-50 is trained on natural images, we propose to use continuous wavelet transform (CWT) to convert our vibration signals to time-frequency spectrums, as has been carried out elsewhere [33]. This brings the data in source and target domains into the same format as the distant domain (i.e., images); see Figure 3 for an example. In addition, CWT conducts multi-scale transform on non-stationary signals, and this retains as much information as possible from the input. For simplicity, we select the complex Morlet wavelet as the mother wavelet, defined as:

$$\psi_{a,\tau}(t) = \frac{1}{\sqrt{a}}\psi\left(\frac{t-\tau}{a}\right) \tag{1}$$

where a represents a scale parameter related to frequency, and τ is a translation parameter related to time.

The wavelet coefficients of a signal are calculated by a convolution operation between the mother wavelet function $\psi_{a,\tau}(t)$ and the signal $x(t)$, as defined by:

$$W(a,\tau) = \frac{1}{\sqrt{a}} \int x(t)\psi^*\left(\frac{t-\tau}{a}\right)dt \quad (2)$$

where $\psi^*(\cdot)$ indicates the complex conjugate of function $\psi(\cdot)$. The signal is decomposed into a series of complex numbers in distinct frequency bands, as shown in Figure 3. The moduli of the complex numbers are calculated and grouped by frequency and time. In this fashion, we get a one-dimensional signal mapping in the time-frequency domain, with the horizontal and vertical axes corresponding to time and frequency, respectively. Next, we conduct three-channel extensions of a time-frequency mapping to produce a time-frequency image.

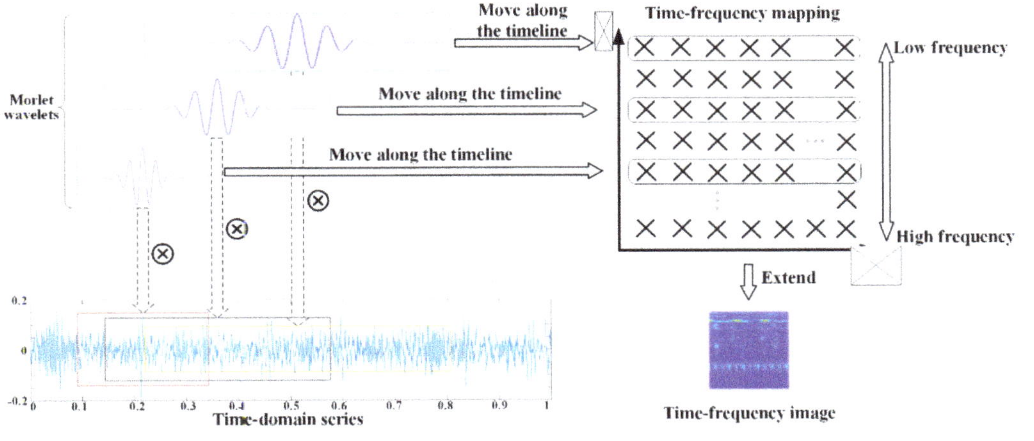

Figure 3. Time-frequency transformation from vibration signals to images using CWT.

The above time-frequency transformation describes the data preprocessing step of our fault diagnosis model. Once this step is completed, the model performs a vanilla transfer from ResNet-50 to the source domain. As the source domain is data-abundant, negative transfer in this step can be avoided. Studies in computer vision have found the convolutional layers of a CNN are essential for automatic feature learning. The deeper the layers, the more abstract the learned features are, and vice versa.

That being said, we propose shallow-layer freezing to ensure feature extractors in the source domain are reused in the target domain. The sharing of feature extractors is supported by assuming the distribution discrepancy between source and target domains is reasonably small. In other words, the weights of those shallow layers in the CNN are fixed to extract features in the target domain (i.e., features shared in common with the source domain).

The last step of our fault diagnosis model is finetuning in the target domain. Concretely, the final classification layer (e.g., a soft-max layer) of the CNN is customized to fit the number of classes in the target domain. Those learnable weights in the deep layers are further trained using the small samples to accomplish the target task. The finetuning step aims to improve the generalization capability in the target task, and it is typically implemented at a small learning rate, depending on the number of samples available. A summary of the above steps appears in Figure 4.

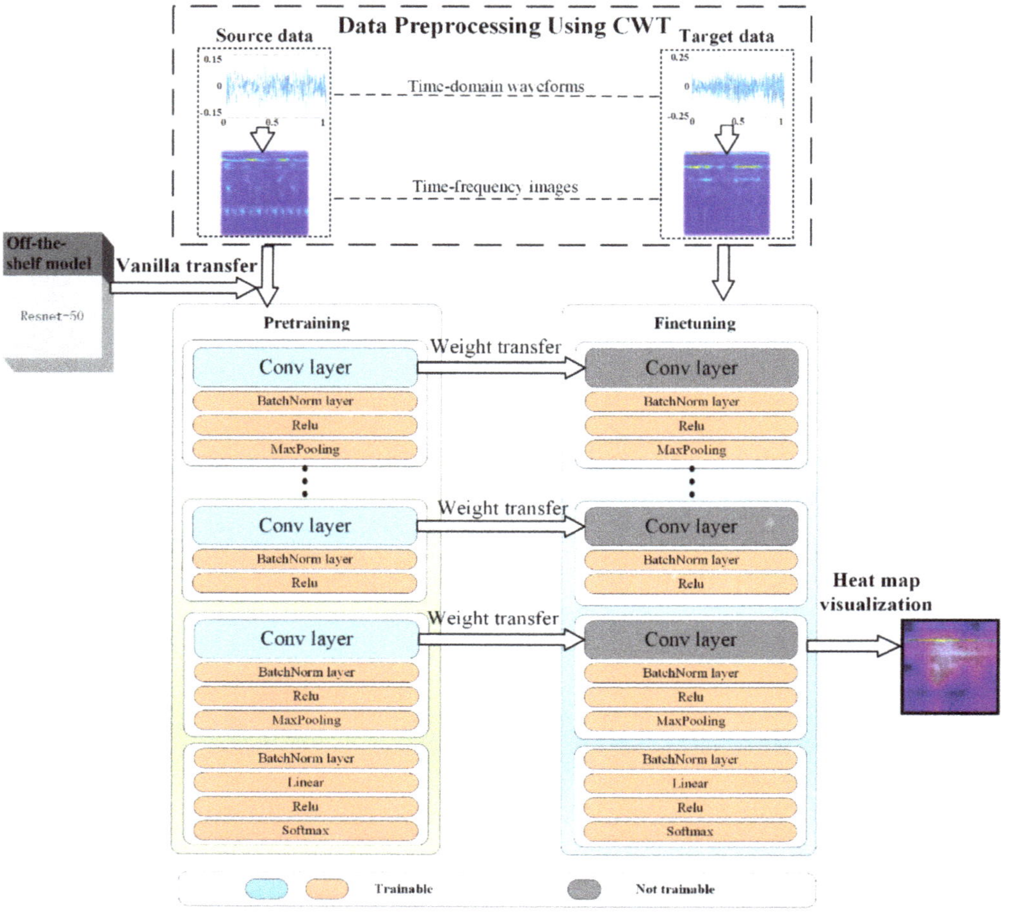

Figure 4. The proposed fault diagnosis model with intra-domain learning strategy.

3. Experiments and Results

To validate the proposed intra-domain transfer learning strategy and the fault diagnosis model, we select a bearing dataset and a gearbox dataset for a case study comparing various transfer strategies. We also carry out extensive experiments to empirically analyze the impact of training sample scarcity on diagnostic accuracy.

3.1. Dataset Introduction

The gearbox dataset, featuring small samples in the target domain, was collected from a two-stage gearbox test rig, as shown in Figure 5. Vibration signals of the gear were measured via an accelerometer attached to the gearbox housing. The signals were recorded using a dSPACE system (DS1006 processor board, dSPACE Inc., Wixom, MI, USA) at a sampling frequency of 20 kHZ. More details about the test rig and data acquisition apparatus are in [8]. The dataset was gathered under nine working conditions: healthy, missing tooth, root crack, spalling, and chipping tip with five different severities. These conditions are shown in Figure 6.

Figure 5. Gearbox test rig of University of Connecticut [6].

Each working condition of the gearbox dataset has 104 samples, each of which has a length of 3600. Table 1 briefly describes some metadata of the dataset. In the case of a nine-class classification problem where the number of samples is far less than the sample dimension, it is challenging to train a reasonably good diagnostic model from scratch. In addition, fault signatures might be submerged by meshing frequencies and noises, thus aggravating the difficulties of feature extraction. Note that when the dataset was released originally in [8], the authors considered utilizing transfer learning to solve the problem, but ultimately chose a different approach.

Figure 6. Nine working conditions, including healthy state and eight types of fault [6].

The Case Western Reserve University (CWRU) bearing dataset is chosen as the source domain for transfer learning in this study, as it is one of the most extensively used datasets for fault diagnosis benchmarking [34]. The dataset contains vibration signals collected at a frequency of 12kHz and 48kHz. We only utilize those with a sampling frequency of 12 kHz to reduce variation in the source domain. Three types of faults (ball, inner race, and outer race) are introduced in the experiments, each with three fault sizes (0.007, 0.014, and 0.021 in).

The monitoring items, sampling frequency, and working conditions are different across source and target domains, but both datasets have high-frequency aperiodicity. Moreover, the source and target domains resemble each other visually after time-frequency transformation, allowing us to transfer feature extractors between intra-domains.

Table 1. Metadata of the gearbox dataset.

Working Conditions	Sampling Frequency (HZ)	Number of Samples	Length of Each Sample
Healthy	20k	104	3600
Missing	20k	104	3600
Crack	20k	104	3600
Spalling	20k	104	3600
Chip5a	20k	104	3600
Chip4a	20k	104	3600
Chip3a	20k	104	3600
Chip2a	20k	104	3600
Chip1a	20k	104	3600

3.2. Data Preparation

Following the proposed fault diagnosis model in Section 2, we convert the time-domain waveforms of the source and target domains into time-frequency images using CWT. Under the condition that Nyquist's sampling theorem is satisfied, the waveforms are first sliced into vibrational snippets of equal length, 1200 points in our experiments. Each vibration snippet is transformed into one time-frequency image of size 224×224 via down-sampling. The images are then extended to three channels to fit the input dimension of ResNet-50. Figure 7 shows an example of the time-frequency images in our target domain.

Using the above approach, we construct two data repositories of time-frequency images: one is the source-domain CWRU dataset for pretraining; the other is the target-domain gearbox dataset for finetuning and model validation. The former consists of nine faulty states (three fault locations and three sizes), each of which has 1203 time-frequency domain images; approximately 20% of them (240) are used for model selection. The latter has 2808 samples, and each of the nine working conditions includes 312 samples. To investigate how the number of samples in the target domain affects the diagnostic accuracy, we split the target domain into a training set and a testing set using different ratios: 90:10, 50:50, 10:90, 5:95, and 1:99. This is summarized in Table 2.

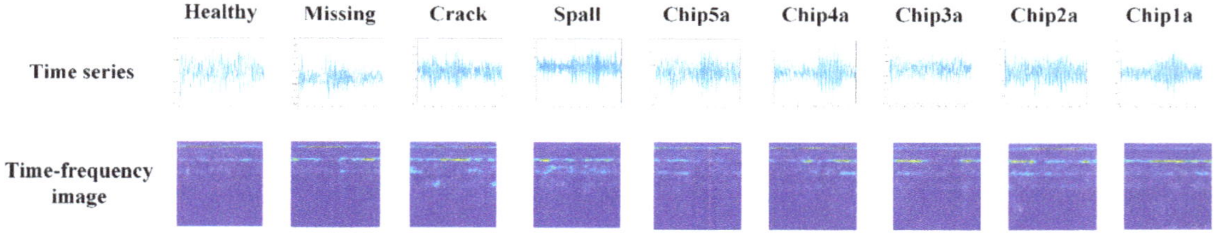

Figure 7. Vibration signals of nine working conditions and their corresponding time-frequency images in the target domain.

Table 2. Target-domain training/testing set division under different splitting ratios.

Splitting Ratios	Total Number of Samples	Number of Samples for Each Working State	
		Training Set	Testing Set
90:10	2808	2529	279
50:50	2808	1404	1404
10:90	2808	279	2529
5:95	2808	144	2664
1:99	2808	27	2781

3.3. Pretraining and Finetuning

To validate the intra-domain transfer learning strategy and the proposed fault diagnosis model, we introduce two more learning schemes for comparison: train from scratch (TFS) and transfer from ImageNet (TFI). The proposed intra-domain transfer learning method is abbreviated as TFC, signifying transfer from the CWRU dataset.

The weights of the ResNet-50 model are randomly reinitialized in TFS. We can imagine the dilemma the TFS scheme ran into with small samples to train such a complex model. All of the convolutional layers of TFI and TFC inherit weights from the off-the-shelf ResNet-50 model, and their batch normalization layers and the fully-connected layers are reinitialized.

With prior knowledge that the target-domain samples differ significantly from natural images, the convolutional layers of TFI are trainable but at a relatively small learning rate. Therefore, the convolutional layers of TFC are frozen to reuse the feature extractors learned from intra-domain samples. The final soft-max layer of each of the above three schemes is customized to match the number of desired classes in our target task. We choose Adam optimizer in our pretraining and finetuning, and the learning rate is set to 1×10^{-4}. All models are trained with mini-batch samples of a size of 16.

Under the settings of the three training schemes and five splitting ratios of target-domain samples, we train the ResNet-50 model for 100 epochs, each repeated five times. The training history is shown in Figure 8. From the figure, we observe the proposed method converges faster than the alternatives, and it consistently plateaus at the highest diagnostic accuracy in our target task. Given adequate training samples, all three learning schemes can achieve a satisfactory result if the model is sufficiently trained. But when training samples are scarce, intra-domain transfer learning can boost the performance in terms of both convergence speed and accuracy in the target task.

3.4. Testing Resultss

Testing accuracy of the three learning schemes in our target task is shown in Table 3. The proposed intra-domain transfer learning strategy exhibits the best diagnostic accuracy, even with small training samples. With proper model training, TFS can achieve a high prediction accuracy in the testing set when training samples are abundant, but it degenerates as the number of training samples decreases. In general, TFI improves the performance compared to TFS, indicating that inter-domain experience helps classify faults in the gearbox dataset. But as shown in Figure 8B,C, negative transfer does occur occasionally, though this is not always evident.

Table 3. Testing accuracy in the target task under three learning schemes and various splitting ratios of target-domain samples (each number is an average of five runs of experiments).

Learning Schemes	Best Accuracy in First 10 Epochs (%)					Best Accuracy in First 50 Epochs (%)					Best Accuracy in First 100 Epochs (%)				
	90:10	50:50	10:90	5:95	1:99	90:10	50:50	10:90	5:95	1:99	90:10	50:50	10:90	5:95	1:99
TFS	97.1	82.4	72.5	51.6	11.1	100	97.0	89.2	69.2	11.6	100	97.0	96.6	79.0	33.6
TFI	98.9	90.1	92.2	74.5	14.9	100	97.9	97.2	82.9	51.2	100	99.0	97.5	82.3	53.1
TFC	**100**	**99.0**	**97.8**	**92.9**	**27.9**	**100**	**99.5**	**98.9**	**97.5**	**83.2**	**100**	**99.8**	**99.7**	**97.8**	**84.3**

As shown in Figure 8 and Table 3, TFC (the proposed learning scheme) consistently outperforms TFI. There is a large gain in accuracy using intra-domain transfer instead of its inter-domain counterpart. This confirms that the feature extractors learned from a different but akin domain can not only avoid negative transfer but also boost prediction accuracy in the target task. Table 3 also demonstrates that TFC has strong generalization capability even when the splitting ratio is set to 1:99, resulting in an accuracy of 84.3% after 100 epochs of training. We note that a nine-class classification problem is not an easy task.

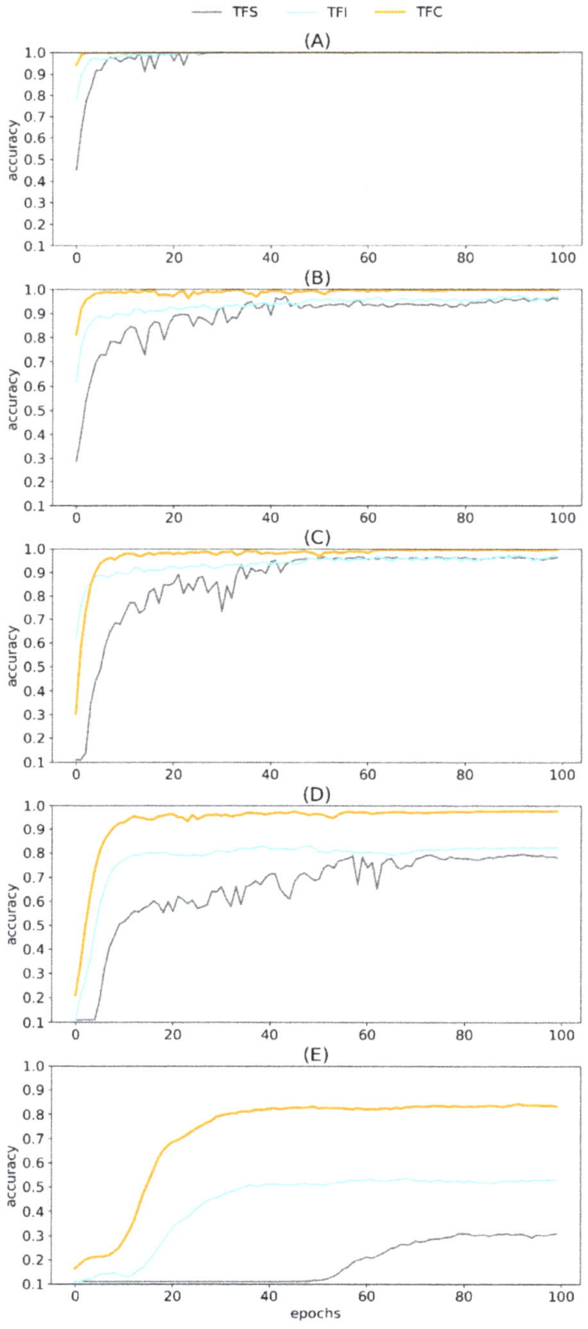

Figure 8. Training history of the three learning schemes in our target task, under various splitting ratios: (**A**) 90:10; (**B**) 50:50; (**C**) 10:90; (**D**) 5:95; (**E**) 1:99.

The confusion matrices of the three learning schemes with small training samples are shown in Figure 9. As the figure indicates, TFS performs slightly better than random guessing, and the most misclassified classes are "chip4a" and "chip2a". TFI obtains a higher diagnostic accuracy, and misclassifications distribute more evenly across different classes. TFC significantly improves the classification of "healthy", "missing", "crack", "spall", and "chip5a", leading to an accuracy of 93.5% in these five states (red rectangle). The remaining four states (blue rectangle) have a relatively higher misclassification rate. An intuitive explanation is that the four faults are of the same type but have different severities. Consequently, their features are comparable, making it hard to distinguish them. Although data size considerably affects the diagnostic accuracy, the proposed method can achieve satisfactory results when training samples are scarce. This proves the efficacy of using the intra-domain transfer learning strategy for fault diagnosis with small samples.

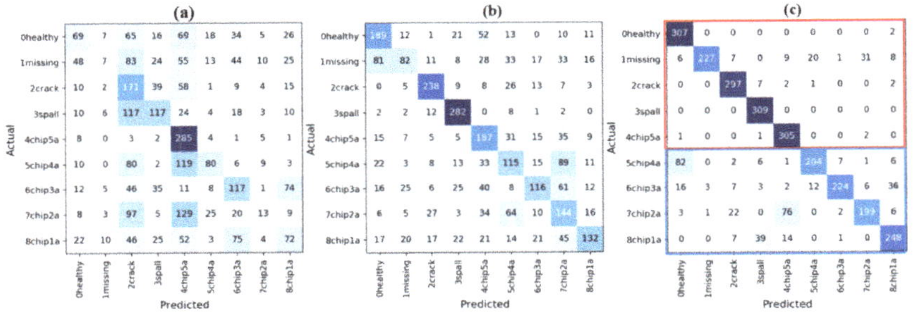

Figure 9. Confusion matrices of the three transfer schemes (**a**) TFS; (**b**) TFI; (**c**) TFC, when the splitting ratio of training/testing samples is 1:99.

4. Demystifying the "Black-Box" via Visualization

This section adopts the heat map method to demystify the learned fault diagnosis model, giving us a way to visualize high-level features of inputs in an intuitive and explainable manner. As a complementary study to the proposed intra-domain transfer learning strategy, we compare the above three learning schemes via heat map visualization.

4.1. Heat Map Visualization

Heat mapping is widely used in computer vision applications [35,36]. It attempts to visualize the most activated area in an input to reveal how the final classification conclusions are drawn by deep learning models. For example, in a face recognition problem, those image patches corresponding to noses, eyes, and mouths will contribute the most to the final decision. Similarly, we hypothesize that fault-related frequencies in our time-frequency images should be recognized by our ResNet-50 model in the target task.

The deeper the layer in a model, the more abstract the extracted features. Therefore, we use the feature maps yielded by the last convolutional layer of the ResNet-50 model to construct a heat map. In the ResNet-50 architecture, the output size of the last convolutional layer is $512 \times 7 \times 7$, corresponding to channels, heights, and widths, respectively. Taking the average along the first axis produces a 7×7 matrix. Note this channel-wise average operation is different from the average pooling layer in the ResNet-50 model, as it averages over the whole feature map.

The magnitude of the above matrix reflects the extent of the activation in the original input but on a much smaller scale. To solve this, we up-sample the matrix to match the input size using bilinear interpolation; this results in an average activation matrix, as shown in Figure 10b. We select bilinear interpolation to make the transition between image blocks smoother. Then, the final heat map can be obtained by overlaying the average activation matrix on the original input. Figure 10 demonstrates an example of heat map visualization taking a time-frequency image of our target sample as input.

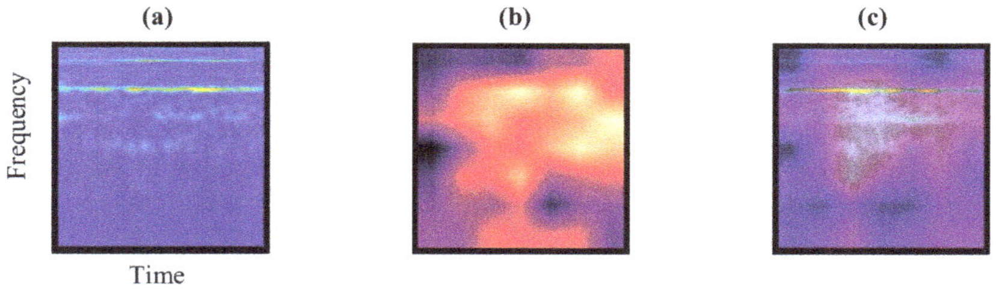

Figure 10. An exemplar illustration of heat map visualization: (**a**) input image, (**b**) average activation matrix, (**c**) heat map.

4.2. Qualitative Assessment of the Heat Maps of the Target Task

The ability to learn high-level abstraction of features using CNN has already been verified in computer vision applications but is relatively underexplored in fault diagnosis. Armed with the above heat map method, we attempt to visualize those high-level features extracted by the transferred ResNet-50 model in our target task and correlate them with the underlying fault characteristics. Hereinafter, to validate the efficacy of our proposed intra-domain transfer learning strategy using small samples, we analyze the model finetuned when the splitting ratio of training and testing samples is set at 1:99.

Figure 11 shows some heat maps generated by the fault diagnosis model at testing time. Each heat map corresponds to one random testing sample in a specific working condition in our target task; see Table 1 for a reference. Three frequency spectrums are presented in Figure 11. Healthy samples are normally characterized by an amplitude located at the shaft rotating frequency and its harmonies, as shown in Figure 11a, while faulty samples can be recognized by anomalous amplitude in their characteristic frequencies.

The activations highlighted in red boxes are consistent with their corresponding fault characteristics. For example, as shown in the frequency spectrum of Figure 11c, the fault "crack" has a characteristic frequency around 5000 Hz, and the energy in this frequency band is correctly identified in its heat map. Similar results can be observed in other fault classes. Notably, the horizontal axis in our heat map is time, the same as in the time-frequency image. Therefore, the horizontal location of the most activated area can also signify the temporal features of non-stationary signals that lead to CNN's final classification.

Another observation is the most activated spots in these heat maps are distinctive, making it easy to separate them. Figure 12 compares the heat maps of the various working conditions obtained via the three learning schemes. Intuitively, it can be deduced that TFC leads to the highest differentiability in these heat maps, TFI ranks second, and TFS comes in last. This coincides with the confusion matrices shown in Figure 9 and the testing accuracy results in Table 3.

From a result-oriented perspective, higher differentiability in the heat maps would yield higher diagnostic accuracy. But it by no means implies that the most activated area in the feature maps must match the underlying physical mechanisms of the corresponding faults. The above qualitative assessment of the heat maps not only proves the efficacy of the proposed intra-domain learning strategy, but also discloses how the transferred ResNet-50 model infers the diagnostic results. Since the feature extractors learned from an intra-domain task are frozen in our target-domain task, the knowledge sharing enabled by our transfer learning strategy is validated in the heat maps. Moreover, the heat maps of misclassification samples can provide new perspectives to improve the fault diagnosis model.

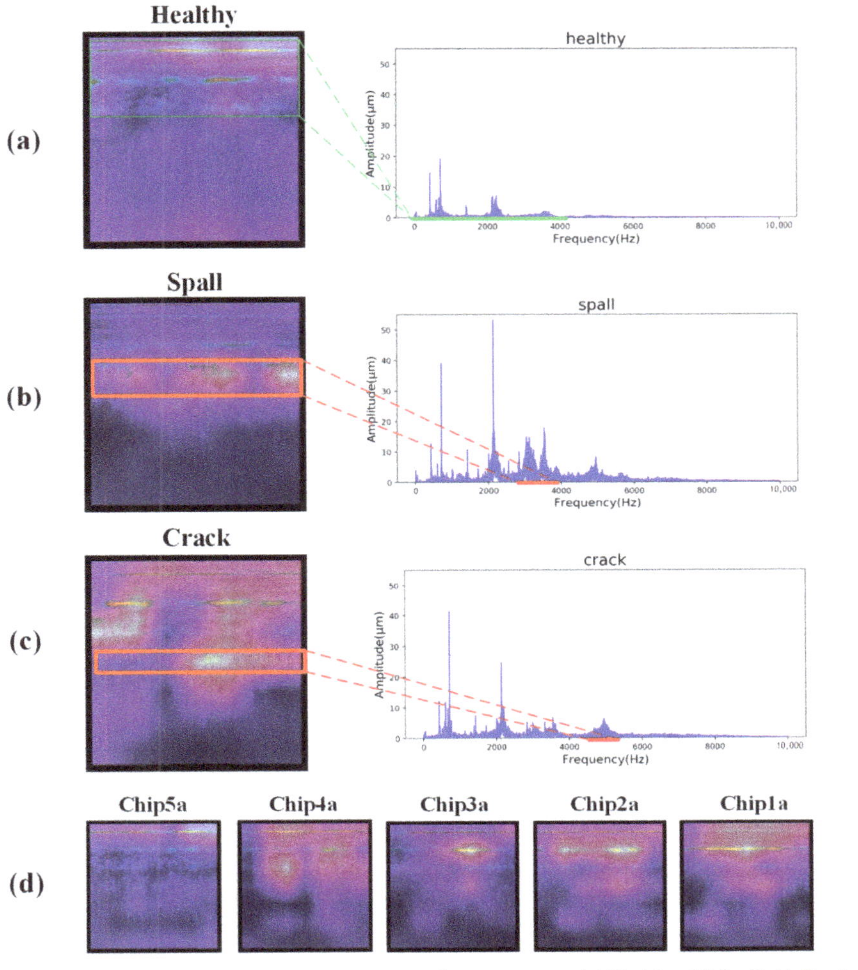

Figure 11. Heat maps and frequency spectrums of various states: (**a**) Healthy; (**b**) Spall; (**c**) Crack; (**d**) Chip.

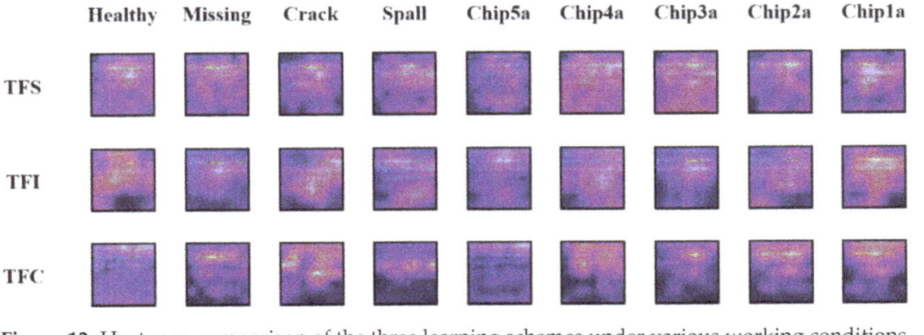

Figure 12. Heat map comparison of the three learning schemes under various working conditions.

5. Conclusions

This study proposed an intra-domain transfer learning strategy to tackle the challenge of insufficient training samples in fault diagnosis applications. To alleviate the impact of negative transfer, the intra-domain transfer learning strategy first uses the vanilla transfer of an off-the-shelf inter-domain model to a data-abundant source domain that is akin to the target domain. The learned feature extractors are then reutilized in the target domain via shallow-layer freezing, followed by a finetuning step with small samples.

We verified the proposed transfer learning strategy in a gearbox fault diagnosis case study and compared it to two other learning schemes. In the case study, we adopted CWT as a preprocessing tool to convert 1D vibrational waveform to 3D time-frequency images, and selected ResNet-50 as the base model. Under various small sample settings (splitting ratio of training and testing samples in the target task), we carried out extensive experiments and observed superior performance of the proposed strategy in both convergence speed and accuracy. Finally, we introduced heat map visualization to demystify the learned deep neural network. We leave the quantitative assessment of a fault diagnosis model using heat maps to future work.

Author Contributions: Conceptualization, L.Z. and J.L.; data curation, J.Z. and Y.P.; investigation, L.Z., Y.P. and J.L.; methodology, L.Z.; validation, J.Z., Y.P., and L.Z.; writing—original draft, L.Z. and J.Z.; writing—review and editing, Y.P., and J.L. All authors have read and agreed to the published version of the manuscript.

Funding: This research was funded in part by the National Science Foundation of China (NSFC) under Grant 71801045, and the research start-up funds of DGUT (GC300502-46).

Institutional Review Board Statement: Not applicable.

Informed Consent Statement: Not applicable.

Conflicts of Interest: The authors declare no conflict of interest.

Nomenclature

Abbreviations	
VAE	Variational Auto-Encoder
GAN	Generative Adversarial Network
DDTL	Distant Domain Transfer Learning
CNN	Convolutional Neural Network
CWT	Continuous Wavelet Transform
CWRU	Case Western Reserve University
TFS	Train from Scratch
TFI	Train from ImageNet
TFC	Train from the CWRU dataset
Superscripts, subscripts, and parameters	
D_S	Source Domain
D_T	Target Domain
$\psi^*(\cdot)$	The complex conjugate of function $\psi(\cdot)$
a	A scale parameter related to frequency
τ	A translation parameter related to time

References

1. Shao, H.; Lin, J.; Zhang, L.; Galar, D.; Kumar, U. A Novel Approach of Multisensory Fusion to Collaborative Fault Diagnosis in Maintenance. *Inf. Fusion* **2021**, *74*, 65–76. [CrossRef]
2. Cai, B.; Huang, L.; Xie, M. Bayesian Networks in Fault Diagnosis. *IEEE Trans. Ind. Inform.* **2017**, *13*, 2227–2240. [CrossRef]
3. Zhang, L.; Lin, J.; Shao, H.; Zhang, Z.; Yan, X.; Long, J. End-to-End Unsupervised Fault Detection Using a Flow-Based Model. *Reliab. Eng. Syst. Saf.* **2021**, *215*, 107805. [CrossRef]
4. Zhang, L.; Lin, J.; Liu, B.; Zhang, Z.; Yan, X.; Wei, M. A Review on Deep Learning Applications in Prognostics and Health Management. *IEEE Access* **2019**, *7*, 162415–162438. [CrossRef]

5. Zhu, Z.; Peng, G.; Chen, Y.; Gao, H. A Convolutional Neural Network Based on a Capsule Network with Strong Generalization for Bearing Fault Diagnosis. *Neurocomputing* **2019**, *323*, 62–75. [CrossRef]
6. Li, C.; Zhang, S.; Qin, Y.; Estupinan, E. A Systematic Review of Deep Transfer Learning for Machinery Fault Diagnosis. *Neurocomputing* **2020**, *407*, 121–135. [CrossRef]
7. Strušnik, D. Integration of Machine Learning to Increase Steam Turbine Condenser Vacuum and Efficiency through Gasket Resealing and Higher Heat Extraction into the Atmosphere. *Int. J. Energy Res.* **2022**, *46*, 3189–3212. [CrossRef]
8. Cao, P.; Zhang, S.; Tang, J. Preprocessing-Free Gear Fault Diagnosis Using Small Datasets with Deep Convolutional Neural Network-Based Transfer Learning. *IEEE Access* **2018**, *6*, 26241–26253. [CrossRef]
9. Suh, S.; Lee, H.; Jo, J.; Lukowicz, P.; Lee, Y.O. Generative Oversampling Method for Imbalanced Data on Bearing Fault Detection and Diagnosis. *Appl. Sci.* **2019**, *9*, 746. [CrossRef]
10. Yan, X.; She, D.; Xu, Y.; Jia, M. Deep Regularized Variational Autoencoder for Intelligent Fault Diagnosis of Rotor–Bearing System within Entire Life-Cycle Process. *Knowl.-Based Syst.* **2021**, *226*, 107142. [CrossRef]
11. Cui, L.; Tian, X.; Shi, X.; Wang, X.; Cui, Y. A Semi-Supervised Fault Diagnosis Method Based on Improved Bidirectional Generative Adversarial Network. *Appl. Sci.* **2021**, *11*, 9401. [CrossRef]
12. Shao, S.; Wang, P.; Yan, R. Generative Adversarial Networks for Data Augmentation in Machine Fault Diagnosis. *Comput. Ind.* **2019**, *106*, 85–93. [CrossRef]
13. Weiss, K.; Khoshgoftaar, T.M.; Wang, D.D. A Survey of Transfer Learning. *J. Big Data* **2016**, *3*, 1–40. [CrossRef]
14. Zhiyi, H.; Haidong, S.; Lin, J.; Junsheng, C.; Yu, Y. Transfer Fault Diagnosis of Bearing Installed in Different Machines Using Enhanced Deep Auto-Encoder. *Measurement* **2020**, *152*, 107393. [CrossRef]
15. Hasan, M.J.; Kim, J.M. Bearing Fault Diagnosis under Variable Rotational Speeds Using Stockwell Transform-Based Vibration Imaging and Transfer Learning. *Appl. Sci.* **2018**, *8*, 2357. [CrossRef]
16. Shao, S.; McAleer, S.; Yan, R.; Baldi, P. Highly Accurate Machine Fault Diagnosis Using Deep Transfer Learning. *IEEE Trans. Ind. Inform.* **2019**, *15*, 2446–2455. [CrossRef]
17. Zhang, W.; Peng, G.; Li, C.; Chen, Y.; Zhang, Z. A New Deep Learning Model for Fault Diagnosis with Good Anti-Noise and Domain Adaptation Ability on Raw Vibration Signals. *Sensors* **2017**, *17*, 425. [CrossRef]
18. Lee, K.; Han, S.; Pham, V.H.; Cho, S.; Choi, H.J.; Lee, J.; Noh, I.; Lee, S.W. Multi-Objective Instance Weighting-Based Deep Transfer Learning Network for Intelligent Fault Diagnosis. *Appl. Sci.* **2021**, *11*, 2370. [CrossRef]
19. Cho, S.H.; Kim, S.; Choi, J.H. Transfer Learning-Based Fault Diagnosis under Data Deficiency. *Appl. Sci.* **2020**, *10*, 7768. [CrossRef]
20. Li, W.; Huang, R.; Li, J.; Liao, Y.; Chen, Z.; He, G.; Yan, R.; Gryllias, K. A Perspective Survey on Deep Transfer Learning for Fault Diagnosis in Industrial Scenarios: Theories, Applications and Challenges. *Mech. Syst. Signal Process.* **2022**, *167*, 108487. [CrossRef]
21. Zhang, W.; Deng, L.; Zhang, L.; Wu, D. Overcoming Negative Transfer: A Survey. *arXiv* **2020**, arXiv:2009.00909.
22. Wang, Z.; Dai, Z.; Carbonell, J. Characterizing and Avoiding Negative Transfer. In Proceedings of the IEEE/CVF Conference on Computer Vision and Pattern Recognition, Long Beach, CA, USA, 15–20 June 2019; pp. 11293–11302.
23. Han, T.; Liu, C.; Wu, R.; Jiang, D. Deep Transfer Learning with Limited Data for Machinery Fault Diagnosis. *Appl. Soft Comput.* **2021**, *103*, 107150. [CrossRef]
24. Zheng, Z.; Fu, J.; Lu, C.; Zhu, Y. Research on Rolling Bearing Fault Diagnosis of Small Dataset Based on a New Optimal Transfer Learning Network. *Measurement* **2021**, *177*, 109285. [CrossRef]
25. Zhang, W.; Li, C.; Peng, G.; Chen, Y.; Zhang, Z. A Deep Convolutional Neural Network with New Training Methods for Bearing Fault Diagnosis under Noisy Environment and Different Working Load. *Mech. Syst. Signal Process.* **2018**, *100*, 439–453. [CrossRef]
26. Tan, B.; Zhang, Y.; Pan, S.J.; Yang, Q. Distant Domain Transfer Learning. In Proceedings of the AAAI Conference on Artificial Intelligence, San Francisco, CA, USA, 4–9 February 2017; pp. 2604–2610.
27. Liu, Y.; Li, J.; Li, Q.; Wang, Q. Transfer Learning with Inception ResNet-Based Model for Rolling Bearing Fault Diagnosis. *J. Adv. Mech. Des. Syst. Manuf.* **2022**, *16*, JAMDSM0023. [CrossRef]
28. Grover, C.; Turk, N. A Novel Fault Diagnostic System for Rolling Element Bearings Using Deep Transfer Learning on Bispectrum Contour Maps. *Eng. Sci. Technol. An Int. J.* **2022**, *31*, 101049. [CrossRef]
29. Deng, M.; Deng, A.; Shi, Y.; Liu, Y.; Xu, M. Intelligent Fault Diagnosis Based on Sample Weighted Joint Adversarial Network. *Neurocomputing* **2022**, *488*, 168–182. [CrossRef]
30. Deng, M.; Deng, A.D.; Shi, Y.; Xu, M. Correlation Regularized Conditional Adversarial Adaptation for Multi-Target-Domain Fault Diagnosis. *IEEE Trans. Ind. Inform.* **2022**, *3203*, 1–11. [CrossRef]
31. Schwendemann, S.; Amjad, Z.; Sikora, A. Bearing Fault Diagnosis with Intermediate Domain Based Layered Maximum Mean Discrepancy: A New Transfer Learning Approach. *Eng. Appl. Artif. Intell.* **2021**, *105*, 104415. [CrossRef]
32. He, K.; Zhang, X.; Ren, S.; Sun, J. Deep Residual Learning for Image Recognition. In Proceedings of the IEEE Computer Society Conference on Computer Vision and Pattern Recognition, Las Vegas, NV, USA, 27–30 June 2016; pp. 770–778.
33. Kankar, P.K.; Sharma, S.C.; Harsha, S.P. Fault Diagnosis of Ball Bearings Using Continuous Wavelet Transform. *Appl. Soft Comput. J.* **2011**, *11*, 2300–2312. [CrossRef]
34. The Case Western Reserve University Bearing Data Center Website. Available online: https://engineering.case.edu/bearingdatacenter (accessed on 30 May 2022).

35. Zhou, B.; Khosla, A.; Lapedriza, A.; Oliva, A.; Torralba, A. Learning Deep Features for Discriminative Localization. In Proceedings of the IEEE Computer Society Conference on Computer Vision and Pattern Recognition, Las Vegas, NV, USA, 27–30 June 2016; pp. 2921–2929.
36. Li, D.; Hu, J.; Wang, C.; Li, X.; She, Q.; Zhu, L.; Zhang, T.; Chen, Q. Involution: Inverting the Inherence of Convolution for Visual Recognition. In Proceedings of the IEEE Computer Society Conference on Computer Vision and Pattern Recognition, Nashville, TN, USA, 20–25 June 2021; pp. 12316–12325.

applied sciences

MDPI

Article

Improved Adversarial Transfer Network for Bearing Fault Diagnosis under Variable Working Conditions

Jun Wang [1,2], Hosameldin Ahmed [2], Xuefeng Chen [1], Ruqiang Yan [1] and Asoke K. Nandi [1,2,*]

1 Department of Mechanical Engineering, Xi'an Jiaotong University, Xi'an 710049, China; jun.wang@brunel.ac.uk (J.W.); chenxf@xjtu.edu.cn (X.C.); yanruqiang@xjtu.edu.cn (R.Y.)

2 Department of Electronic and Electrical Engineering, Brunel University London, London UB8 3PH, UK; hosameldin.abdallahahmed@brunel.ac.uk

* Correspondence: asoke.nandi@brunel.ac.uk

Abstract: Bearings are one of the critical components of rotating machinery, and their failure can cause catastrophic consequences. In this regard, previous studies have proposed a variety of intelligent diagnosis methods. Most existing bearing fault diagnosis methods implicitly assume that the training and test sets are from the same distribution. However, in real scenarios, bearings have been working in complex and changeable working environments for a long time. The data during their working processes and the data used for model training cannot meet this condition. This paper proposes an improved adversarial transfer network for fault diagnosis under variable working conditions. Specifically, this paper combines an adversarial transfer network with a short-time Fourier transform to obtain satisfactory results with the lighter network. Then, this paper employs a channel attention module to enhance feature fusion. Moreover, this paper designs a novel domain discrepancy hybrid metric loss to improve model transfer learning performance. Finally, this paper verifies the method's effectiveness on three datasets, including dual-rotor, a Case Western Reserve University dataset and the Ottawa dataset. The proposed method achieves average accuracy, surpassing other methods, and shows better domain alignment capabilities.

Keywords: bearing intelligent diagnosis; adversarial learning; transfer learning; improved adversarial transfer network

1. Introduction

Bearings are critical rotating machinery components and one of these machines' weakest links [1,2]. Their performance directly affects the system's stable operation and production efficiency [3,4]. Taking an aerospace engine as an example, as shown in Figure 1, rolling bearings are easy to start at low temperatures and have small friction losses, wide operating ranges and strong resistance to oil cutoffs; engines use rolling bearings as main bearings [5,6]. The main bearings support the engine rotor in the engine and transfer rotor loads. The main bearing is divided into the thrust bearing and the traveling bearing. Thrust bearings carry axial loads and control the axial clearance of the rotor and stator. Travel bearings bear radial loads and control radial clearance. When any bearing fails, the aircraft may be forced to land; otherwise, the aircraft may be destroyed, and personnel may lose their lives [7,8]. Therefore, developing efficient and accurate bearing fault diagnosis methods has become necessary to ensure equipment reliability and reduce maintenance costs.

In the past decades, previous studies have proposed various methods for intelligent bearing fault diagnosis [9,10], which can be divided into two categories: model-based methods and data-driven methods [11]. The model-based methods extract features from raw data according to prior knowledge and construct mathematical models to evaluate the health state of bearings [12,13]. However, model-based methods rely too much on prior knowledge, so their robustness could be better. Data-driven methods refer to establishing mappings from fault space to feature space by using a large amount of training data without any knowledge and experience [14]. With the recent development of sensing and intelligent computing technology, it has become more accessible and easier to collect massive amounts of data, and data-driven intelligent diagnosis methods have attracted increased attention [15].

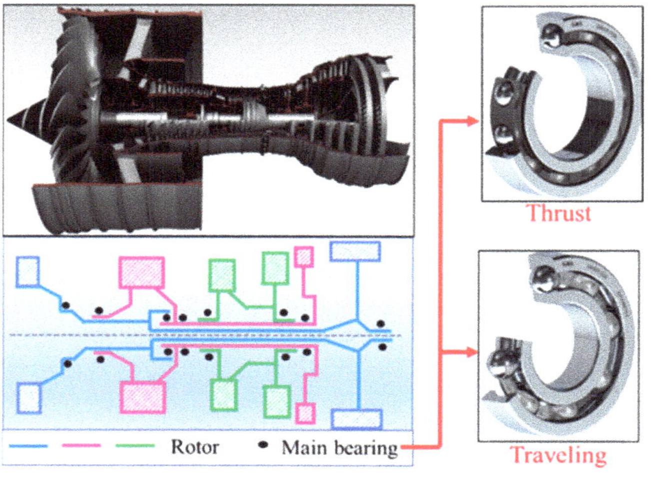

Figure 1. Two kinds of main bearings in aeroengine.

Data-driven methods have swept through the field of intelligent bearing fault diagnosis due to their end-to-end diagnosis mode [16]. For instance, Ahmed et al. [17] used a sparse autoencoder to learn over-complete sparse representations of datasets compressed by compressed sampling. Linshan et al. [18] proposed a novel convolution neural network (CNN) model named the Gramian Time-Frequency Enhancement Network (GTFE-Net) for bearing fault diagnosis. Kaicheng et al. [19] combined Improve Complete Ensemble Empirical Mode Decomposition with Adaptive Noise (CEEMDAN) and CNN to enhance recognition of the state of rolling bearings. Diwang et al. [20] improved the performance of CNN by guiding the CNN design with the physical characteristics of bearing acceleration signals. Wang et al. [21] presented an improved spiking neural network for inter-shaft bearing fault diagnosis. Sinitsin et al. [22] designed a novel hybrid CNN-MPL model-based bearing fault-diagnosis method. He et al. [23] proposed a new framework based on small labeled infrared thermal images and an enhanced convolutional neural network transferred from a convolutional autoencoder.

Although data-driven methods have been applied to bearing fault diagnosis, most assume that the source dataset, which indicates the dataset used for the training model, and the target dataset, which indicates the datasets used for the testing model, are from the same distribution [24,25]. However, the bearing usually works under complex and changing conditions in actual scenarios, so the assumption cannot be achieved [26,27]. To address this problem, transfer learning (TL) emerged [28,29]. Transfer learning improves performance in the target domain by transferring knowledge from different but related source domains [30,31]. In this way, the reliance on copious amounts of target domain data for building learners can be reduced [32,33]. A feasible solution is to find standard latent features through feature transformation and use them as a bridge for knowledge transfer, converting each original feature into a new feature representation for knowledge transfer [34,35]. Wen et al. [36] reduced the distribution discrepancy across the two domains by minimizing the maximum mean discrepancy (MMD). Sinno et al. [37] designed a novel metric to measure the discrepancy between source and target datasets. Chunran et al. [38] proposed a class-level matching TL network to match source and target domain data.

Wang et al. [39] used domain-adversarial training based on the Wasserstein distance to learn domain-invariant features from the raw signal. Cui et al. [40] combined distance-based domain adaptation and adversarial-based domain adaptation to propose a multi-adversarial joint distribution adaptation network.

This paper proposes an improved adversarial transfer network (IATN) for bearing fault diagnosis under variable working conditions, inspired by the effectiveness of distance-based and adversarial-based domain adaptation. Specifically, this paper combines an adversarial transfer network with a short-time Fourier transform (STFT) [41] to obtain satisfactory results with lighter networks. Then, this paper employs an attention module to enhance feature fusion. Moreover, this paper designs a novel domain discrepancy hybrid metric loss to improve model transfer learning performance. Finally, this paper verifies the method's effectiveness on a private dual-rotor dataset, a Case Western Reserve University (CWRU) dataset [42] and the Ottawa dataset [43]. The main contributions are as follows:

(1) This paper combines adversarial TL with the traditional signal processing method, STFT, which allows us to obtain satisfactory results with a much lighter network.
(2) This paper employs a channel attention module to enhance feature fusion.
(3) This paper designs a novel domain discrepancy hybrid metric to measure the discrepancy between features of the source dataset and target dataset features.
(4) This paper evaluates the proposed approach on three test benches' single-source tasks and multi-source tasks. The results demonstrate that the proposed method performs better than other methods.

The structure of this paper is as follows: In Section 2, related knowledge is introduced. Section 3 presents the proposed method in detail. Section 4 provides information about the experiment and analysis of the results. Section 5 provides further discussion, and Section 6 provides conclusions.

2. Preliminary Information

2.1. Related Works

In industrial applications, bearings are key rotating machinery components, and monitoring and fault diagnosis of their condition is crucial to ensuring the stable operation of equipment. Especially under changing working conditions, traditional fault diagnosis methods are often difficult to adapt to different operating environments, resulting in a decline in diagnostic performance. In recent years, deep learning, as a powerful end-to-end strategy, has been widely used to diagnose bearing faults intelligently. This section reviews significant recent research progress in the intelligent diagnosis of bearing faults under variable operating conditions.

In recent years, multi-scale feature learning has shown strong potential in bearing fault diagnosis. For example, Chen et al. [44] proposed a method based on multi-scale convolutional neural networks to effectively extract bearing fault features for fault diagnosis. Hu et al. [45] proposed a novel fault diagnosis method based on enhanced multi-scale sample entropies and balanced adaptation regularization for fault diagnosis needs under different working conditions. To solve the problem of inconsistent training and test data distribution, Guo et al. [46] explored the application of transfer learning technology in bearing fault diagnosis. They proved its effectiveness in cross-domain fault diagnosis. Saha et al. [47] employed transfer learning and random forest classification to exemplify the potential of combining deep learning techniques with traditional algorithms in bearing fault diagnosis. Moreover, researchers have also widely discussed data enhancement methods based on generative adversarial networks (GANs). Wang et al. [48] developed a data augmentation strategy based on GANs to improve the model's generalization ability under varying working conditions. Shi et al. [49] encapsulated a shared feature extractor, a label predictor and a series of domain discriminators to propose an adversarial multi-source data subdomain adaptation (AMDSA) model. Zhu et al. [50] present a deep subdomain adaptation network (DSAN) by aligning the relevant subdomain distributions of domain-specific layer activations across domains based on a local maximum mean.

Meng et al. [51] explore the application of graph neural networks in bearing fault diagnosis. In addition, ensemble learning methods improve diagnostic accuracy by combining the prediction results of multiple learners and have also received increasing attention. Tong et al. [52] proposed a multi-sensor information fusion method based on ensemble learning to automatically learn fault-related information from multi-sensor signals and provide accurate diagnosis results. Chen et al. [53] proposed an explainable learning framework to further improve the accuracy and robustness of bearing fault diagnosis under variable working conditions.

Through these studies, one can see the rapid development of intelligent diagnostic technology in bearing fault detection. These techniques improve the accuracy of fault diagnosis and enhance the model's adaptability in the face of changing operating conditions.

2.2. Domain-Adversarial Adaptation Network

DAT maps two different domains into a common subspace to eliminate the differences between the domains [54]. As Figure 2 shows, a domain-adversarial adaptation network consists of three parts—feature extractor θ_f, classifier θ_c and domain discriminator θ_D [55].

Given a labeled source dataset D_s and an unlabeled target dataset D_T, both are input into θ_f, and then the features of D_s are input into θ_c and obtain the predicted class. According to the predicted label and the true label, class loss L_C is calculated. Moreover, the features of D_s and D_T are input into θ_D to obtain the predicted domain, followed by the domain loss L_D.

In the training process of domain-adversarial adaptation networks, all the parameters are updated by Equation (1):

$$\begin{gathered}\theta_f \leftarrow \theta_f - \alpha\left(\frac{\partial L_C}{\partial \theta_f} - \beta\frac{\partial L_D}{\partial \theta_f}\right)\\ \theta_C \leftarrow \theta_C - \alpha\frac{\partial L_C}{\partial \theta_C}\\ \theta_D \leftarrow \theta_D - \alpha\frac{\partial L_D}{\partial \theta_D}\end{gathered} \tag{1}$$

where α and β are positive constants. When this paper updates the parameters, this paper multiplies the gradient of L_D to θ_f by a negative constant $-\beta$, which prevents θ_f from extracting features for domain discrimination.

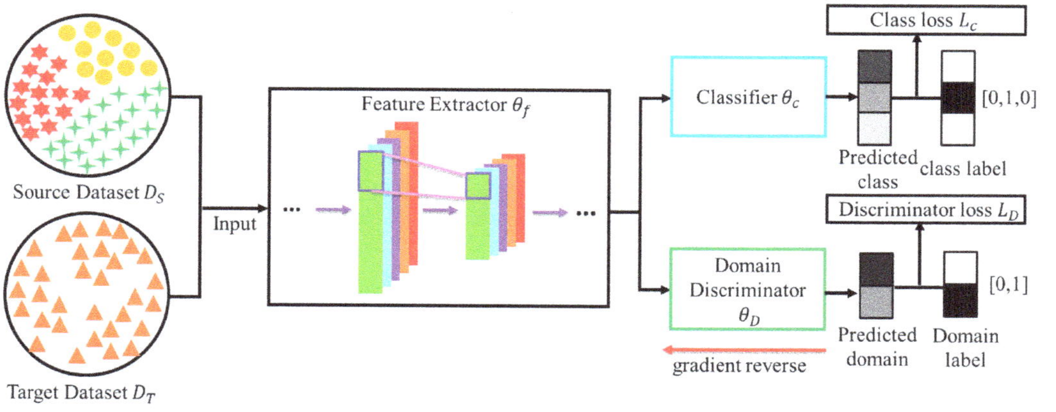

Figure 2. The structure of a domain-adversarial adaptation network. The lighter the output color, the closer the value is to 0; the darker the output color, the closer the value is to 1.

In general, domain-adversarial adaptation networks can be thought of as a two-player game. In this game, the players are the domain discriminator θ_D and the feature extractor

θ_f [56]. The target of θ_D is to identify the domain of each sample, but θ_f strives to deceive θ_D. After long-term adversarial training, θ_f specifically extracts domain-invariant features.

2.3. Maximum Mean Difference (MMD)

The purpose of TL is to apply the knowledge learned in the source domain to a different but related target domain. Essentially, it is to find a loss function that minimizes the distance between domain features and target domain features. Therefore, MMD is used to measure the difference in data distribution between the two domains [57]. Given two distributions, X and Y, MMD is defined as follows:

$$\mathrm{MMD}(X,Y) = \left\| \frac{1}{n}\sum_{i=1}^{n} \phi(x_i) - \frac{1}{m}\sum_{j=1}^{m} \phi(y_j) \right\|_H^2 \quad (2)$$

where H means that this distance is measured by mapping the data into the regenerative Hilbert space using the kernel function $\phi(\mathrm{x})$. m, n are the sample numbers of two distributions, X and Y.

In a word, the meaning of MMD is to map variables onto a high-dimensional space through a kernel function and then find the expected difference between two distributed random variables after mapping.

3. IATN for Bearing Fault Diagnosis under Variable Working Conditions

Like domain-adversarial adaptation networks, the proposed IATN also consists of three parts—feature extractor θ_f, classifier θ_c and domain discriminator θ_D. Meanwhile, there are some improvements in the proposed approach.

3.1. Preprocess

Compared with the raw time-domain signal, the time-frequency image has both time and frequency domain information and contains more fault information. In addition, the features of time-frequency images are more prominent and require lower feature extraction capabilities of the network. In fault diagnosis research, relevant studies have proved that time-frequency domain input is better than time-domain input [58]. By converting the raw 1D time-domain signal into a 2D time-frequency diagram, the features of the signal are enhanced, allowing us to use lighter downstream networks for classification.

Therefore, this paper performs STFT on the collected raw signals. STFT selects a time-frequency localized window function, assumes that the signal is stationary (pseudo-stationary) within a brief time interval and moves the window function to calculate the frequency spectrum at different moments [59]. STFT is defined as follows:

$$STFT_x(t,f) = \int_{-\infty}^{+\infty} x(\tau)g^*(\tau - t)e^{-j2\pi f\tau}d\tau \quad (3)$$

where $x(t)$ is the raw signal, and $g^*(\tau - t)$ is the window function.

Ideally, STFT can provide time and frequency information at the same time so that the frequency distribution of the signal over time can be observed, which is crucial for fault diagnosis. By converting a signal into its time-frequency representation, STFT can reveal structural features that may not be apparent in the original time-domain signal. This conversion can improve the separability between different categories, such as inner rings, outer rings, rolling elements and different degrees, such as standard, minor and major faults, so shallow neural networks can achieve high-accuracy classification.

However, in actual scenarios, various factors often interfere with bearing signals, including background noise, structural resonance, amplitude modulation, etc. At this time, the classification of time-frequency diagrams may involve highly nonlinear and complex data structures, especially when there are multiple fault types and changing work conditions. Shallow neural network performance cannot model such complexity and nonlinearity. In contrast, deep neural networks could be better at extracting high-level features from complex data. Deep neural networks extract abstract features that are difficult to observe directly through multi-layer nonlinear transformations to achieve high-precision classification of time-frequency images.

3.2. Feature Extractor

As shown in Figure 3, differently from previous studies, the input of IATN is a 2D time-frequency diagram rather than a 1D vibration signal. Therefore, the feature extractor consists of a 2D CNN. Since this paper performs STFT on the raw signal, the feature extractor can use a lighter network. The structure of the feature extractor of the model is shown in Table 1.

Table 1. The structure of feature extractor.

Layer	Kernel	Channel
Conv2D	3×3	16
BatchNorm	/	16
MaxPool	/	16
Relu	/	16
Conv2D	3×3	32
BatchNorm	/	32
MaxPool	/	32
Relu	/	32

Moreover, the hyperparameters of IATN for all tasks are shown in Table 2.

Table 2. Hyperparameters used in IATN.

Hyperparameter	First Appeared	Value
d	(4)	32
w	(4)	64
β	(14)	0.04
γ	(13)	0.1

In addition, since convolution only operates in a local space, obtaining enough information to extract the relationship between channels takes much work. Therefore, this paper introduces the channel attention mechanism to determine the importance of each feature channel, thereby enhancing the weight of essential features and reducing the weight of irrelevant information [60].

The channel attention mechanism includes three steps: squeezing, excitation and scaling. As shown in Figure 3, the number of channels of the feature X_2 of the input channel attention layer is C_2, and the data are expressed as $H_2 \times W_2$. The squeeze operation applies global average pooling to convert the data into Z, with a shape of $1 \times 1 \times C_2$. This process expands the receptive field and encodes the entire spatial feature on a channel into a global feature, as shown in (4).

$$Z_c = F_{sq}(u_c) = \frac{1}{H \times W} \sum_{i=1}^{H} \sum_{j=1}^{W} u_c(i,j) \tag{4}$$

where z_c denotes the c-th element of matrix Z, and u_c denotes the c-th channel of C_2.

Subsequently, excitation obtains the feature weight of each channel through two fully connected layers, as shown in (5):

$$S = F_{ex}(Z, W) = \sigma(g(Z, W)) = \sigma(W_2 \delta(W_1 Z)) \tag{5}$$

where $\sigma(x)$ denotes sigmoid function, $\delta(x)$ denotes the Relu function.

Finally, scaling multiplies the obtained feature weights S with the input X_2 channel by channel, as shown in (6):

$$X_3^c = F_{\text{scale}}(u_c, s_c) = s_c u_c \quad (6)$$

where X_3^c denotes the c-th channel of X_3.

After that, in the training, both the samples from source datasets and the target dataset are input into the feature extractor, and then the feature extractor outputs source features F_S and F_T.

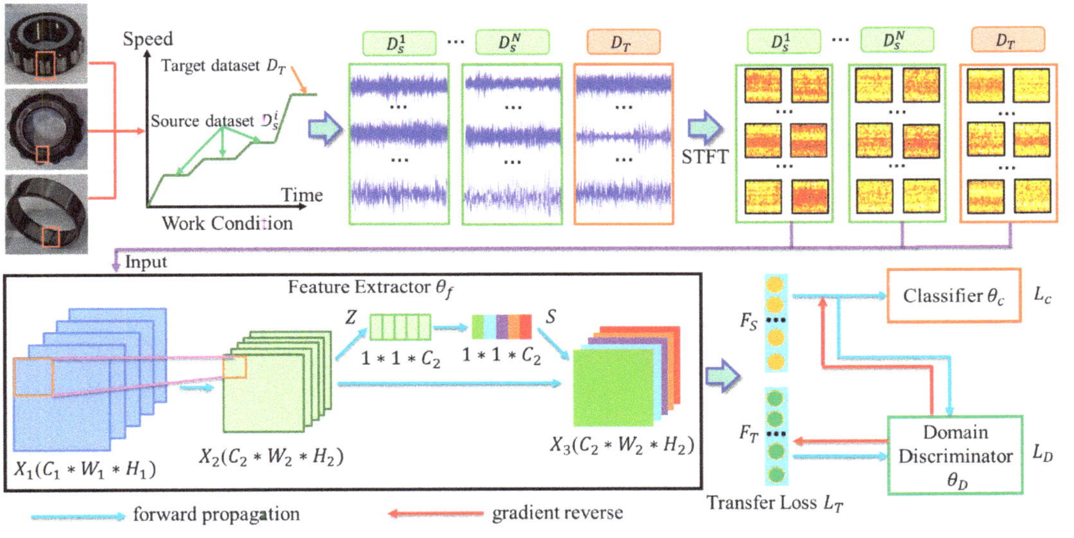

Figure 3. IATN for bearing fault diagnosis under variable working conditions.

3.3. Transfer Loss

As mentioned in Section 2, the most used transfer loss function is currently MMD. One of the most important concepts of MMD is the kernel function $\phi(x)$. In MMD, $\phi(x)$ is fixed. The Gaussian kernel function is usually chosen as the kernel function because the Gaussian kernel function can map data onto an infinite-dimensional space:

$$k(x, x') = \exp\left(-\| x - x' \|^2 / \left(2\sigma^2\right)\right) \quad (7)$$

Theoretically, any selection of a kernel function can determine the MMD of two different distributions. But in some extreme cases, this kernel function is determined to make the MMD distance of two different distributions exceedingly small. To avoid this situation, MK-MMD is proposed.

$$\mathcal{K} \triangleq \left\{k = \sum_{u=1}^{m} \beta_u k_u : \sum_{u=1}^{m} \beta_u = 1, \beta_u \geqslant 0, \forall u\right\} \quad (8)$$

By using multi-kernel functions, this paper can better represent the differences in data distribution in high-dimensional space and improve the representation ability of model features.

$$\text{MK} - \text{MMD}(X, Y) = \left\| \frac{1}{n}\sum_{i=1}^{n} \mathcal{K}(x_i) - \frac{1}{m}\sum_{j=1}^{m} \mathcal{K}(y_j) \right\|_H^2 \quad (9)$$

Moreover, to further enhance the domain adaptation performance of the transfer loss, this paper designs a novel domain difference measure combining MK-MMD and feature center distance (FCD). FCD refers to calculating the distance between the target domain feature clustering center and the source domain feature clustering center. Assuming that there are m samples $X^k = [X_1, X_2 \cdots X_m]$ of the k-th category in the source domain, and there are n samples $Y^k = [Y_1, Y_2 \cdots Y_m]$ in the target domain judged to be the k-th category by the model, the definition of FCM is shown in (10):

$$L_{FCM} = \frac{1}{k}\sum_{k=1}^{K} \dot{s}_k = \frac{1}{k}\sum_{k=1}^{K} \left|\overline{X^k} - \overline{Y^k}\right| \tag{10}$$

where K is the number of categories. By combining (9) and (10), the novel transfer loss is defined as (11):

$$L_T = L_{\text{MK-MMD}} + L_{FCM} \tag{11}$$

The combination of MMD and FCD enables the model to use different feature alignment strategies for feature transfer. FCD promotes local feature alignment, and MMD is used for global feature alignment. The integration of these two loss functions combines local and global feature alignment, thereby enhancing domain adaptation performance.

3.4. Training Process

As Figure 3 shows, the total loss function consists of three parts, transfer loss L_T, class loss L_C and domain loss L_D. They are defined as follows:

$$\begin{aligned} L_C &= \text{MSE}\left(Y_C, Y_C^{label}\right) \\ L_D &= \text{MSE}\left(Y_D, Y_D^{label}\right) \\ L_{total} &= \gamma L_T + L_C + L_D \end{aligned} \tag{12}$$

where MSE is the mean squared error, Y_C and Y_D are the output of the classifier and domain discriminators and Y_C^{label} and Y_D^{label} are the class label and domain label.

Since there is gradient reverse in the backpropagation update process of domain loss L_D, the model parameters are updated as shown in (13):

$$\begin{aligned} \theta_f &\leftarrow \theta_f - \alpha\left(\frac{\partial L_C}{\partial \theta_f} - \beta\frac{\partial L_D}{\partial \theta_f} + \gamma\frac{\partial L_T}{\partial \theta_f}\right) \\ \theta_c &\leftarrow \theta_c - \alpha\left(\frac{\partial L_C}{\partial \theta_c}\right) \\ \theta_D &\leftarrow \theta_D - \alpha\left(\frac{\partial L_D}{\partial \theta_D}\right) \end{aligned} \tag{13}$$

3.5. Dataset

To evaluate the performance of the proposed method more comprehensively, this paper verifies its effectiveness on three datasets.

The first dataset is private. As Figure 4 shows, it is a dual-rotor test bench to simulate dual-rotor engine operation. The low-pressure rotor (LR) and the high-pressure rotor (HR) are driven by different motors and rotate in the same direction at different speeds. The LR is supported by the No. 1 and No. 4 bearings, and the high-pressure rotor HR is supported by the No. 2 and 3 bearings. The inter-shaft bearing is No. 3, and the experimental bearing is between the two rotors. The outer ring of the inter-shaft bearing rotates with the HR at a higher speed, and the inner ring rotates with the LR at a lower speed.

This paper collects vibration data under four working conditions (WDs), as shown in Table 3. Each working condition includes bearing data in nine different health states, including normal state, three degrees of inner-ring failure, three degrees of rolling-element failure and two different degrees of outer-ring failure. The sampling frequency is 20.48 k Hz. In addition, this paper uses a sliding window of a length of 1024 to sample the raw signal without overlap and collect 100 samples for each health state under each operating condition.

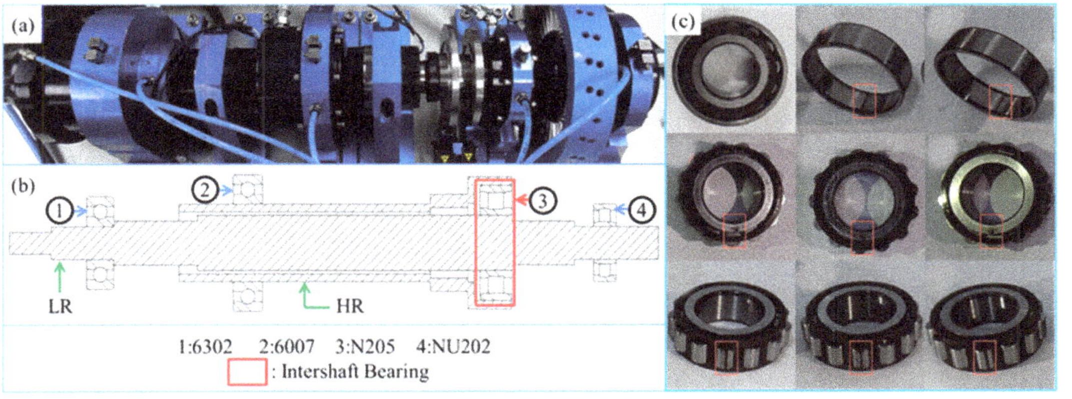

Figure 4. Dual-rotor test bench. (**a**) The test rig; (**b**) engineering sketch; (**c**) bearing failure mode.

Table 3. The working conditions of dual-rotor test bench.

Number	LR Speed (rpm)	HR Speed (rpm)
D0	2000	4000
D1	3000	6000
D2	4000	8000
D3	5500	10,000

The second dataset is the public Case Western Reserve University dataset (CWRU). The CWRU dataset has four working conditions: 1797 rpm (R0), 1772 rpm (R1), 1750 rpm (R2) and 1730 rpm (R3). Under all working conditions, this paper selects normal data and three different degrees of fault data of the inner ring, outer ring and rolling elements collected at the driving end as experimental data. The sampling frequency is 12,000 Hz.

The third dataset is the public Ottawa dataset. The Ottawa dataset has four working conditions: increasing speed (O0), decreasing speed (O1), increasing then decreasing speed (O2) and decreasing then increasing speed (O3). Under all working conditions, this paper selects normal data and fault data of the inner ring and outer ring. The sampling frequency is 200,000 Hz.

The training epoch is uniformly set to 300 using the Adams optimizer, and the learning rate is 0.001. The code is written through the Pytorch 2.2 framework and runs on the GPU of GPTX1650.

4. Results

Here this paper compares the performance of the IATN with CNN [61], DANN [62], AMDSA, DASN, ISAE-CSDF [63] and CDGATLN [64] models on the three datasets, and the results are shown in Tables 4–6. In addition, the experiment in this paper is repeated 20 times to eliminate the influence of randomness.

One can generally make three observations: (1) the IATN has the highest accuracy on most tasks. (2) the IATN has the most minor variances. (3) The more similar the working conditions of the source and target domain data are, the better the transfer learning effect will be.

In Table 4, the diagnosis accuracy of the IATN is significantly higher than the other six methods on most tasks. In the task of D0→D3, all the methods have the lowest accuracy. The accuracies of the CNN, DANN, AMDSA, DASN, ISAE-CSDF and SA-SN-DCGAN are 58.75%, 61.82%, 65.28%, 60.32%, 65.64% and 66.82%, respectively, but the accuracy of the IATN is 70.62%. In the task of D0, D1, D3→D2, the accuracies of the CNN, DANN, TL and DSAN are 95.12%, 96.16%, 95.44%, 93.66%, 96.08% and 96.06%, respectively, but the accuracy of the IATN is 98.44%. In the last row, this paper calculates the average accuracy of the five methods on all tasks, and the average accuracy of the IATN is 91.89%, which is 6.82%, 4.77%, 1.06%, 2.52%, 2.58% and 1.23% higher than the CNN, DANN, ISAE-CSDF and SA-SN-DCGAN, respectively. As for variance, the average variance of the IATN is 1.10%, which is significantly lower than the averages of the other six methods, which are 2.29%, 1.72%, 1.52%, 1.91%, 1.35% and 1.49%, respectively.

Table 4. Performance comparison of dual-rotor test bench.

Task	Model Performance (%)						
	CNN	DANN	AMDSA	DSAN	ISAE-CSDF	CDGATLN	IATN
D1→D0	84.22 ± 1.21	85.22 ± 1.08	89.24 ± 0.98	89.20 ± 1.04	89.78 ± 1.22	89.92 ± 1.14	90.76 ± 1.06
D2→D0	68.80 ± 2.42	72.80 ± 3.42	79.26 ± 1.82	76.84 ± 2.42	78.24 ± 1.79	79.94 ± 1.62	82.82 ± 1.58
D3→D0	68.67 ± 1.46	72.76 ± 3.18	79.32 ± 1.74	78.75 ± 2.36	76.58 ± 1.58	78.75 ± 2.36	80.76 ± 1.38
D1, D2→D0	84.11 ± 1.82	87.14 ± 1.69	91.42 ± 1.58	91.26 ± 1.78	90.68 ± 1.58	91.36 ± 1.68	92.44 ± 1.28
D1, D3→D0	86.44 ± 2.05	89.84 ± 2.16	90.72 ± 1.46	91.13 ± 1.68	89.94 ± 1.02	91.24 ± 1.08	92.31 ± 0.76
D2, D3→D0	78.89 ± 2.28	81.86 ± 1.06	89.92 ± 1.88	88.98 ± 2.56	86.73 ± 1.42	90.28 ± 1.52	91.89 ± 1.26
D1, D2, D3→D0	87.33± 3.79	89.67 ± 0.96	92.56 ± 1.24	93.12 ± 1.36	90.54 ± 1.06	92.02 ± 1.12	93.42 ± 0.88
D0→D1	88.89 ± 1.67	89.16 ± 1.56	92.12 ± 1.72	92.02 ± 1.92	89.49 ± 1.82	92.12 ± 1.26	92.82 ± 1.08
D2→D1	88.26 ± 1.88	89.06 ± 1.68	95.86 ± 1.94	95.46 ± 2.12	94.62 ± 1.58	95.46 ± 2.12	97.24 ± 1.32
D3→D1	80.33 ± 1.24	81.33 ± 1.12	81.48 ± 1.68	81.86 ± 1.94	81.28 ± 1.26	81.86 ± 1.94	82.62 ± 1.12
D0, D2→D1	97.49 ± 1.40	97.89 ± 0.62	98.02 ± 1.72	98.08 ± 1.64	97.77 ± 0.98	97.08 ± 1.24	98.10 ± 0.86
D0, D3→D1	90.67 ± 1.38	92.04 ± 1.28	94.32 ± 1.72	96.28 ± 1.98	94.76 ± 1.44	96.58 ± 1.78	97.52 ± 1.28
D2, D3→D1	90.72 ± 2.38	92.06 ± 1.42	92.58 ± 1.20	96.38 ± 1.04	94.82 ± 0.94	96.28 ± 1.02	97.13 ± 0.80
D0, D2, D3→D1	97.67 ± 1.24	98.06 ± 0.82	98.00 ± 0.52	98.02 ± 0.66	97.78 ± 0.58	98.02 ± 0.76	98.12 ± 0.46
D0→D2	73.42 ± 3.42	75.72 ± 3.12	98.24 ± 2.08	79.36 ± 2.36	78.24 ± 1.96	79.06 ± 2.46	79.26 ± 1.82
D1→D2	91.11 ± 2.14	92.48 ± 2.08	95.62 ± 0.75	95.46 ± 0.92	95.26 ± 0.68	94.62 ± 0.90	96.78 ± 0.76
D3→D2	87.18 ± 1.04	87.78 ± 0.94	92.12 ± 1.52	90.16 ± 2.02	91.20 ± 1.16	92.76 ± 1.42	94.58 ± 1.06
D0, D1→D2	92.24 ± 1.58	93.92 ± 1.76	96.88 ± 1.06	96.84 ± 1.38	95.58 ± 0.92	96.82 ± 0.98	97.12 ± 0.76
D0, D3→D2	89.94 ± 1.88	91.72 ± 1.06	94.52 ± 1.74	93.24 ± 2.12	93.56 ± 1.42	94.28 ± 1.52	96.13 ± 1.08
D1, D3→D2	95.12 ± 0.88	96.16 ± 1.42	95.44 ± 0.98	93.66 ± 1.38	96.02 ± 0.88	97.68 ± 0.68	98.44. ± 0.44
D0, D1, D3→D2	96.76 ± 1.02	96.72 ± 1.42	96.06 ± 1.42	96.16 ± 1.98	96.08 ± 0.82	96.06 ± 1.58	97.36 ± 1.02
D0→D3	58.78 ± 5.68	61.82 ± 3.28	65.28 ± 2.92	60.32 ± 3.88	65.64 ± 2.96	66.82 ± 3.28	70.62 ± 2.18
D1→D3	75.12 ± 2.66	77.24 ± 3.08	77.42 ± 2.40	74.78 ± 3.18	79.54 ± 2.62	80.18 ± 2.78	81.68 ± 1.88
D2→D3	86.08 ± 2.14	90.06 ± 2.02	93.64 ± 1.48	93.82 ± 2.26	91.54 ± 1.28	94.42 ± 0.76	97.89 ± 0.48
D0, D1→D3	70.24 ± 5.98	75.24 ± 2.46	84.28 ± 1.62	76.17 ± 2.14	84.21 ± 1.56	88.12 ± 1.64	90.16 ± 1.18
D0, D2→D3	92.12 ± 3.56	94.66 ± 1.06	97.16 ± 1.04	97.26 ± 1.24	93.53 ± 0.96	95.16 ± 1.04	97.62 ± 0.64
D1, D2→D3	89.33 ± 2.16	91.53 ± 1.48	95.68 ± 0.88	95.92 ± 1.88	93.52 ± 0.92	94.02 ± 1.08	96.74 ± 0.74
D0, D1, D2→D3	92.24 ± 3.68	95.48 ± 1.09	96.12 ± 1.68	96.18 ± 2.26	93.88 ± 1.42	96.78 ± 1.16	97.33 ± 1.08
Average	85.07 ± 2.29	87.12 ± 1.72	90.83 ± 1.52	89.37 ± 1.91	89.31 ± 1.35	90.66 ± 1.49	91.89 ± 1.10

Table 5. Performance comparison of CWRU.

Task	Model Performance (%)						
	CNN	DANN	AMDSA	DSAN	ISAE-CSDF	SA-SN-DCGAN	IATN
R1→R0	93.10 ± 0.38	94.35 ± 0.98	96.02 ± 1.20	95.62 ± 1.32	94.68 ± 0.52	95.22 ± 0.72	97.76 ± 0.36
R2→R0	86.30 ± 0.46	88.82 ± 0.82	90.82 ± 1.52	90.42 ± 1.88	90.26 ± 0.89	91.22 ± 0.88	94.82 ± 0.58
R3→R0	83.76 ± 1.22	86.86 ± 1.68	88.72 ± 1.98	84.26 ± 2.14	90.13 ± 1.76	87.06 ± 1.54	93.76 ± 1.28
R0→R1	97.25 ± 0.36	97.66 ± 0.62	98.40 ± 0.92	97.88 ± 1.04	98.26 ± 0.72	97.28 ± 0.32	99.40 ± 0.26
R2→R1	97.22 ± 0.48	99.02 ± 0.46	99.30 ± 0.65	99.02 ± 0.78	98.52 ± 0.48	99.22 ± 0.78	99.60 ± 0.18
R3→R1	94.76 ± 0.82	94.96 ± 0.58	98.24 ± 0.95	97.76 ± 1.14	96.28 ± 0.78	97.06 ± 0.72	97.86 ± 0.62
R0→R2	91.22 ± 0.66	91.76 ± 1.66	96.28 ± 1.65	95.64 ± 1.68	93.58 ± 0.92	94.64 ± 1.14	97.64 ± 0.84
R1→R2	99.96 ± 0.04	99.98 ± 0.02	99.90 ± 0.04	99.96 ± 0.04	99.98 ± 0.02	99.96 ± 0.04	100.00 ± 0.00
R3→R2	99.94 ± 0.06	99.96 ± 0.04	99.95 ± 0.04	99.92 ± 0.06	99.98 ± 0.02	99.92 ± 0.08	100.00 ± 0.00
R0→R3	81.56 ± 1.24	86.26 ± 1.36	89.24 ± 1.53	90.26 ± 1.62	87.38 ± 0.92	90.06 ± 1.42	92.56 ± 1.06
R1→R3	95.32 ± 0.48	94.82 ± 0.68	98.38 ± 0.92	98.82 ± 1.08	97.66 ± 0.68	98.22 ± 0.88	99.52 ± 0.48
R2→R3	99.90 ± 0.08	99.90 ± 0.10	99.90 ± 0.10	99.92 ± 0.08	99.95 ± 0.05	99.92 ± 0.08	100.00 ± 0.00
Average	93.36 ± 0.52	94.52 ± 0.75	96.26 ± 0.96	95.79 ± 1.07	95.55 ± 0.64	95.81 ± 0.72	97.74 ± 0.47

In Tables 5 and 6, one could also find similar discoveries by comparing the performance of the IATN and the other methods. The average accuracy of the IATN on all CWRU tasks is 97.74%, which is 4.38%, 3.22%, 1.48%, 1.95%, 2.19% and 1.93% higher than the CNN, DANN, AMDSA, DASN, ISAE-CSDF and SA-SN-DCGAN, respectively. The average variance of the IATN on all CWRU tasks is 0.47%, which is lower than the other six methods, which are 0.52%, 0.75%, 0.96%, 1.07%, 0.64% and 0.72%, respectively. The average accuracy of the IATN on all Ottawa dataset tasks is 81.11%, which is 6.77%, 4.23%, 1.42%, 3.78%, 3.28% and 0.84% higher than the CNN, DANN, AMDSA, DASN, ISAE-CSDF and SA-SN-DCGAN, respectively. The average variance of the IATN on all Ottawa dataset tasks is 1.24%, which is lower than those of the other six methods, which are 1.56%, 1.97%, 2.10%, 2.38%, 1.28% and 1.34%, respectively.

Table 6. Performance comparison of Ottawa dataset.

Task	Model Performance (%)						
	CNN	DANN	AMDSA	DSAN	ISAE-CSDF	SA-SN-DCGAN	IATN
O1 → O0	64.20 ± 1.82	65.38 ± 2.48	66.81 ± 2.57	65.58 ± 2.95	65.76 ± 1.86	66.64 ± 1.72	67.53 ± 1.42
O2→O0	58.30 ± 1.86	62.82 ± 2.52	67.84 ± 2.62	72.26 ± 3.01	64.86 ± 1.42	68.52 ± 1.78	70.36 ± 1.46
O3→O0	70.26 ± 1.52	71.86 ± 1.68	73.79 ± 1.86	63.82 ± 2.04	72.58 ± 1.26	73.36 ± 1.14	74.76 ± 1.08
O0→O1	60.25 ± 1.76	63.76 ± 2.62	66.94 ± 2.68	64.04 ± 3.14	64.36 ± 1.62	65.78 ± 1.68	68.53 ± 1.56
O2→O1	84.32 ± 1.28	85.08 ± 1.46	87.98 ± 1.56	85.82 ± 1.71	86.92 ± 0.78	87.12 ± 1.08	88.33 ± 0.76
O3→O1	85.26 ± 1.22	86.42 ± 1.56	87.24 ± 1.72	86.72 ± 1.98	87.08 ± 1.18	97.76 ± 1.14	88.76 ± 1.28
O0→O2	48.76 ± 2.56	54.36 ± 2.68	61.9 ± 3.04	55.75 ± 3.29	57.18 ± 1.62	62.54 ± 1.88	65.67 ± 1.84
O1→O2	76.26 ± 1.24	80.88 ± 1.82	84.84 ± 1.84	81.33 ± 2.14	81.88 ± 1.02	83.76 ± 0.94	86.83 ± 0.98
O3→O2	93.94 ± 1.28	94.98 ± 1.54	95.91 ± 1.64	95.25 ± 1.82	95.58 ± 0.92	95.92 ± 0.88	96.38 ± 0.86
O0→O3	70.56 ± 1.84	73.56 ± 1.98	76.84 ± 2.32	73.92 ± 2.57	74.38 ± 1.82	76.26 ± 1.82	78.48 ± 1.78
O1→O3	85.32 ± 1.28	87.72 ± 1.88	89.54 ± 1.94	87.64 ± 2.27	87.56 ± 1.29	88.88 ± 1.28	90.46 ± 1.18
O2→O3	94.70 ± 1.12	95.76 ± 1.46	96.76 ± 1.52	95.87 ± 1.72	95.85 ± 0.68	96.42 ± 0.82	97.26 ± 0.78
Average	74.34 ± 1.56	76.88 ± 1.97	79.69 ± 2.10	77.33 ± 2.38	77.83 ± 1.28	80.27 ± 1.34	81.11 ± 1.24

The comparison results demonstrate that the IATN significantly improves bearing fault diagnosis under variable WDs. Moreover, when the WDs of the source and target domains are similar, the test accuracy obtained is also higher. For example, the accuracy of all methods on the D2→D3 task is higher than that on the D0→D3 task. Data similarity may be higher when operating conditions are similar, making it easier to transfer fault diagnosis knowledge. This suggests that data similarity may be higher when WDs are similar, making it easier to transfer fault diagnosis knowledge.

5. Discussion

5.1. Running Process Comparison

To evaluate the performance of the model more comprehensively, this paper compares the running process of 300 epochs of different methods on the D0→D1 task, the D0, D1→D2 task and the D0, D1, D2→D3 task, the results are shown in the Table 7 and Figure 5.

Table 7. Comparison of running-times of dual-rotor dataset.

Task	Running Time (s)				
	CNN	DANN	ISAE-CSDF	CDGATLN	IATN
D0→D1	276	280	508	526	410
D0, D1→D2	284	291	564	602	417
D0, D1, D2→D3	287	295	765	826	601

According to Table 7, it is obvious that the running time of the IATN is more than those of the CNN and DANN but less than those of the ISAE-CSDF and CDGATLN. By comparing the running time of the IATN with those of the CNN and DANN, it can be concluded that the transfer process is much more time-consuming than the adversarial process.

Moreover, the training curves of the IATN, CNN and DANN are presented in Figure 5. In Figure 5, there are three subfigures. Figure 5a presents the training curves of the D0→D1 task; Figure 5b presents the training curves of the D0, D1→D2 task; and Figure 5c presents the training curves of the D0, D1, D2→D3 task.

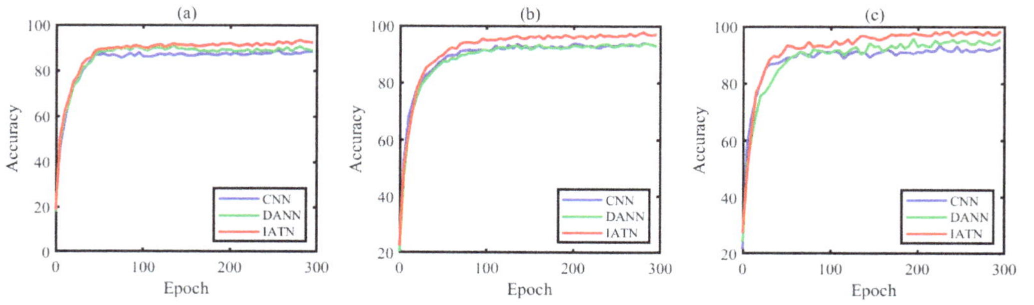

Figure 5. Training curves. (**a**) D0→D1, (**b**) D0, D1→D2, (**c**) D0, D1, D2→D3.

In Figure 5, the blue lines represent the accuracy of the CNN; the green lines represent the accuracy of the DANN, and the red lines represent the accuracy of the IATN. According to Figure 5, it is obvious that although the calculation process of the IATN is more complex, this does not affect its convergence. On the three tasks with different numbers of source domains, the convergence speed of the IATN is like that of the CNN and DANN.

5.2. Feature Visualization

According to Tables 4–6, it is obvious that the IATN performs better than the other methods in extracting domain-invariant features. To demonstrate the superiority of the IATN more intuitively in extracting domain-invariant features, this paper performs feature visualization of the D2→D3 task. Specifically, the source domain features and target domain features extracted by each model are reduced to two dimensions for visualization with t-distributed stochastic neighbor embedding (t-SNE) [65]. The results are shown in Figure 6.

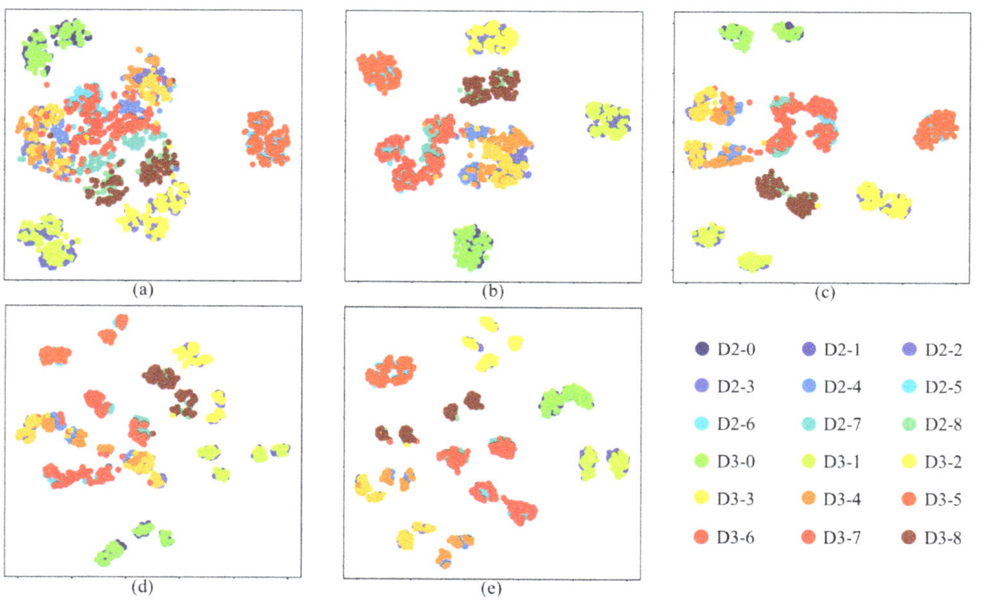

Figure 6. t-SNE feature visualization. (**a**) CNN. (**b**) DANN. (**c**) ISAE-CSDF. (**d**) CDGATLN. (**e**) IATN. D2-0 indicates data collected for category "0" fault bearings in D2 working conditions.

In Figure 6, there are five subfigures. Figure 6a is the feature visualization of the CNN, Figure 6b is the feature visualization of the DANN, Figure 6c is the feature visualization of the ISAE-CSDF, Figure 6d is the feature visualization of the CDGATLN and Figure 6e is the feature visualization of the IATN. In each subfigure, there are 18 clusters; 9 belong to D2 tasks and 9 belong to D3 tasks. There are two criteria for judging model performance. One is the degree of separation of the nine fault clusters in each WD, and the other is the degree of overlap of the same fault clusters in the two working conditions.

According to Figure 6, it is obvious that there is significant overlap between different fault clusters of source domain features and target domain features extracted by the CNN, DANN, ISAE-CSDF and CDGATLN, which initially have completely separate clusters, but there is still aliasing at the boundaries of some categories. In Figure 6e, all the clusters of each WD are separated entirely, and all the same fault clusters of D2 and D3 are basically overlapped, which indicates that the IATN has strong domain-invariant feature extraction capabilities.

5.3. Data Analysis

Based on the observations in Tables 4 and 5, the closer the working conditions are, the more similar the distribution of the collected data will be. This paper calculates the MK-MMD and MMD between the data collected under different working conditions to verify the hypothesis. The results are shown in Table 8.

Table 8. The distribution connections of different domains.

Task	MK-MMD	MMD	Mean Acc
D0↔D1	0.438	0.122	91.79
D0↔D2	1.085	0.312	90.03
D0↔D3	1.782	0.511	75.74
D1↔D2	0.316	0.087	97.01
D1↔D3	1.175	0.253	82.15
D2↔D3	0.321	0.084	96.23
R0↔R1	0.061	0.011	98.58
R0↔R2	0.072	0.010	96.23
R0↔R3	0.071	0.014	93.16
R1↔R2	0.052	0.008	99.80
R1↔R3	0.065	0.013	98.69
R2↔R3	0.054	0.012	100
O0↔O1	0.854	0.198	68.03
O0↔O2	0.927	0.209	68.01
O0↔O3	0.833	0.194	76.62
O1↔O2	0.207	0.045	87.58
O1↔O3	0.029	0.006	89.61
O2↔O3	0.021	0.004	96.82

According to Table 8, it is evident that the more similar the working conditions of the source domain data and the target domain data are, the lower the MK-MMD and MMD between them are. To express the connection between the distribution difference between source and target domain data and the model performance more intuitively, we perform correlation analysis on them according to (14):

$$\rho_{x,y} = \frac{\mathrm{Cov}(X,Y)}{\sqrt{D(X)D(Y)}} = \frac{E(XY) - E(X)E(Y)}{\sqrt{E(X^2) - E^2(X)}\sqrt{E(Y^2) - E^2(Y)}} \tag{14}$$

The correlation coefficient between the MK-MMD of dual-rotor tasks and their transfer learning accuracy is defined as ρ_1; the correlation coefficient between the MMD of dual-rotor tasks and their transfer learning accuracy is defined as ρ_2; the correlation coefficient between the MK-MMD of CWRU tasks and their transfer learning accuracy is defined as ρ_3; the correlation coefficient between the MMD of CWRU tasks and their transfer learning accuracy is defined as ρ_4; the correlation coefficient between the MK-MMD of Ottawa dataset tasks and their transfer learning accuracy is defined as ρ_5; the correlation coefficient between the MMD of Ottawa dataset tasks and their transfer learning accuracy is defined as ρ_6. According to (11), it can be calculated that $\rho_1 = -0.9523$, $\rho_2 = -0.9044$, $\rho_3 = -0.8456$, $\rho_4 = -0.5045$, $\rho_5 = -0.9574$ and $\rho_6 = -0.9549$.

It is obvious that there exists a strong negative correlation. Therefore, it can be summarized that the more similar the working conditions of the source domain data and the target domain data are, the lower the MK-MMD and MMD between them are, and the better the transfer learning results will be.

5.4. Hyperparameter Optimization

The IATN has two main hyperparameters, the inverse gradient coefficient β and the transfer learning coefficient γ. To study the impact of their values on model performance, this paper adjusts k and ϑ to find the optimal values. Here, this paper sets β and γ to different values on D1→D2 and D2→D3 tasks and conducts 10 repeated experiments to obtain their average accuracy. The results are shown in Figure 7. In Figure 7, the abscissa of each graph represents the value of the hyperparameter, the ordinate represents the accuracy, the blue line represents the experimental results of D1→D2 and the red line represents the experimental results of D2→D3.

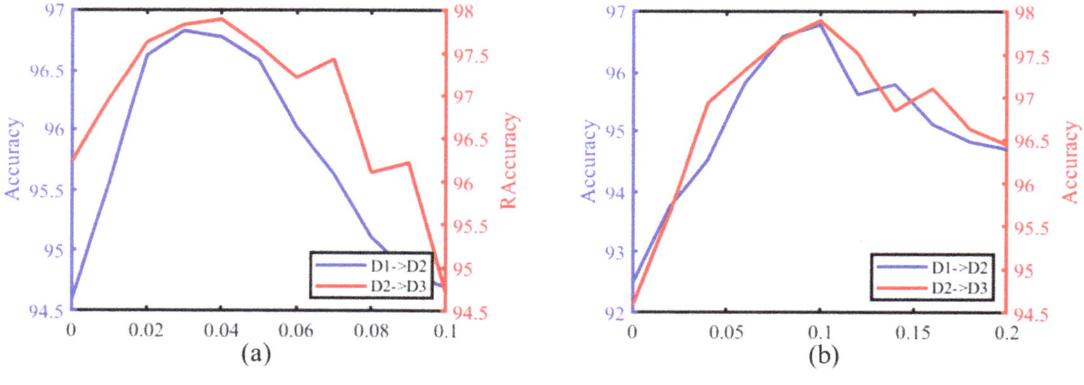

Figure 7. The testing accuracy of β and γ on D1→D2 and D2→D3 tasks. (**a**) The impact of different values of β on D1→D2 and D2→D3 tasks; (**b**) the impact of different values of γ on D1→D2 and D2→D3 tasks.

In Figure 6a, it is obvious that the optimal parameters are different for different tasks. For the D1→D2 task, the best β is 0.03; for the D2→D3 task, the best β is 0.04. For the D1→D2 task, the best β is 0.03; for the D2→D3 task, the best β is 0.04. In Figure 6b, the optimal γ for both tasks is 0.1.

In summary, this paper chooses $\beta = 0.04$ and $\gamma = 0.1$ as the experimental parameters.

6. Conclusions

This paper proposes an IATN for bearing fault diagnosis under variable WDs. Specifically, this paper combines adversarial transfer networks with an STFT to obtain satisfactory results with lighter networks. Then, this paper employs a channel attention module to enhance feature fusion, allowing the model to emphasize useful features and suppress less-useful features selectively. Moreover, this paper designs a novel domain discrepancy hybrid metric loss by combining MK-MMD and FCD. FCD promotes local feature alignment, and MMD is used for global feature alignment. The integration of these two loss functions combines local and global feature alignment. Finally, this paper verifies the method's effectiveness on three datasets, a private dual-rotor dataset, a CWRU dataset and the Ottawa dataset. The IATN achieves an average accuracy that is significantly higher than those of the other six methods, CNN, DANN, AMDSA, DASN, ISAE-CSDF and SA-SN-DCGAN. The results indicate that the IATN has strong domain-invariant feature extraction capabilities.

In the future, we will pay more attention to the interpretability of the transfer process and study what knowledge is transferred during the transfer learning process.

Author Contributions: J.W. contributed to conceptualization, formal analysis, investigation, methodology and writing—original draft; H.A. contributed to methodology, validation and writing—review and editing; X.C. contributed to funding acquisition, project administration and writing—review and editing; R.Y. contributed to funding acquisition, project administration and data curation; and A.K.N. contributed to supervision, validation and writing—review and editing. All authors have read and agreed to the published version of the manuscript.

Funding: This work was supported in part by the Natural Science Foundation of China (No. 52175116), Major Research Programs of the Natural Science Foundation of China (No. 92060302), the Research Foundation of the Higher Educational Key Laboratory for Flexible Manufacturing Equipment Integration of Fujian Province, the Xiamen Institute of Technology, the National Key Science and Technology Infrastructure Opening Project Fund for Research and Evaluation facilities for Service Safety of Major Engineering Materials and the Aeronautical Science Foundation (No. 2019ZB070001). Also, this work was supported in part by the Royal Society award (number IEC\NSFC\223294) to Asoke K. Nandi. Jun Wang acknowledges the financial support from the Innovative Leading Talents Scholarship and Brunel University London.

Institutional Review Board Statement: Not applicable.

Informed Consent Statement: Not applicable.

Data Availability Statement: Data are contained within the article.

Conflicts of Interest: The authors declare no conflict of interest.

References

1. Ahmed, H.; Nandi, A.K. *Condition Monitoring with Vibration Signals: Compressive Sampling and Learning Algorithms for Rotating Machines*; John Wiley & Sons: Chichester, UK, 2020; ISBN 978-1-119-54462-3.
2. Desnica, E.; Ašonja, A.; Radovanović, L.; Palinkaš, I.; Kiss, I. Selection, Dimensioning and Maintenance of Roller Bearings. In Proceedings of the 31st International Conference on Organization and Technology of Maintenance (OTO 2022), Osijek, Croatia, 12 December 2022; Blažević, D., Ademović, N., Barić, T., Cumin, J., Desnica, E., Eds.; Lecture Notes in Networks and Systems. Springer: Cham, Switzerland, 2022; Volume 592, pp. 210–227. [CrossRef]
3. Liu, S.; Chen, S.; Chen, Z.; Gong, Y. Fault Diagnosis Strategy Based on BOA-ResNet18 Method for Motor Bearing Signals with Simulated Hydrogen Refueling Station Operating Noise. *Appl. Sci.* **2024**, *14*, 157. [CrossRef]
4. Patil, A.A.; Desai, S.S.; Patil, L.N.; Patil, S.A. Adopting Artificial Neural Network for Wear Investigation of Ball Bearing Materials Under Pure Sliding Condition. *Appl. Eng. Lett.* **2022**, *7*, 81–88. [CrossRef]
5. Xu, Z.; Ji, F.; Ding, S.; Zhao, Y.; Zhou, Y.; Zhang, Q.; Du, F. Digital Twin-Driven Optimization of Gas Exchange System of 2-Stroke Heavy Fuel Aircraft Engine. *J. Manuf. Syst.* **2021**, *58*, 132–145. [CrossRef]
6. Mu, X.; Wang, Y.; Yuan, B.; Sun, W.; Liu, C.; Sun, Q. A New Assembly Precision Prediction Method of Aeroengine High-Pressure Rotor System Considering Manufacturing Error and Deformation of Parts. *J. Manuf. Syst.* **2021**, *61*, 112–124. [CrossRef]
7. He, Y.; Hu, M.; Feng, K.; Jiang, Z. An Intelligent Fault Diagnosis Scheme Using Transferred Samples for Intershaft Bearings under Variable Working Conditions. *IEEE Access* **2020**, *8*, 203058–203069. [CrossRef]
8. Berghout, T.; Benbouzid, M. Diagnosis and Prognosis of Faults in High-Speed Aeronautical Bearings with a Collaborative Selection Incremental Deep Transfer Learning Approach. *Appl. Sci.* **2023**, *13*, 10916. [CrossRef]
9. Ahmed, H.O.A.; Nandi, A.K. Connected Components-Based Colour Image Representations of Vibrations for a Two-Stage Fault Diagnosis of Roller Bearings Using Convolutional Neural Networks. *Chin. J. Mech. Eng.* **2021**, *34*, 37. [CrossRef]
10. Ahmed, H.; Nandi, A.K. Three-Stage Hybrid Fault Diagnosis for Rolling Bearings with Compressively Sampled Data and Subspace Learning Techniques. *IEEE Trans. Ind. Electron.* **2019**, *66*, 5516–5524. [CrossRef]
11. Zuo, L.; Zhang, L.; Zhang, Z.-H.; Luo, X.-L.; Liu, Y. A Spiking Neural Network-Based Approach to Bearing Fault Diagnosis. *J. Manuf. Syst.* **2021**, *61*, 714–724. [CrossRef]
12. Li, Y.; Wang, S.; Deng, Z. Intelligent Fault Identification of Rotary Machinery Using Refined Composite Multi-Scale Lempel-Ziv Complexity. *J. Manuf. Syst.* **2020**, *61*, 725–735. [CrossRef]
13. Ahmed, H.; Nandi, A.K. Compressive Sampling and Feature Ranking Framework for Bearing Fault Classification with Vibration Signals. *IEEE Access* **2018**, *6*, 44731–44746. [CrossRef]

14. Wang, Z.; Liang, B.; Guo, J.; Wang, L.; Tan, Y.; Li, X. Fault Diagnosis Based on Residual–Knowledge–Data Jointly Driven Method for Chillers. *Eng. Appl. Artif. Intell.* **2023**, *125*, 106768. [CrossRef]
15. Pan, J.; Zi, Y.; Chen, J.; Zhou, Z.; Wang, B. LiftingNet: A Novel Deep Learning Network with Layerwise Feature Learning from Noisy Mechanical Data for Fault Classification. *IEEE Trans. Ind. Electron.* **2018**, *65*, 4973–4982. [CrossRef]
16. Wen, L.; Li, X.; Gao, L.; Zhang, Y. A New Convolutional Neural Network-Based Data-Driven Fault Diagnosis Method. *IEEE Trans. Ind. Electron.* **2018**, *65*, 5990–5998. [CrossRef]
17. Ahmed, H.O.A.; Wong, M.L.D.; Nandi, A.K. Intelligent Condition Monitoring Method for Bearing Faults from Highly Compressed Measurements Using Sparse Over-Complete Features. *Mech. Syst. Signal Process.* **2018**, *99*, 459–477. [CrossRef]
18. Jia, L.; Chow, T.W.S.; Yuan, Y. GTFE-Net: A Gramian Time Frequency Enhancement CNN for Bearing Fault Diagnosis. *Eng. Appl. Artif. Intell.* **2023**, *119*, 105794. [CrossRef]
19. Zhao, K.; Xiao, J.; Li, C.; Xu, Z.; Yue, M. Fault Diagnosis of Rolling Bearing Using CNN and PCA Fractal-Based Feature Extraction. *Measurement* **2023**, *223*, 113754. [CrossRef]
20. Ruan, D.; Wang, J.; Yan, J.; Gühmann, C. CNN Parameter Design Based on Fault Signal Analysis and Its Application in Bearing Fault Diagnosis. *Adv. Eng. Inform.* **2023**, *55*, 101877. [CrossRef]
21. Wang, J.; Li, T.; Sun, C.; Yan, R.; Chen, X. Improved Spiking Neural Network for Intershaft Bearing Fault Diagnosis. *J. Manuf. Syst.* **2022**, *65*, 208–219. [CrossRef]
22. Sinitsin, V.; Ibryaeva, O.; Sakovskaya, V.; Eremeeva, V. Intelligent Bearing Fault Diagnosis Method Combining Mixed Input and Hybrid CNN-MLP Model. *Mech. Syst. Signal Process.* **2022**, *180*, 109454. [CrossRef]
23. Zhiyi, H.; Haidong, S.; Xiang, Z.; Yu, Y.; Junsheng, C. An Intelligent Fault Diagnosis Method for Rotor-Bearing System Using Small Labeled Infrared Thermal Images and Enhanced CNN Transferred from CAE. *Adv. Eng. Inform.* **2020**, *46*, 101150. [CrossRef]
24. Huo, C.; Jiang, Q.; Shen, Y.; Qian, C.; Zhang, Q. New Transfer Learning Fault Diagnosis Method of Rolling Bearing Based on ADC-CNN and LATL under Variable Conditions. *Measurement* **2022**, *188*, 110587. [CrossRef]
25. Zheng, B.; Huang, J.; Ma, X.; Zhang, X.; Zhang, Q. An Unsupervised Transfer Learning Method Based on SOCNN and FBNN and Its Application on Bearing Fault Diagnosis. *Mech. Syst. Signal Process.* **2024**, *208*, 111047. [CrossRef]
26. Li, C.; Zhang, S.; Qin, Y.; Estupinan, E. A Systematic Review of Deep Transfer Learning for Machinery Fault Diagnosis. *Neurocomputing* **2020**, *407*, 121–135. [CrossRef]
27. Liu, Z.; Su, S.; Cao, J.; Yu, P. Transfer Learning for Bearing Fault Diagnosis Based on Improved Residual Network. In Proceedings of the 2023 IEEE 2nd International Conference on Electrical Engineering, Big Data and Algorithms (EEBDA), Changchun, China, 24–26 February 2023; pp. 1214–1216. [CrossRef]
28. Liu, Y.; Wang, Y.; Chow, T.W.S.; Li, B. Deep Adversarial Subdomain Adaptation Network for Intelligent Fault Diagnosis. *IEEE Trans. Ind. Inform.* **2022**, *18*, 6038–6046. [CrossRef]
29. Öztürk, C.; Taşyürek, M.; Türkdamar, M.U. Transfer Learning and Fine-Tuned Transfer Learning Methods' Effectiveness Analyse in the CNN-Based Deep Learning Models. *Concurr. Comput. Pract. Exp.* **2023**, *35*, e7542. [CrossRef]
30. Nguyen, C.T.; Van Huynh, N.; Chu, N.H.; Saputra, Y.M.; Hoang, D.T.; Nguyen, D.N.; Pham, Q.-V.; Niyato, D.; Dutkiewicz, E.; Hwang, W.-J. Transfer Learning for Wireless Networks: A Comprehensive Survey. *Proc. IEEE* **2022**, *110*, 1073–1115. [CrossRef]
31. Chen, K.-F.; Lee, M.-C.; Lin, C.-H.; Yeh, W.-C.; Lee, T.-S. Multi-Fault and Severity Diagnosis for Self-Organizing Networks Using Deep Supervised Learning and Unsupervised Transfer Learning. *IEEE Trans. Wirel. Commun.* **2024**, *23*, 141–157. [CrossRef]
32. Azari, M.S.; Flammini, F.; Santini, S.; Caporuscio, M. A Systematic Literature Review on Transfer Learning for Predictive Maintenance in Industry 4.0. *IEEE Access* **2023**, *11*, 12887–12910. [CrossRef]
33. Bu, K.; He, Y.; Jing, X.; Han, J. Adversarial Transfer Learning for Deep Learning Based Automatic Modulation Classification. *IEEE Signal Process. Lett.* **2020**, *27*, 880–884. [CrossRef]
34. Zhuang, F.; Qi, Z.; Duan, K.; Xi, D.; Zhu, Y.; Zhu, H.; Xiong, H.; He, Q. A Comprehensive Survey on Transfer Learning. *Proc. IEEE* **2021**, *109*, 43–76. [CrossRef]
35. Liu, D.; Zhang, J.; Wu, H.; Liu, S.; Long, J. Multi-Source Transfer Learning for EEG Classification Based on Domain Adversarial Neural Network. *IEEE Trans. Neural Syst. Rehabil. Eng.* **2023**, *31*, 218–228. [CrossRef]
36. Wen, L.; Gao, L.; Li, X. A New Deep Transfer Learning Based on Sparse Auto-Encoder for Fault Diagnosis. *IEEE Trans. Syst. Man Cybern. Syst.* **2019**, *49*, 136–144. [CrossRef]
37. Pan, S.J.; Kwok, J.T.; Yang, Q. Transfer Learning via Dimensionality Reduction. In Proceedings of the 23rd National Conference on Artificial Intelligence—Volume 2, Chicago, IL, USA, 13–17 July 2008; AAAI Press: Washington, DC, USA, 2008; pp. 677–682.
38. Huo, C.; Jiang, Q.; Shen, Y.; Lin, X.; Zhu, Q.; Zhang, Q. A Class-Level Matching Unsupervised Transfer Learning Network for Rolling Bearing Fault Diagnosis under Various Working Conditions. *Appl. Soft Comput.* **2023**, *146*, 110739. [CrossRef]
39. Wang, Y.; Sun, X.; Li, J.; Yang, Y. Intelligent Fault Diagnosis with Deep Adversarial Domain Adaptation. *IEEE Trans. Instrum. Meas.* **2021**, *70*, 1–9. [CrossRef]
40. Cui, Z.; Cao, H.; Ai, Z.; Wang, J. A Multi-Adversarial Joint Distribution Adaptation Method for Bearing Fault Diagnosis under Variable Working Conditions. *Appl. Sci.* **2023**, *13*, 10606. [CrossRef]
41. Guo, Y.; Zhou, X.; Li, J.; Ba, R.; Xu, Z.; Tu, S.; Chai, L. A Novel and Optimized Sine–Cosine Transform Wavelet Threshold Denoising Method Based on the sym4 Basis Function and Adaptive Threshold Related to Noise Intensity. *Appl. Sci.* **2023**, *13*, 10789. [CrossRef]

42. Case Western Reserve University. Bearing Data Center. Available online: https://engineering.case.edu/bearingdatacenter/apparatus-and-procedures (accessed on 25 January 2024).
43. Huang, H.; Baddour, N. Bearing Vibration Data Collected under Time-Varying Rotational Speed Conditions. *Data Brief* **2018**, *21*, 1745–1749. [CrossRef] [PubMed]
44. Chen, J.; Huang, R.; Zhao, K.; Wang, W.; Liu, L.; Li, W. Multiscale Convolutional Neural Network With Feature Alignment for Bearing Fault Diagnosis. *IEEE Trans. Instrum. Meas.* **2021**, *70*, 1–10. [CrossRef]
45. Hu, Q.; Si, X.; Qin, A.; Lv, Y.; Liu, M. Balanced Adaptation Regularization Based Transfer Learning for Unsupervised Cross-Domain Fault Diagnosis. *IEEE Sens. J.* **2022**, *22*, 12139–12151. [CrossRef]
46. Guo, L.; Lei, Y.; Xing, S.; Yan, T.; Li, N. Deep Convolutional Transfer Learning Network: A New Method for Intelligent Fault Diagnosis of Machines with Unlabeled Data. *IEEE Trans. Ind. Electron.* **2019**, *66*, 7316–7325. [CrossRef]
47. Saha, D.; Hoque, M.E.; Chowdhury, M.E.H. Enhancing Bearing Fault Diagnosis Using Transfer Learning and Random Forest Classification: A Comparative Study on Variable Working Conditions. *IEEE Access* **2024**, *12*, 5986–6000. [CrossRef]
48. Wang, H.; Li, P.; Lang, X.; Tao, D.; Ma, J.; Li, X. FTGAN: A Novel GAN-Based Data Augmentation Method Coupled Time–Frequency Domain for Imbalanced Bearing Fault Diagnosis. *IEEE Trans. Instrum. Meas.* **2023**, *72*, 1–14. [CrossRef]
49. Shi, J.; Wang, X.; Lu, S.; Zheng, J.; Dong, H.; Zhang, J. An Adversarial Multisource Data Subdomain Adaptation Model: A Promising Tool for Fault Diagnosis of Induction Motor Under Cross-Operating Conditions. *IEEE Trans. Instrum. Meas.* **2023**, *72*, 1–14. [CrossRef]
50. Zhu, Y.; Zhuang, F.; Wang, J.; Ke, G.; Chen, J.; Bian, J.; Xiong, H.; He, Q. Deep Subdomain Adaptation Network for Image Classification. *IEEE Trans. Neural Netw. Learn. Syst.* **2021**, *32*, 1713–1722. [CrossRef] [PubMed]
51. Meng, Z.; Zhu, J.; Cao, S.; Li, P.; Xu, C. Bearing Fault Diagnosis Under Multisensor Fusion Based on Modal Analysis and Graph Attention Network. *IEEE Trans. Instrum. Meas.* **2023**, *72*, 1–10. [CrossRef]
52. Tong, J.; Liu, C.; Bao, J.; Pan, H.; Zheng, J. A Novel Ensemble Learning-Based Multisensor Information Fusion Method for Rolling Bearing Fault Diagnosis. *IEEE Trans. Instrum. Meas.* **2023**, *72*, 1–12. [CrossRef]
53. Chen, Z.; Qin, W.; He, G.; Li, J.; Huang, R.; Jin, G.; Li, W. Explainable Deep Ensemble Model for Bearing Fault Diagnosis under Variable Conditions. *IEEE Sens. J.* **2023**, *23*, 17737–17750. [CrossRef]
54. Sun, S.; Yeh, C.-F.; Hwang, M.-Y.; Ostendorf, M.; Xie, L. Domain Adversarial Training for Accented Speech Recognition. In Proceedings of the 2018 IEEE International Conference on Acoustics, Speech, and Signal Processing (ICASSP), Calgary, AB, Canada, 15–20 April 2018; pp. 4854–4858. [CrossRef]
55. Li, Y.; Wu, H.; Kang, Y.; Guo, Y.; Cui, Z.; Xing, J.; Wang, Q.; Meng, J. Channel Discrepancies Adaptive Modulation Recognition Using Domain Adversarial Training. In Proceedings of the 2021 Asia-Pacific International Symposium on Electromagnetic Compatibility (APEMC), Nusa Dua, Bali, Indonesia, 27–30 September 2021; pp. 1–4. [CrossRef]
56. Ma, X.; Mou, X.; Wang, J.; Liu, X.; Geng, J.; Wang, H. Cross-Dataset Hyperspectral Image Classification Based on Adversarial Domain Adaptation. *IEEE Trans. Geosci. Remote Sens.* **2021**, *59*, 4179–4190. [CrossRef]
57. Qin, R.; Lu, C. Research on Measurement Methods of Transferability between Different Domains in Transfer Learning. In Proceedings of the 2019 CAA Symposium on Fault Detection, Supervision and Safety for Technical Processes (SAFEPROCESS), Xiamen, China, 5–7 July 2019; pp. 926–931. [CrossRef]
58. Zhang, Q.; Deng, L. An Intelligent Fault Diagnosis Method of Rolling Bearings Based on Short-Time Fourier Transform and Convolutional Neural Network. *J. Fail. Anal. Prev.* **2023**, *23*, 795–811. [CrossRef]
59. Wang, X.; Ying, T.; Tian, W. Spectrum Representation Based on STFT. In Proceedings of the 2020 13th International Congress on Image and Signal Processing, BioMedical Engineering, and Informatics (CISP-BMEI), Chengdu, China, 17–19 October 2020; pp. 435–438. [CrossRef]
60. Hu, J.; Shen, L.; Albanie, S.; Sun, G.; Wu, E. Squeeze-and-Excitation Networks. *IEEE Trans. Pattern Anal. Mach. Intell.* **2020**, *42*, 2011–2023. [CrossRef] [PubMed]
61. Zhang, W.; Li, C.; Peng, G.; Chen, Y.; Zhang, Z. A Deep Convolutional Neural Network with New Training Methods for Bearing Fault Diagnosis under Noisy Environment and Different Working Load. *Mech. Syst. Signal Process.* **2018**, *100*, 439–453. [CrossRef]
62. Di, Y.; Yang, R.; Huang, M. Fault Diagnosis of Rotating Machinery Based on Domain Adversarial Training of Neural Networks. In Proceedings of the 2021 IEEE 30th International Symposium on Industrial Electronics (ISIE), Kyoto, Japan, 20–23 June 2021; pp. 1–6. [CrossRef]
63. Luo, S.; Huang, X.; Wang, Y.; Luo, R.; Zhou, Q. Transfer Learning Based on Improved Stacked Autoencoder for Bearing Fault Diagnosis. *Knowl.-Based Syst.* **2022**, *256*, 109846. [CrossRef]
64. Wu, Z.; Jiang, H.; Liu, S.; Liu, Y.; Yang, W. Conditional Distribution-Guided Adversarial Transfer Learning Network with Multi-Source Domains for Rolling Bearing Fault Diagnosis. *Adv. Eng. Inform.* **2023**, *56*, 101993. [CrossRef]
65. Senanayake, D.A.; Wang, W.; Naik, S.H.; Halgamuge, S. Selforganizing Nebulous Growths for Robust and Incremental Data Visualization. *IEEE Trans. Neural Netw. Learn. Syst.* **2021**, *32*, 4588–4602. [CrossRef]

Disclaimer/Publisher's Note: The statements, opinions and data contained in all publications are solely those of the individual author(s) and contributor(s) and not of MDPI and/or the editor(s). MDPI and/or the editor(s) disclaim responsibility for any injury to people or property resulting from any ideas, methods, instructions or products referred to in the content.

applied sciences

MDPI

Article

Time Series Recovery Using Adjacent Channel Data Based on LSTM: A Case Study of Subway Vibrations

Tao Xin [1,2,*], Yi Yang [1], Xiaoli Zheng [3], Jing Lin [4,5], Sen Wang [1,6] and Pengsong Wang [1,7]

1 School of Civil Engineering, Beijing Jiaotong University, Beijing 100044, China
2 Beijing Key Laboratory of Track Engineering, Beijing Jiaotong University, Beijing 100044, China
3 Beijing Municipal Institute of City Planning and Design, Beijing 100045, China
4 Department of Civil, Environmental and Natural Resources Engineering, Luleå University of Technology, 97187 Luleå, Sweden
5 School of Innovation, Design and Engineering, Mälardalen University, 63220 Eskilstuna, Sweden
6 Frontiers Science Center for Smart High-Speed Railway System, Beijing 100044, China
7 Collaborative Innovation Center of Railway Traffic Safety, Beijing 100044, China
* Correspondence: xint@bjtu.edu.cn

Abstract: Multi-sensor technology has been widely applied in the condition monitoring of rail transit. In practice, the data of some channels in the high channel counts are often abnormal or lost due to the abnormality and damage of the sensors, thus resulting in a large amount of data waste. A method for the data recovery of lost channels by using adjacent channel data is proposed to solve this problem. Based on the LSTM network algorithm, a data recovery model is established based on the "sequence-to-sequence" regression analysis of adjacent channel data. Taking the measured vibration data of a subway as an example, the network is trained with multi-channel measured data to recover the lost channel data of time-series characteristics. The results show that this multi-channel data recovery model is feasible, and the accuracy is up to 98%. This method can also further reduce the number of channels that need to be collected.

Keywords: multi-channel data; time-series recovery; neural network; regression analysis; data recovery; time domain; frequency domain

1. Introduction

With the continuous development of urban rail transit, subway lines have developed rapidly in many cities. With the increasing demand for transportation facilities construction, a large amount of measured data in terms of subway evaluation, field testing, experimental verification, and simulation have been collected. Due to the sensor used for data acquisition being affected by vibration, some sensors are damaged, and the data is lost during the collection process, leading to a large amount of data waste. The data waste can be significantly reduced if the lost data can be recovered based on the existing normal data from adjacent channels.

Research on lost data recovery is usually based on historical data to build a data recovery model [1]. At present, the research methods for data recovery can be roughly divided into three categories: The first is traditional mathematical models, including the moving average method, historical trend method, residual grey model, and interpolation method [2–6]. The second is intelligent recovery methods, including nonparametric regression and neural networks [7,8]. The third is combination models [9,10], which refers to a combination of two or more data processing methods. Although conventional methods can achieve the accurate recovery of single missing datasets, important information in long-term time series data is often ignored in addition to a prolonged estimation time, which causes missing data and distortion and has a greater impact on subsequent data analysis. The intelligent recovery method, especially the use of a neural network, can identify the characteristics, trends, and development laws of the variable changes from the time series, thereby achieving the effective prediction and recovery of the missing data. The intelligent recovery method mainly employs machine learning algorithms to identify and recover abnormal and lost data, and it is currently widely used for data recovery.

With the advancement of machine learning, it is possible to predict future multi-epoch data by quantifying the relationship between current data and historical data through the characteristic trend in the data over time. Kamat et al. and Sanakkayala et al. [11,12] tried to use deep learning to monitor abnormal bearing data, and the results showed that deep learning achieved very high accuracy and robustness in monitoring the future state of things with the historical data. As the data have characteristics in the format of time series, learning temporal dependencies from sequence data poses challenges to classical machine learning techniques and standard neural networks. The long short-term memory (LSTM) network addresses this problem by introducing memory units into the network architecture. The LSTM network has great advantages in analyzing large-scale data [13,14], and it has high prediction accuracy for time-series data such as atmospheric and ocean temperature, traffic flow, and vibration signals [15,16]. For example, Milad et al. applied the LSTM model to the temperature prediction of pavement roads [17]. Based on the excellent performance of the LSTM model, Li et al. [18] considered the influence factors of historical air quality and meteorological data and realized the fast and accurate prediction of PM2.5 by analyzing the long-term historical process of time-series data. Han et al. [19] applied a one-dimensional convolutional neural network (1-D CNN) and an LSTM to analyze vibration signals for bearing fault diagnosis. Ma et al. [20] proposed a CNN-LSTM model for the accurate point-by-point prediction of high-speed railway car body vibration. LSTM networks perform well in extracting knowledge from sequence data over time spans, as has been shown in previous studies. Nowadays, an LSTM neural network is mainly used for the prediction of single-channel time-series data based on historical data. However, in a scenario where multi-channel data is collected, especially when there is a certain correlation of data between channels, there is inadequate research on the use of neural networks in which the complete data of multiple channels is used for restoring missing data on a neighboring channel. The accuracy of recovering the data of one channel using neighboring channels should be further studied. To be specific, a set of data from a total of eight channels was collected, but the data from Channel 7 were missing. Then, the neural network was used to train the remaining complete data of multiple channels, thus recovering the missing data of Channel 7.

2. LSTM Neural Network Modeling

In this study, an LSTM neural network is proposed to analyze time-series data based on the characteristic trend of multi-channel data collected over time.

2.1. LSTM Basic Algorithm

A traditional RNN network can retain the effective information of the previous sequence samples and impact the current output. However, traditional RNN network training often has the phenomenon of gradient disappearance or gradient explosion when processing long input sequence samples, resulting in a poor network training effect. To solve this problem and improve the performance of an RNN network, some scholars have introduced gate operations on the hidden layer nodes of a traditional RNN network, resulting in some RNN variants, such as an LSTM network and a gated recurrent unit (GRU) network [21–23]. In practice, LSTM networks have achieved good results. An LSTM network can selectively memorize sequence information, with advantages such as the satisfactory applicability to measurement data, high accuracy, and short response time. The research of Zhen et al. [24] proved that the recognition rate of LSTM neural networks in diagnosis of different faults of a single or multiple sensors is over 96% and that the response time is less than 1 millisecond. Wang et al. [25] applied an LSTM to the fault detection and recovery of UAV flight data, and they found that satisfactory performance was achieved. Three types of gates in LSTM networks are as follows: input gate, forget gate, and output gate. It is possible to control the flow of information between the hidden layers of an LSTM network by adjusting the state of these gates. The hidden layer of the original RNN has only one state and is very sensitive to short-term inputs. An LSTM adds one cell state to the RNN to keep the long-term state. An LSTM neural network structure is shown in Figure 1.

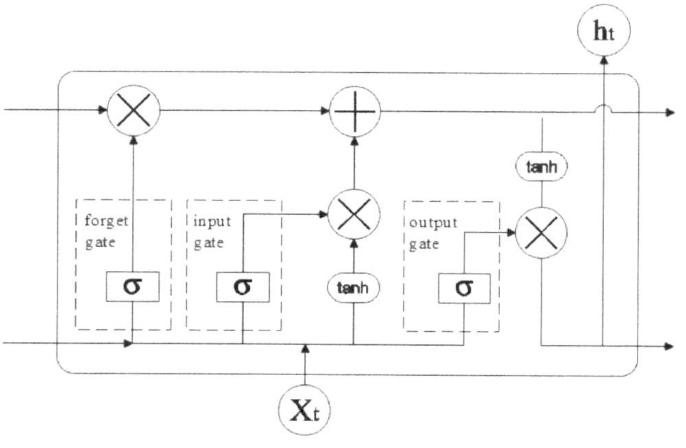

Figure 1. LSTM neural network structure.

The forget gate f_t determines the degree to which the long-term characteristics of preorder information are preserved or discarded in the cell state, which can be expressed as:

$$\mu = \frac{1}{m}\sum_{i=1}^{m} x_i$$

where f_t is the activation value of output h_{t-1} at $t-1$, the input value at t, and the bias term of b_f. The function is sigmoid, and f_t is scaled down between zero (complete forgetting) and one (complete memory). W_f is the weight term and includes W_{fx} and W_{fh}.

The input gate determines how to update the cell state, namely, the new information is selectively memorized to the cell state C_t. The formula is as follows:

$$\begin{cases} i_t = \sigma\ (W_i \cdot [h_{t-1}, x_t] + b_i) \\ \widetilde{C}_t = \tanh\ (W_C \cdot [h_{t-1}, x_t] + b_C) \\ C_t = f_t{}^* C_{t-1} + i_t{}^* \widetilde{C}_t \end{cases}$$

where h_{t-1} and x_t are passed to the sigmoid function to obtain its value; h_{t-1} and x_t are passed to the tanh activation function to obtain the candidate value $\widetilde{C}_t$ of the cell state, and it and $\widetilde{C}_t$ jointly determine the new cell state; W_i, and W_C are weight terms, including W_{ix} and W_{ih} and W_{Cx} and W_{Ch}, respectively; b_i and b_C are bias terms; and "*" refers to multiplication with matrix elements.

The output gate O_t determines the input at the present moment. Similar to the input gate, the information carried by h_t is determined by O_t and C_t, and the formula is as follows:

$$\begin{cases} O_t = \sigma(W_O \cdot [h_{t-1}, x_t] + b_O) \\ h_t = O_t{}^* \tanh(C_t) \end{cases}$$

where W_O is the weight term, which includes W_{ox} and W_{ho}, and b_O is the bias term.

The LSTM method diagram is shown in Figure 2.

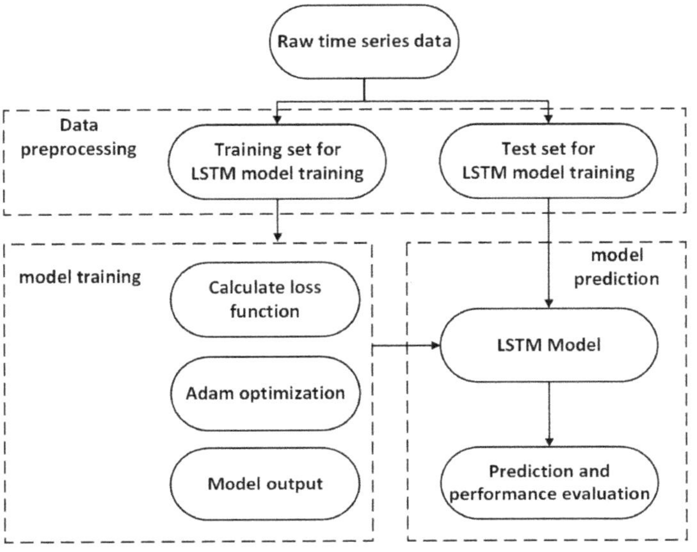

Figure 2. LSTM method diagram.

2.2. Sample Data Normalization

In order to make the data more convenient for processing with the model and avoid the influence of different dimensions between each feature and the target value on the prediction performance, as well as to speed up the gradient descent, the Z-score method is used to normalize the sample data, where the training predictor variable is normalized into a variable with a zero mean and unit variance. The main steps are as follows:

1. Obtain the mean μ of the output data x as Equation (1):

$$\mu = \frac{1}{m}\sum_{i=1}^{m} x_i. \quad (1)$$

2. Obtain the variance σ of the output data as Equation (2):

$$\sigma^2 = \frac{1}{m}\sum_{i=1}^{m}(x_i - \mu)^2. \quad (2)$$

3. Obtain the output data normalization as Equation (3):

$$x_i = \frac{x_i - \mu}{\sqrt{\sigma^2}}, \quad (3)$$

where m represents the number of samples in one batch of the input batch training and x_i represents the input value of sample i.

2.3. Evaluation Index

In this paper, the root mean square error (RMSE) and mean absolute error (MAE) evaluation indicators are used to aid the evaluation of the restoration effectiveness of the abnormal data. The foregoing means the deviation between the restored value and true value, reflecting the restoration capacity. The lower the error value, the better the restoration effect and the higher the accuracy of the restored data. At the same time, the coefficient of determination R^2 is adopted to evaluate the effectiveness of the model. The coefficient of determination R^2 is a value between zero and one. The closer it is to zero, the poorer the fitting effect of the prediction model for the real data, and the closer it is to one, the better the fitting effect of the prediction model for the real data. The calculation formula is:

$$\mathrm{RMSE} = \sqrt{\frac{\sum_{i=1}^{n}(v_r - v_c)^2}{n}}, \tag{4}$$

$$\mathrm{MAE} = \frac{1}{N}\sum_{i=1}^{N}|y_i - \hat{y}_i|, \text{ and} \tag{5}$$

$$R^2 = 1 - \frac{\sum_{i=1}^{N}(y_i - \hat{y}_i)^2}{\sum_{i=1}^{N}(y_i - y)^2}, \tag{6}$$

where v_c is the real value, v_r is the restored value, and n is the amount of data.

3. Case Study

Taking the vibration and noise level test in a subway tunnel as an example, INV3062V network distributed acquisition instrumentation and LC0161BG piezoelectric acceleration sensors are used for the data acquisition, with a frequency response of 0.1–1000 Hz, a measuring range of 5 g, and a testing accuracy of 0.0002 g. The arrangement of measuring points is shown in Figure 3. The measuring points of the tunnel section are arranged and include the rail, the track slab, and the tunnel wall. Eight channel data are selected, and each channel has 12,800 sampling points for analysis to ensure the validity of the network model.

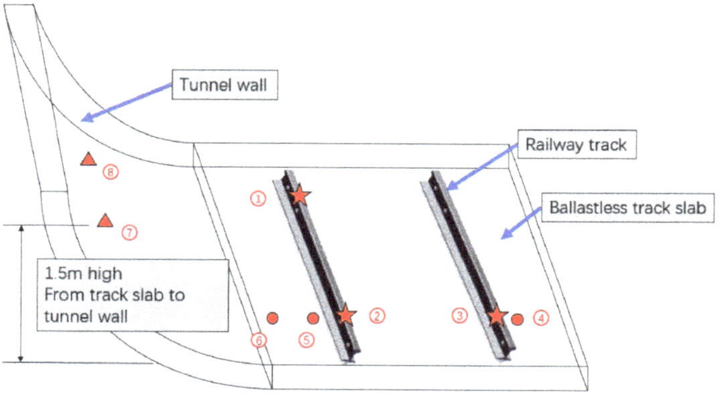

Figure 3. Location of the acceleration sensors. Points 1–3 are on the rail, points 4–6 are on the track slab, and points 7–8 are on the tunnel wall. The circle is the point on the track slab, the star is the point on the railway track, and the triangle is the point on the tunnel wall.

In order to verify the accuracy of this prediction model, two different data processing methods will be used to predict the missing channel data. The first method is to directly train and verify the neural network with the raw data (time-domain data) collected by sensors. The second method is to perform a Fourier transform on the original data (frequency domain data) and then use the transformed data to train and verify the neural network.

3.1. Training the Network with Time-Domain Data

3.1.1. Neural Network Training Process

This study builds a multi-channel data recovery model and defines the network architecture. The LSTM network that contains an LSTM layer with a specified number of hidden units and a fully connected layer is created. The solver 'Adam' is used to train epochs and batch size based on user-specified size. The learning rate is specified as 0.03.

When a subway train passed, a set of 12.8 s vibration signals were generated, and then the collected eight-channel vibration signals were used as the training network.

Channels 1–6 were used as test predictor variables (XTest), and Channels 7–8 were used as the response sequence (YTest). The LSTM network was used to predict the incomplete sequences that lacked data from Channels 7 and 8. The prediction results from the training network are shown in Figures 4 and 5. Figure 4 showcases the prediction of the missing data for Channel 7, while Figure 5 showcases that for Channel 8. It can be observed from the figures that this training network can correctly demonstrate the relationships between the data from Channels 1–6 and those from Channels 7 and 8.

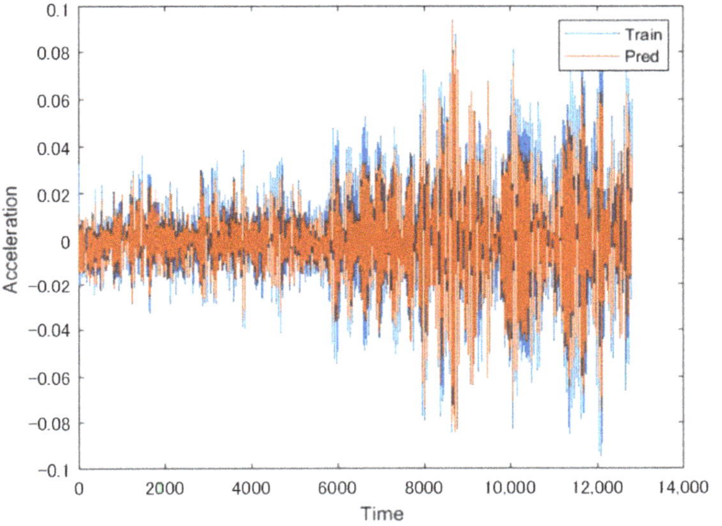

Figure 4. Training results from Channel 7.

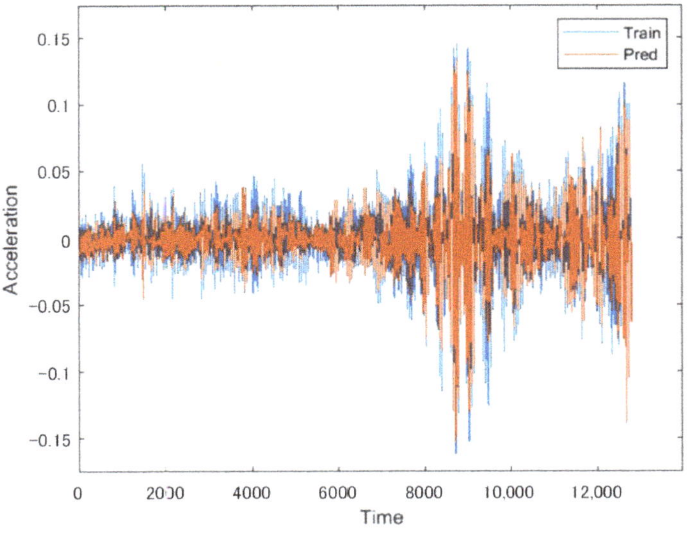

Figure 5. Training results from Channel 8.

The user-specified parameters in the above neural network are optimized to maintain the model's training accuracy and speed. The optimization results are as follows:

Effect of Epoch on Model Accuracy

As shown in Figure 6, when the epoch value is less than 50, the RMSE value drops rapidly. The decline rate of the RMSE value slows down with the continuous increase in epoch value, and the RMSE region stabilizes when the epoch value exceeds 100. Therefore, 100 is selected as the epoch value for this model by considering the computational accuracy and efficiency.

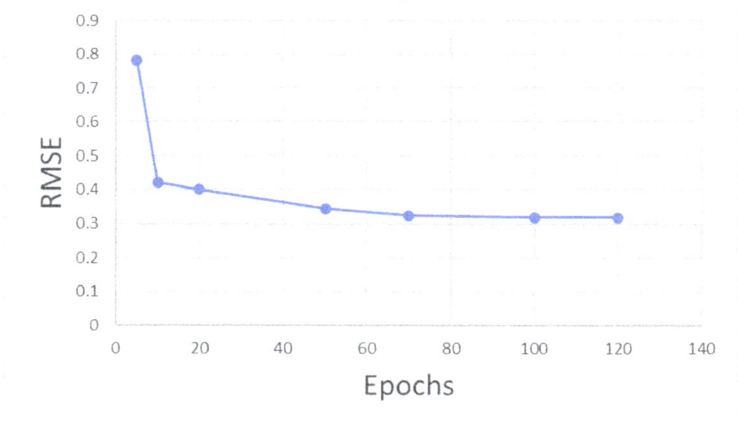

Figure 6. Effect of different epochs on model accuracy.

Effect of Batch Size on Model Accuracy

An appropriate batch size selected in an LSTM can increase the model accuracy and reduce the model's training time. As shown in Figure 7, the value of the RMSE shows a trend of first decreasing and then increasing with the increase in batch size in this model. When the batch size is selected as 250, the value of the RMSE is the minimum, and so 250 is used as the value of the batch size for this model.

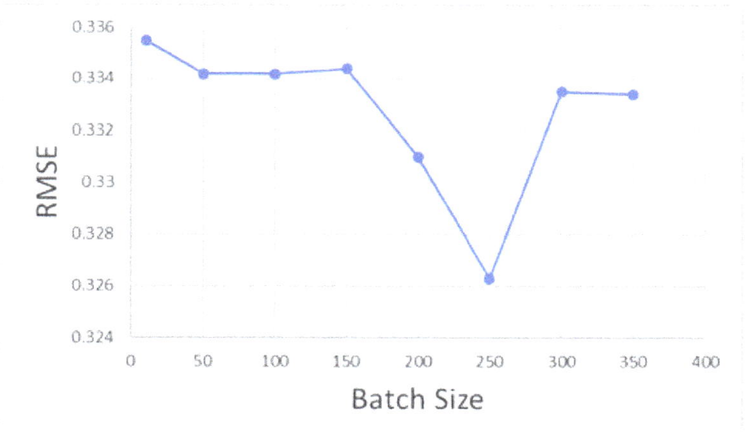

Figure 7. Effect of different batch sizes on model accuracy.

Effect of the Number of Hidden Units in the LSTM Layer on Model Accuracy

When the number of hidden units is less than 30, the RMSE of the training model drops rapidly, as shown in Figure 8. When the number of hidden units is up to 50, the RMSE value is the minimum. As the number of hidden units continues to increase, the RMSE value tends to increase slowly. Therefore, 50 is selected as the value of the number of hidden units in the LSTM layer for this model.

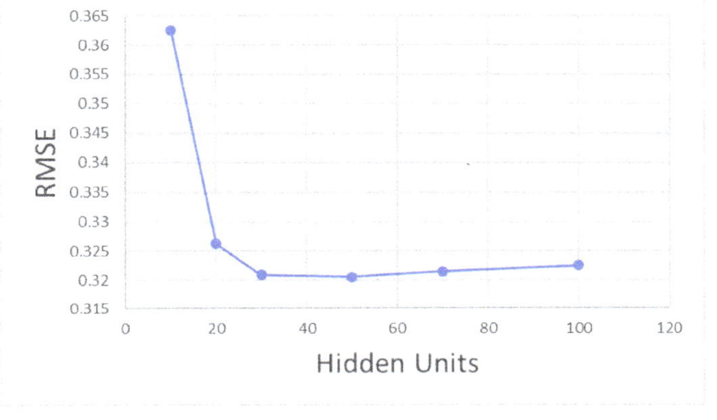

Figure 8. Effect of different hidden units on model accuracy.

3.1.2. Analysis of Training Results

Another set of 12.8 s vibration signals collected from the same subway at the same location but at different times was used as a test network, and the LSTM network as trained above was used to predict the incomplete sequences that lacked data from Channels 7 and 8. Figure 9 indicates the prediction of the lost data in Channel 7, and Figure 10 represents the prediction of the lost data in Channel 8. As shown in the above figures, the vibration signal predicted by the multi-channel data recovery model proposed in this paper based on the lost data recovery model of adjacent channel time series is largely consistent with the real vibration signal curve.

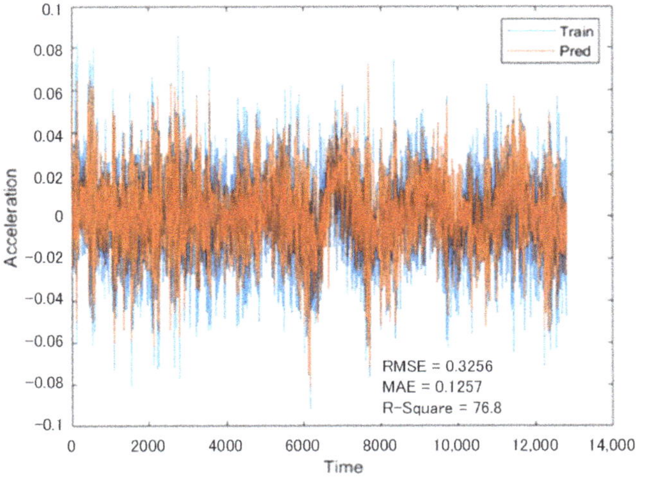

Figure 9. Comparison of vibration acceleration from Channel 7.

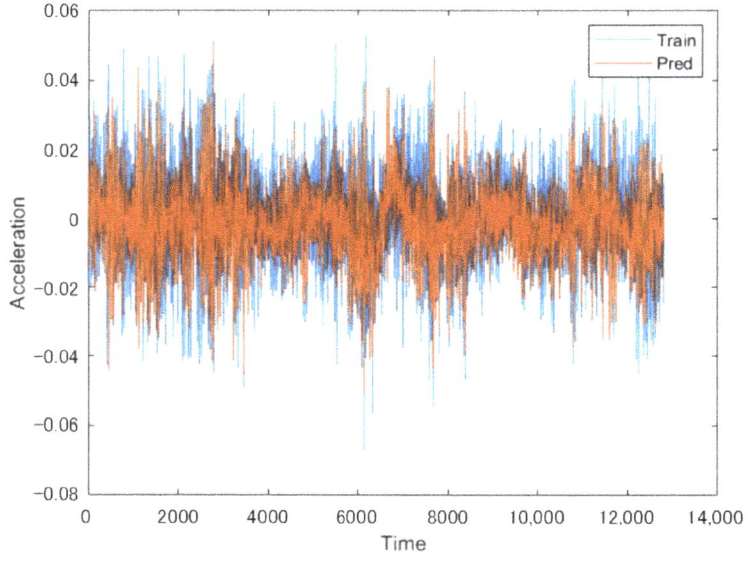

Figure 10. Comparison of vibration acceleration from Channel 8.

The vibration acceleration signal is further wavelet-transformed to obtain a more intuitive time-frequency image of Channel 7's vibration acceleration frequency over time. Figure 11 shows the wavelet-transformed time-frequency image of the measured value, and Figure 12 shows the wavelet-transformed time-frequency image of the predicted value. It can be seen from the figures that the change trend in signal frequency of the missing data's predicted value and measured value over time is largely the same, and that the peak frequency appears at approximately the same position.

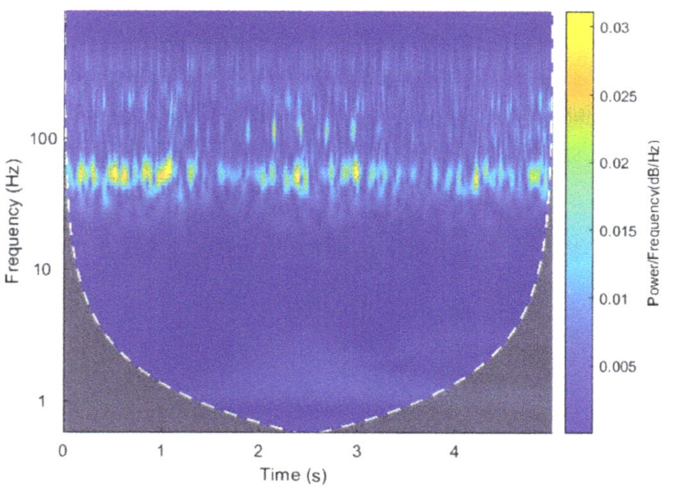

Figure 11. Wavelet-transformed time-frequency image of the measured value from Channel 7.

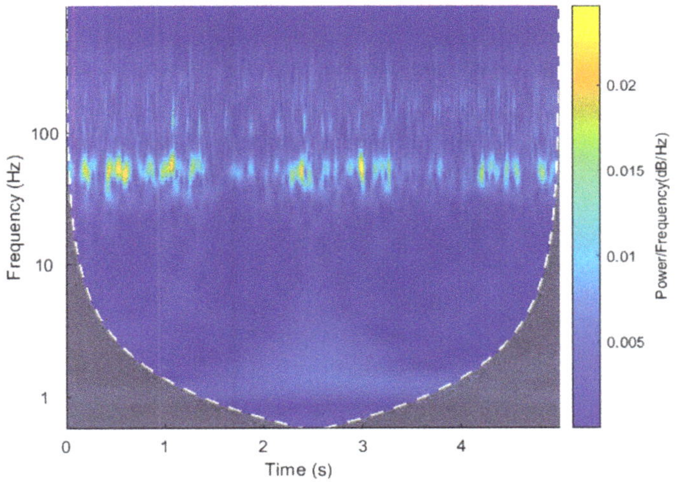

Figure 12. Wavelet-transformed time-frequency image of the predicted value from Channel 7.

A Fourier transform is performed on the vibration acceleration signal to obtain a more intuitive comparison diagram of the vibration acceleration amplitude loss data prediction and actual measurement of Channel 7, as shown in Figure 13, and the predicted value of the lost data is consistent with the actual value at all frequencies.

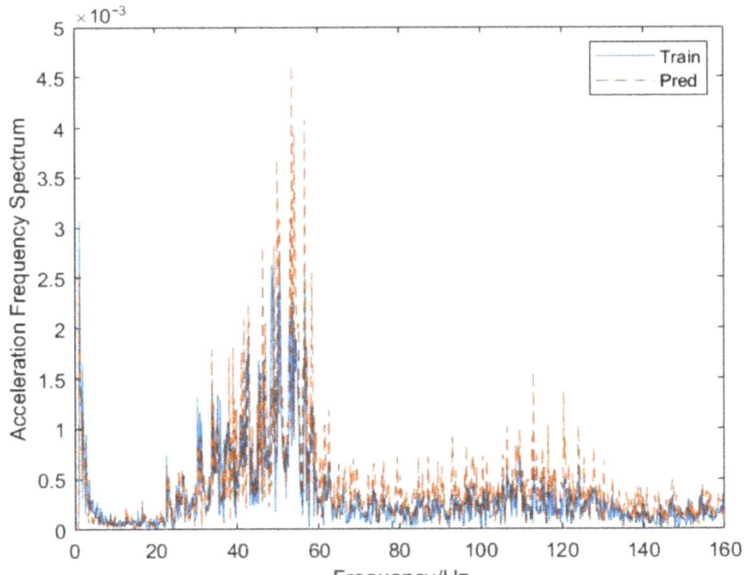

Figure 13. Comparison of acceleration frequency spectrum from Channel 7.

Figure 14 shows the comparison between the predicted data of the LSTM neural network and the actual vibration acceleration data at different frequencies. It can be seen that the predicted data and the actual data have the same trend at all frequencies. Since Channel 7 is from the sensor on the tunnel wall, the accuracy of the predicted data between

10 and 100 Hz is mainly analyzed. The relative error between the predicted data and the actual data is calculated, as shown in Figure 15, and the error values at all frequencies are less than 10%, the overall accuracy rate is up to 95%, and for the 10–100 Hz frequency of interest, the accuracy rate reaches 97%.

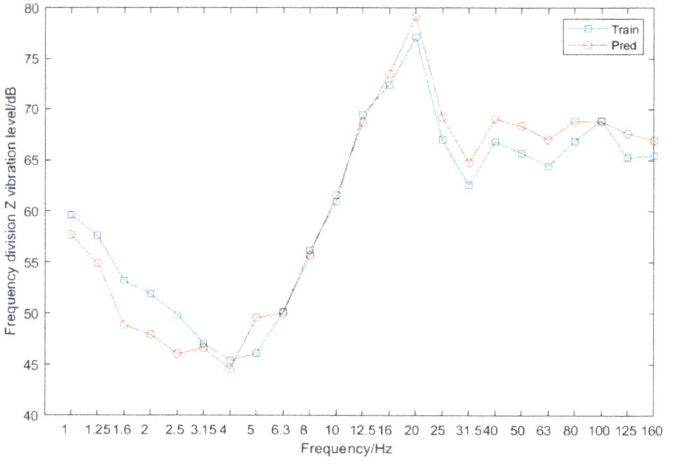

Figure 14. Comparison of vibration level at the one-third octave from Channel 7.

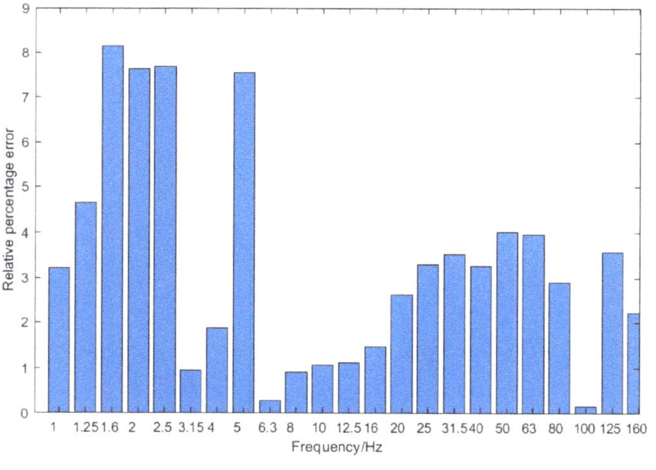

Figure 15. Relative error of the one-third octave spectrum from Channel 7.

3.2. *Training the Network by Using Frequency-Domain Data*

3.2.1. Training Process of the Neural Network

By using the same set of 12.8 s vibration signals in Section 3.1.1, a Fourier transform was performed on the Channel 8 data. After the transform, Channels 1–6 were used as the test predictor variables (XTest2) and Channels 7–8 as the response sequence (YTest2), while keeping the other parameters consistent with those in Section 3.1.1, and the LSTM network was used to predict the incomplete sequence that lacked data from Channels 7 and 8. The prediction results from the training network for the missing data from Channel 7 are shown in Figure 16.

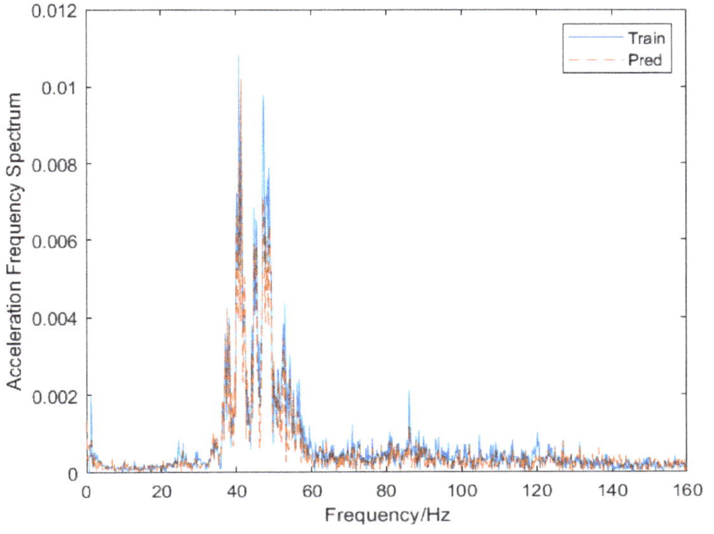

Figure 16. Training results from Channel 7 using processed data.

The user-specified parameters in the above neural network are optimized to maintain the model's training accuracy and speed. The optimization results are as follows:

Effect of Epoch on Model Accuracy

As shown in Figure 17, when the epoch value is less than 20, the RMSE value drops rapidly. The decline rate of the RMSE value slows down with the continuous increase in epoch value, and the RMSE region stabilizes when the epoch value exceeds 50. Therefore, 50 is selected as the epoch value for this model by considering the computational accuracy and efficiency.

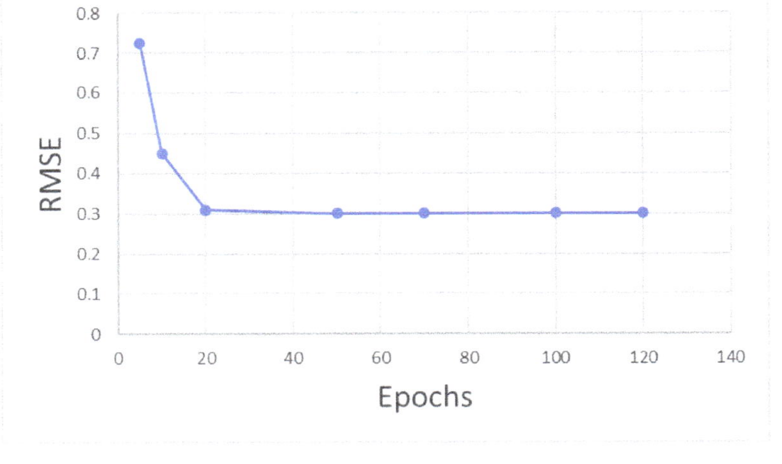

Figure 17. Effect of different epochs on model accuracy using frequency-domain data.

Effect of Batch Size on Model Accuracy

An appropriate batch size selected in an LSTM can increase the model's accuracy and reduce the model's training time. As shown in Figure 18, the value of the RMSE shows a trend of first decreasing and then increasing with the increase in batch size in this model. When the batch size is selected as 250, the value of the RMSE is the minimum, and so 250 is used as the value of the batch size for this model.

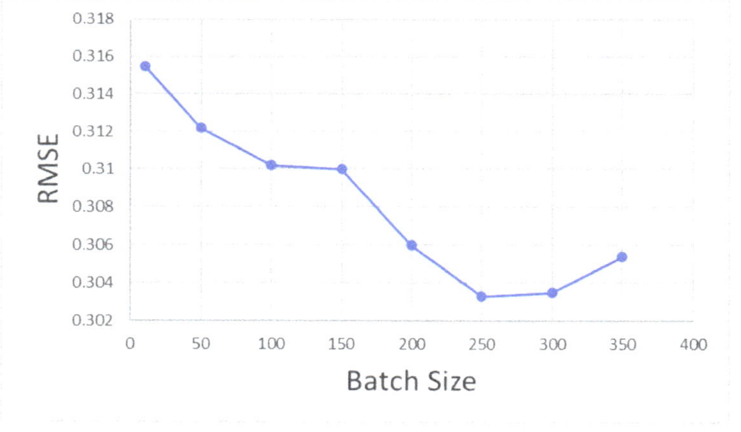

Figure 18. Effect of different batch sizes on model accuracy using frequency-domain data.

Effect of the Number of Hidden Units in the LSTM Layer on Model Accuracy

When the number of hidden units is less than 30, the RMSE of the training model drops rapidly, as shown in Figure 19. When the number of hidden units is up to 50, the RMSE value is the minimum. As the number of hidden units continues to increase, the RMSE value tends to increase slowly. Therefore, 50 is selected as the value of the number of hidden units in the LSTM layer for this model.

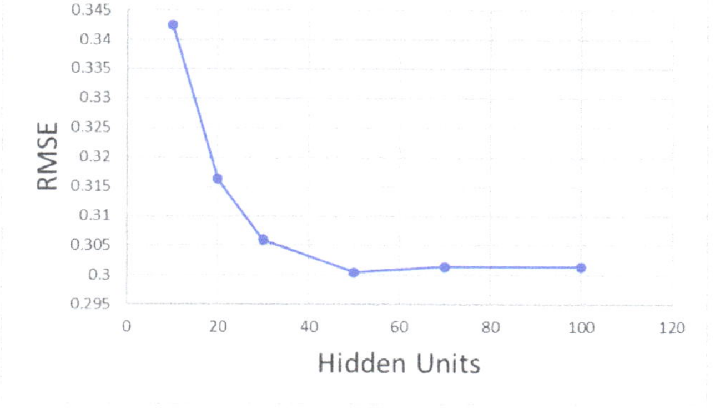

Figure 19. Effect of different hidden units on model accuracy using frequency-domain data.

In general, the model built with frequency-domain data achieves faster iterative efficiency in the training process than that built with time-domain data. The overall time spent on training is shorter, with lower losses and RMSE, MEA, and R2 values.

3.2.2. Analysis of Training Results

By using another set of 12.8 s vibration signals in Section 3.1.2 at the same location, a Fourier transform was performed on the vibration signals. The LSTM network as trained above was used to predict the incomplete sequences that lacked data from Channels 7 and 8, with the prediction results shown in Figure 20. Compared with prediction using raw time domain data, where a Fourier transform was performed after the original data were used to make a prediction, Figure 20 shows that the predicted signal curves fit more closely to the vibration curves.

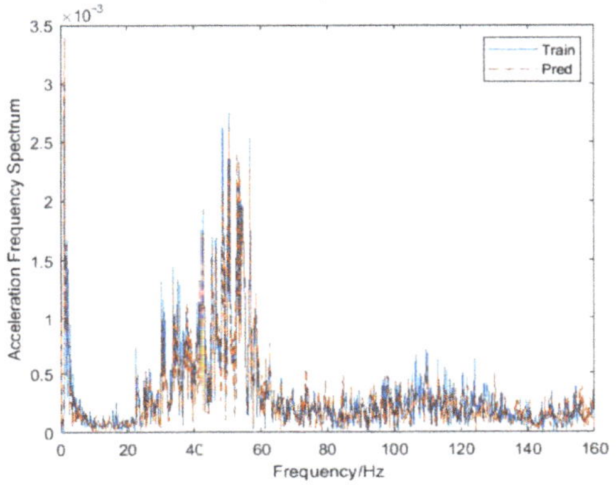

Figure 20. Comparison of acceleration frequency spectrum from Channel 7 using processed data.

After the inverse Fourier transform, the vibration acceleration signal is further wavelet-transformed to obtain a more intuitive time-frequency image of Channel 7's vibration acceleration frequency over time. Figure 21 shows the wavelet-transformed time-frequency image of the measured value, and Figure 22 shows the wavelet-transformed time-frequency image of the predicted value. It can be seen from the figures that the height of the signal frequency of the missing data's predicted value and measured value over time is approximately the same, and that the peak frequency appears at approximately the same height.

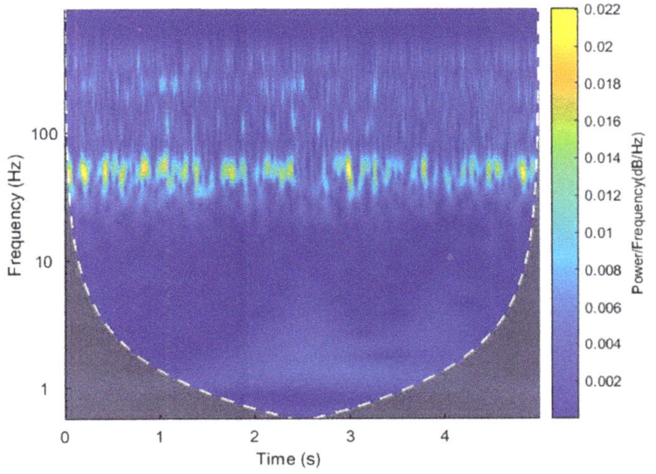

Figure 21. Wavelet-transformed time-frequency image of the measured value from Channel 7 using frequency-domain data.

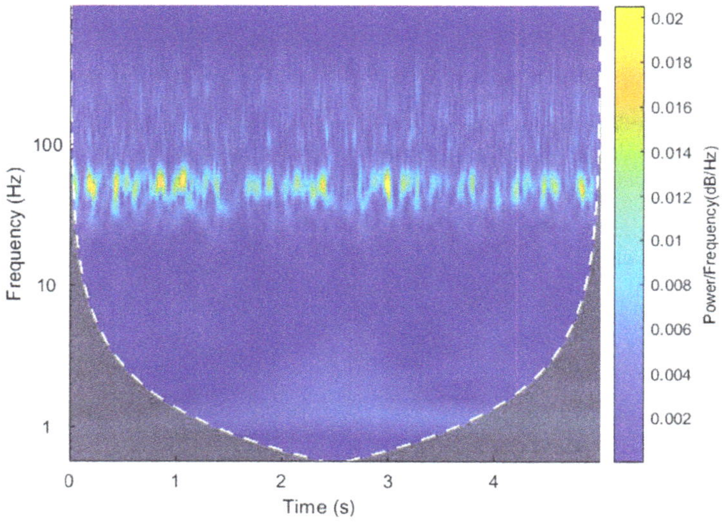

Figure 22. Wavelet-transformed time-frequency image of the predicted value from Channel 7 using frequency-domain data.

After an inverse Fourier transform was further performed on the predicted frequency-domain data, the corresponding time-domain data could be obtained, as shown in Figure 23. Then, the vibration acceleration data acquired were processed in the one-third octave. Figure 24 shows the one-third octave center frequency spectrum of Channel 7's vibration acceleration value as predicted with the Fourier transform data, as well as that spectrum of Channel 7's real value. Figure 25 shows the percentage difference between actual and predicted values. Figure 25 demonstrates smaller errors at all frequencies than those shown in Figure 15, especially between 1 and 10 Hz. The overall accuracy rates of as high as 98% indicate that the method can deliver higher accuracy rates.

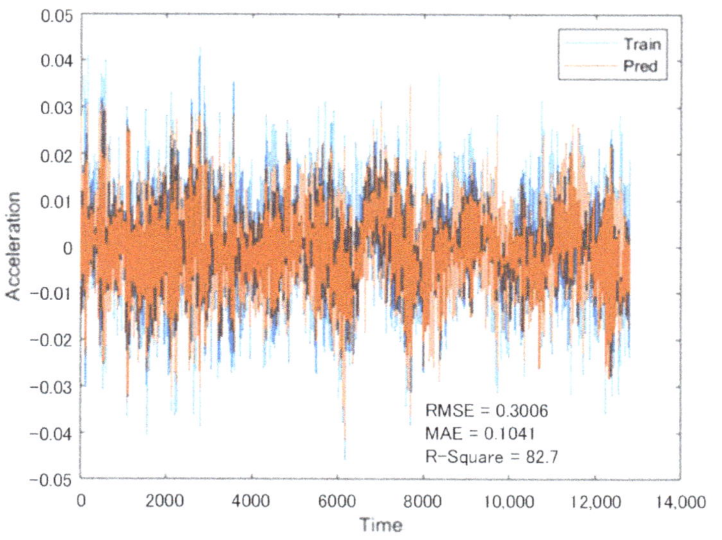

Figure 23. Vibration acceleration after inverse Fourier transform from Channel 7.

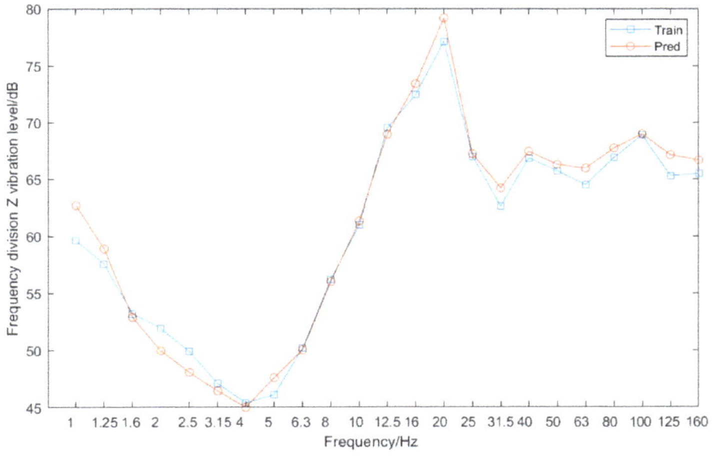

Figure 24. Comparison of acceleration of the one-third octave spectrum from Channel 7 using processed data.

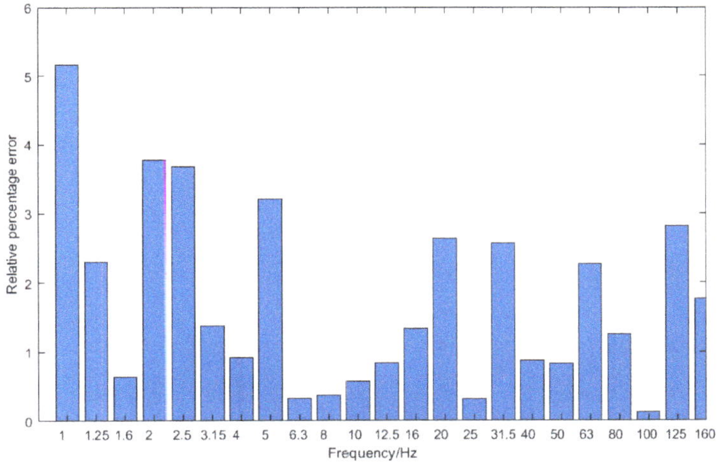

Figure 25. Relative error of the one-third octave spectrum from Channel 7 using processed data.

In summary, in the scenarios of collecting multi-channel data, this model can use the data from adjacent channels to restore the data of the lost channels and ensure that the restored data have high accuracy. This method can reduce the cases of discarding whole data groups due to the loss of individual channels' data. Primarily, if the original data are applied to the LSTM network after performing a Fourier transform, the model can deliver higher accuracy because the processed data have higher correlations than the original data.

4. Conclusions

This study proposes a time-series recovery method using adjacent multi-channel data based on an LSTM neural network. In the case of collecting multi-channel data, the adjacent channel data is used to recover the lost time-series data. Based on the measured data of a subway, a multi-channel data recovery model is established, and the accuracy of the training data is guaranteed by optimizing the parameters of the neural network. This study uses time-domain and frequency-domain data to train the LSTM neural network. The results show that the network trained with time-domain data is 95% accurate for time-series data recovery, while the network trained with frequency-domain data is as high as 98% accurate. Time-domain data can better reflect the internal relationship between adjacent channels so that the trained neural network has higher accuracy. This data recovery method is generally feasible and provides a reference value for further exploration of multi-channel time-series data recovery using machine learning. Currently, this method's internal connection of multi-channel vibration signals is data-driven. Therefore, it is necessary to further study the internal relationship of data among channels and optimize the model's algorithm by adding physical information to the neural network. On this basis, weak mechanism modeling theory is used to derive the laws and summarize weak mechanisms. There are two ways to obtain weak mechanisms: first, they can be obtained through derivation from a stronger mechanism, which may be more useful than mathematical analysis, and second, they can be obtained through summarization of the data mainly by means of statistics and machine learning. Then, some weak mechanisms are obtained. Next, proceeding from one or more weak mechanisms, a computable model can be established through a specific combination/fusion, which helps to understand the inherent connection of the data learned by the neural network and further expands the interpretability of the neural network.

Author Contributions: Conceptualization, T.X. and X.Z.; methodology, T.X. and Y.Y.; software, T.X. and Y.Y.; data curation, T.X. and S.W.; writing—original draft preparation, Y.Y. and P.W.; writing—review and editing, T.X. and J.L. All authors have read and agreed to the published version of the manuscript.

Funding: This study is sponsored by the Beijing Nova Program (Z191100001119126), the Shanghai Key Laboratory of Rail Infrastructure Durability and System Safety (R202101), and the Fundamental Research Funds for the Central Universities (2020JBM049). The support from the 111 Project (B20040) is also greatly appreciated.

Institutional Review Board Statement: Not applicable.

Informed Consent Statement: Not applicable.

Data Availability Statement: Not applicable.

Conflicts of Interest: The authors declare no conflict of interest.

References

1. Lu, H.-P.; Sun, Z.-Y.; Qu, W.-C. Repair method of traffic flow malfunction data based on temporal-spatial model. *Jiaotong Yunshu Gongcheng Xuebao/J. Traffic Transp. Eng.* **2015**, *15*, 92–100.
2. Barabadi, A.; Ayele, Y.Z. Post-Disaster Infrastructure Recovery: Prediction of Recovery Rate Using Historical Data. *Reliab. Eng. Syst. Saf.* **2018**, *169*, 209–223. [CrossRef]
3. Anderssen, R.; Davies, A. Simple Moving-Average Formulae for the Direct Recovery of the Relaxation Spectrum. *J. Rheol.* **2001**, *45*, 1–27. [CrossRef]
4. Wang, S.; Hu, J.; Shan, H.; Shi, C.-X.; Huang, W. Temperature Field Data Reconstruction Using the Sparse Low-Rank Matrix Completion Method. *Adv. Meteorol.* **2019**, *2019*, 3676182. [CrossRef]
5. Zhao, T.; Nehorai, A.; Porat, B. K-Means Clustering-Based Data Detection and Symbol-Timing Recovery for Burst-Mode Optical Receiver. *IEEE Trans. Commun.* **2006**, *54*, 1492–1501. [CrossRef]
6. Jon, Y.; Wen, Y.; Lee, T.; Cho, H. Missing Data Treatment on Travel Time Estimation for ATIS. In Proceedings of the 2003 IEEE International Conference on Systems, Man and Cybernetics, Washington, DC, USA, 8 October 2003; pp. 102–107.
7. Chen, M.; Jiang, H.; Liao, W.; Zhao, T. Nonparametric Regression on Low-Dimensional Manifolds Using Deep ReLU Networks: Function Approximation and Statistical Recovery. *Inf. Inference J. IMA* **2022**. [CrossRef]
8. Oh, B.K.; Glisic, B.; Kim, Y.; Park, H.S. Convolutional Neural Network-Based Data Recovery Method for Structural Health Monitoring. *Struct. Health Monit.-Int. J.* **2020**, *19*, 1821–1838. [CrossRef]
9. Fang, W.; Fu, L.; Liu, S.; Li, H. Dealiased Seismic Data Interpolation Using a Deep-Learning-Based Prediction-Error Filter. *Geophysics* **2021**, *86*, V317–V328. [CrossRef]
10. Wisdom, S.; Powers, T.; Pitton, J.; Atlas, L. Building recurrent networks by unfolding iterative thresholding for sequential sparse recovery. In Proceedings of the 2017 IEEE International Conference on Acoustics, Speech and Signal Processing (ICASSP), New Orleans, LA, USA, 5–9 March 2017; pp. 4346–4350.
11. Kamat, P.V.; Sugandhi, R.; Kumar, S. Deep learning-based anomaly-onset aware remaining useful life estimation of bearings. *PeerJ Comput. Sci.* **2021**, *7*, e795. [CrossRef]

12. Sanakkayala, D.C.; Varadarajan, V.; Kumar, N.; Soni, G.; Kamat, P.; Kumar, S.; Kotecha, K. Explainable AI for Bearing Fault Prognosis Using Deep Learning Techniques. *Micromachines* **2022**, *13*, 1471. [CrossRef]
13. Chung, H.; Shin, K. Genetic Algorithm-Optimized Long Short-Term Memory Network for Stock Market Prediction. *Sustainability* **2018**, *10*, 3765. [CrossRef]
14. Guo, A.; Jiang, A.; Lin, J.; Li, X. Data Mining Algorithms for Bridge Health Monitoring: Kohonen Clustering and LSTM Prediction Approaches. *J. Supercomput.* **2020**, *76*, 932–947. [CrossRef]
15. Wu, T.; Liu, C.; He, C. Prediction of Egional Temperature Change Trend Based on LSTM Algorithm. In Proceedings of the 2020 IEEE 4th Information Technology, Networking, Electronic and Automation Control Conference (ITNEC), Chongqing, China, 12–14 June 2020; pp. 62–66.
16. Liu, J.; Zhang, T.; Han, G.; Gou, Y. TD-LSTM: Temporal Dependence-Based LSTM Networks for Marine Temperature Prediction. *Sensors* **2018**, *18*, 3797. [CrossRef]
17. Milad, A.; Adwan, I.; Majeed, S.A.; Yusoff, N.I.M.; Al-Ansari, N.; Yaseen, Z.M. Emerging Technologies of Deep Learning Models Development for Pavement Temperature Prediction. *IEEE Access* **2021**, *9*, 23840–23849. [CrossRef]
18. Li, T.; Hua, M.; Wu, X. A Hybrid CNN-LSTM Model for Forecasting Particulate Matter (PM2.5). *IEEE Access* **2020**, *8*, 26933–26940. [CrossRef]
19. Hao, S.; Ge, F.-X.; Li, Y.; Jiang, J. Multisensor Bearing Fault Diagnosis Based on One-Dimensional Convolutional Long Short-Term Memory Networks. *Measurement* **2020**, *159*, 107802. [CrossRef]
20. Ma, S.; Gao, L.; Liu, X.; Lin, J. Deep Learning for Track Quality Evaluation of High-Speed Railway Based on Vehicle-Body Vibration Prediction. *IEEE Access* **2019**, *7*, 185099–185107. [CrossRef]
21. Gers, F.; Schmidhuber, J.; Cummins, F. Learning to Forget: Continual Prediction with LSTM. *Neural Comput.* **2000**, *12*, 2451–2471. [CrossRef]
22. Hochreiter, S.; Schmidhuber, J. Long Short-Term Memory. *Neural Comput.* **1997**, *9*, 1735–1780. [CrossRef]
23. Ma, X.; Tao, Z.; Wang, Y.; Yu, H.; Wang, Y. Long Short-Term Memory Neural Network for Traffic Speed Prediction Using Remote Microwave Sensor Data. *Transp. Res. Part C-Emerg. Technol.* **2015**, *54*, 187–197. [CrossRef]
24. Zheng, M.; Yang, K.; Shang, C.; Luo, Y. A PCA-LSTM-Based Method for Fault Diagnosis and Data Recovery of Dry-Type Transformer Temperature Monitoring Sensor. *Appl. Sci.* **2022**, *12*, 5624. [CrossRef]
25. Wang, B.; Liu, D.; Peng, Y.; Peng, X. Multivariate regression-based fault detection and recovery of UAV flight data. *IEEE Trans. Instrum. Meas.* **2019**, *69*, 3527–3537. [CrossRef]

applied sciences

MDPI

Article

Outlier Detection for Permanent Magnet Synchronous Motor (PMSM) Fault Detection and Severity Estimation

Konstantinos Koutrakos and Epameinondas Mitronikas *

Department of Electrical and Computer Engineering, University of Patras, 26504 Patras, Greece; k.koutrakos@ac.upatras.gr
* Correspondence: e.mitronikas@upatras.gr

Abstract: Today, Permanent Magnet Synchronous Motors (PMSMs) are a dominant choice in industry applications. During operation, different possible faults in the system can occur, so early and automated fault detection and severity estimation are required to ensure smooth operation and optimal maintenance planning. In this direction, outlier detection methods are employed in this paper. The motor's current signals are used to extract useful indicators of the fault, along with d-q transform. Statistical indicators in both time and frequency domains are selected to describe fault-related patterns. Based on the extracted features, three outlier detection methods are investigated: the Isolation Forest, the One Class Support Vector Machine, and the Robust Covariance Ellipse. Each method is investigated through different model parameters to evaluate fault detection and severity estimation capabilities. Finally, an ensemble approach is proposed based on decisions and outlier score ensemble. The proposed methodology is verified through different operating conditions in a PMSM test bench.

Keywords: PMSM; fault detection; severity estimation; Isolation Forest; One Class SVM; Robust Covariance Ellipse; ensemble learning

1. Introduction

Today, Permanent Magnet Synchronous Motors (PMSMs) are increasingly used in more applications in industry due to their high efficiency, high power density, dynamic performance, and control capabilities. However, condition monitoring (CM) is necessary to avoid anomalies and fault conditions. Different faults may appear during PMSM operation and can be related to the rotor [1], bearings [2], magnets [3], or stator windings [4]. Continuous and online condition monitoring of the system offers the possibility for early fault detection and thus better maintenance planning [5], avoidance of catastrophic faults, and reduction in downtime [6]. Various tools are used for motor health assessment, such as vibration analysis, acoustic emission analysis, temperature monitoring, or Motor Current Signature Analysis. Based on the requirements for continuous monitoring of the system, without interruptions, along with integrating possibilities into the existing system, Motor Current Signature Analysis (MCSA) is one of the prevailing options. Indeed, it has been shown that different possible faults can be detected through MCSA [7,8].

However, to automate fault diagnosis processes and enhance their capabilities, various AI and machine learning approaches must be utilized. In recent years, various approaches have been proposed for PMSM fault diagnosis. In [9], MCSA and Wavelets were used to extract eccentricity-related features and then Principal Component Analysis (PCA), k-NN, and Support Vector Machine (SVM) were used for feature reduction, eccentricity type identification, and eccentricity degree estimation, respectively. In [10], a similar approach is used to detect open circuits. FFT was used to extract frequency domain features from the measured signals, PCA was used to reduce dimensionality of the data, and then Bayesian Network (BN) was employed for open circuit fault diagnosis. Moreover, in [11], interturn short-circuit fault diagnosis is addressed through Recurrent Neural Networks (RNNs). In that work, negative sequence and positive sequence currents are extracted from three-phase current measurements and fed into an Attention-Based RNN along with the PMSM's operating speed. The authors achieve the evaluation of the severity of the fault under various operating conditions. In [12], raw current signals from the PMSM are fed directly to a Convolutional Neural Network (CNN) to avoid the features extraction and signal processing stage. The proposed method achieves the identification of damages in permanent magnets of a PMSM. An extensive overview of PMSM fault diagnosis, including artificial intelligence and machine learning applications, can also be found in [6–8].

In outlier detection, the goal is to identify and isolate outliers in the data. Outliers can be categorized into Point Outliers and Collective Outliers [13]. In the first category, a single point in a dataset deviates from the rest, while in the second one, there is a collection of points that deviate from the rest of the dataset. For fault detection and severity estimation of a PMSM, the Collective Outlier case is considered, as with the occurrence of a fault, samples that deviate from the normal operation will appear, especially with increasing severity. Three different widely used methods for outlier detection are investigated in this work. The first one is Isolation Forest, proposed by Liu et al. in [14]. The second one is One Class Support Vector Machine (SVM) [15] and the last one is Robust Covariance Ellipse [16]. The three methods are investigated separately and then an ensemble approach is proposed to address the methods' limitations and further PMSM fault diagnosis challenges.

Several challenges exist in the integration of AI and machine learning for outlier detection and fault diagnosis of electrical machines, and particularly for PMSM. A key challenge is related to the Signal Processing and Feature Extraction stage [17]. The features to be used by the diagnostic model must reflect as close as possible the existence of faults. In addition, they must be as independent as possible of load and speed operating conditions. For this reason, this paper proposes the utilization of the Park transform and the transformation of currents in the d-q rotational reference system. The presence of a fault in the motor leads to the distortion of d-q current signals in both time and frequency domains, so statistical indicators are used to describe these changes. An additional challenge in diagnosing PMSM faults has to do with the estimation of the severity of the fault. Other than the prediction of a fault, the diagnostic system should also give an estimate of how serious it is for the appropriate action to be taken. For this reason, the so-called Anomaly or Outlier Scores extracted by the selected methods are exploited. Moreover, for a specific diagnostic task, many models may fit and perform well, so a combination of them may yield to better results, improving the overall performance of the diagnostic system. For this reason, simple ensemble approaches, like Majority Voting and Mean Ensemble [18], which combine independently the predictions and scores from the used models, are employed.

2. PMSM Mathematical Model

The voltage equations of the PMSM in the *abc* reference system are the following:

$$\begin{bmatrix} V_a(t) \\ V_b(t) \\ V_c(t) \end{bmatrix} = \begin{bmatrix} R_s & 0 & 0 \\ 0 & R_s & 0 \\ 0 & 0 & R_s \end{bmatrix} \cdot \begin{bmatrix} I_a(t) \\ I_b(t) \\ I_c(t) \end{bmatrix} + \begin{bmatrix} \frac{d\psi_a(t)}{dt} \\ \frac{d\psi_b(t)}{dt} \\ \frac{d\psi_c(t)}{dt} \end{bmatrix} \tag{1}$$

where R_s is the stator's winding resistance, I_a, I_b, I_c are the stator's phase currents, and ψ_a, ψ_b, ψ_c are the stator's flux linkages, which are given by the following equations:

$$\begin{bmatrix} \psi_a(t) \\ \psi_b(t) \\ \psi_c(t) \end{bmatrix} = \begin{bmatrix} L_{aa} & L_{ab} & L_{ac} \\ L_{ba} & L_{bb} & L_{bc} \\ L_{ca} & L_{cb} & L_{cc} \end{bmatrix} \cdot \begin{bmatrix} I_a(t) \\ I_b(t) \\ I_c(t) \end{bmatrix} + \begin{bmatrix} \psi_{apm}(t) \\ \psi_{bpm}(t) \\ \psi_{cpm}(t) \end{bmatrix} \tag{2}$$

where L_{xx} is the self-inductance of the x phase, L_{xy} is the mutual inductance of the x and y phases, and ψ_{apm}, ψ_{bpm}, ψ_{cpm} are the flux linkages associated to the windings abc and the permanent magnets. A schematic representation of the PMSM model is shown in Figure 1.

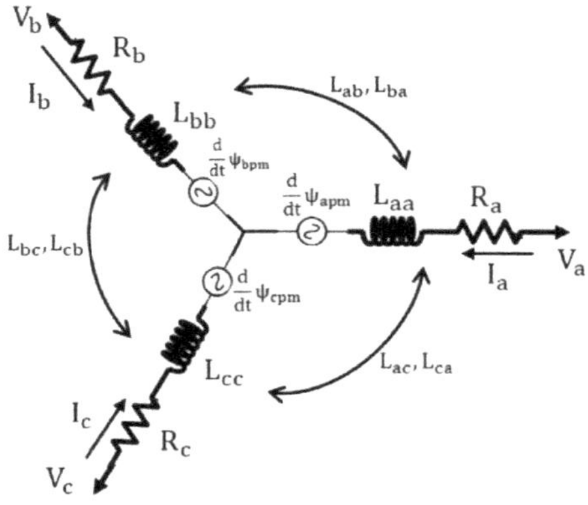

Figure 1. PMSM mathematical model.

To simplify the above equations, the Park transform is used, where the Park matrix is obtained as:

$$T^{\theta} = \frac{2}{3}\begin{bmatrix} sin(\theta) & sin(\theta - \frac{2\pi}{3}) & sin(\theta + \frac{2\pi}{3}) \\ cos(\theta) & cos(\theta - \frac{2\pi}{3}) & cos(\theta + \frac{2\pi}{3}) \\ 0.5 & 0.5 & 0.5 \end{bmatrix} \quad (3)$$

with respect to an arbitrary reference frame with reference angle θ and q-axis alignment. By selecting $\theta = \theta_e$, where θ_e is the rotational reference angle, the simplified voltage equations are the following:

$$\begin{aligned} V_q &= R_s I_q + L_q \frac{dI_q}{dt} + p\omega(I_d L_d + \psi_{pm}) \\ V_d &= R_s I_d + L_d \frac{dI_q}{dt} - p\omega I_q L_q \end{aligned} \quad (4)$$

where p is the number of pole pairs, L_q, L_d are the d-q reference frame inductances, ψ_{pm} is the flux linkage associated with permanent magnets. The zero sequence Vo is neglected here.

3. Motor d-q Current Signature Analysis

Motor Current Signature Analysis (MCSA) relies on examining a motor's stator current to find specific patterns associated with faults. The stator's current is analyzed in the frequency domain in the interest of extracting specific frequencies associated with motor's faults. For a healthy motor, the form of stator currents is as follows:

$$\begin{aligned} I_{A_h}(t) &= Asin(2\pi f_s t) \\ I_{B_h}(t) &= Asin(2\pi f_s t - \frac{2\pi}{3}) \\ I_{C_h}(t) &= Asin(2\pi f_s t + \frac{2\pi}{3}) \end{aligned} \quad (5)$$

Using the Park transform matrix in the synchronous rotational reference system $(\theta = \theta_e)$:

$$\begin{bmatrix} I_q \\ I_d \end{bmatrix} = \frac{2}{3}\begin{bmatrix} cos(\theta_e) & cos(\theta_e - \frac{2\pi}{3}) & cos(\theta_e + \frac{2\pi}{3}) \\ sin(\theta_e) & sin(\theta_e - \frac{2\pi}{3}) & sin(\theta_e + \frac{2\pi}{3}) \end{bmatrix} \cdot \begin{bmatrix} I_a \\ I_b \\ I_c \end{bmatrix} \quad (6)$$

It follows that:

$$\begin{aligned} I_{q_h} &= \frac{A}{3} \\ I_{d_h} &= 0 \end{aligned} \quad (7)$$

With the presence of a fault in the motor, additional harmonics appear in the current's spectrum. The frequency patterns for MCSA related to eccentricity, demagnetization, and bearings faults are presented in Table 1.

Table 1. Fault-related frequency patterns.

Faults	Expression
Eccentricity	$f_{ecc} = \left(1 \pm \frac{2k-1}{p}\right) \cdot f_s$
Demagnetization	$f_{dmg} = \left(1 \pm \frac{k}{p}\right) \cdot f_s$
Bearings	$f_{bear} = f_s \pm k \cdot f_{bch}$

fs: fundamental supply frequency, p: pole pairs, f_{bch}: bearing characteristic frequency, k: integer.

Currents under faulty PMSM state can be expressed as follows:

$$\begin{aligned} I_A(t) &= Asin(2\pi f_s t) + \sum_{K} A_K sin(2\pi f_k t) = I_{A_h}(t) + I_{A_f}(t) \\ I_B(t) &= Asin\left(2\pi f_s t - \tfrac{2\pi}{3}\right) + \sum_{K} A_K sin\left(2\pi f_k t - \tfrac{2\pi}{3}\right) = I_{B_h}(t) + I_{B_f}(t) \\ I_c(t) &= Asin\left(2\pi f_s t + \tfrac{2\pi}{3}\right) + \sum_{K} A_K sin\left(2\pi f_k t + \tfrac{2\pi}{3}\right) = I_{C_h}(t) + I_{C_f}(t) \end{aligned} \quad (8)$$

The above-mentioned quantities are expressed in the synchronous rotational reference system as:

$$\begin{aligned} I_q &= I_{q_h} + I_{q_f} = \tfrac{A}{3} + I_{q_f} \\ I_d &= I_{d_h} + I_{d_f} = I_{d_f} \end{aligned} \quad (9)$$

From the above equations, we can observe that currents in the d-q rotating system contain two quantities, the dc quantities I_{q_h}, I_{d_h}, which refer to the healthy operational state, and the oscillating quantities I_{q_f}, I_{d_f}, which refer to additional harmonics associated with the fault. By analyzing the above quantities in the frequency domain, the frequencies that occur beyond dc can be used to detect and identify PMSM faults.

4. Features Extraction

To extract useful indicators of fault conditions from d-q currents, multiple features can be exploited. The occurrence of a fault affects both the form of d-q current in time and frequency domains. The fault's effect on the current's time waveform can be described by features such as standard deviation, variance, skewness, and kurtosis. In the current spectrum, the appearance of fault harmonics results in the distortion of the shape with the appearance of new peaks. Depending on the fault and its severity, the amplitudes and position of the harmonics may vary. To describe the spectrum, various spectral descriptors can be used. The spectral features shown in Table 2 are used for this task. It is important to note that the total features to be used should be kept as few as possible so that computational complexity remains low, but they must also be sufficient to describe the desired fault characteristics.

Table 2. Spectral features.

Features	Expression
Spectral Centroid	$S_{centroid}(t) = \frac{\sum_{k=b1}^{b2} f_k s_k}{\sum_{k=b1}^{b2} s_k}$
Spectral Spread	$S_{spread}(t) = \sqrt{\frac{\sum_{k=b1}^{b2} (f_k - S_{centroid})^2}{\sum_{k=b1}^{b2} s_k}}$
Spectral Skewness	$S_{skewness}(t) = \sqrt{\frac{\sum_{k=b1}^{b2} (f_k - S_{centroid})^3 \cdot s_k}{S_{spread}^3 \cdot \sum_{k=b1}^{b2} s_k}}$
Spectral Kurtosis	$S_{kurtosis}(t) = \sqrt{\frac{\sum_{k=b1}^{b2} (f_k - S_{centroid})^4 \cdot s_k}{S_{spread}^4 \cdot \sum_{k=b1}^{b2} s_k}}$

f_k is the k-th frequency bin, s_k is the k-th spectral magnitude value.

5. Outlier Detection

5.1. Isolation Forest (iForest)

Isolation Forest (iForest) is an unsupervised outlier detection method that was proposed by Liu et al. [14]. iForest creates a forest of decision trees and assigns anomaly scores to every data point in the forest according to the length from the root node. The basic idea is to assign high anomaly scores to the data points with the least length and thus identify outliers.

iForest consists of two phases, the training and evaluating phases. Considering a dataset X with n samples and m features, training starts by selecting a subset of data and building the first Isolation Tree. The procedure is repeated until a certain number of Isolation Trees are built. Every tree is built by splitting the selected samples. Splitting is performed by randomly selecting a feature and a value between the maximum and minimum value of the selected feature. The splitting procedure is repeated until a certain depth in the tree is achieved. In the second phase, the evaluating phase, each data point receives a score based on its position in the created forest. The score of every data point is calculated as:

$$s(x,n) = 2^{-\frac{E(h(x))}{c(n)}} \tag{10}$$

The path of every sample from the root node is denoted as $h(x)$ and the average value of the paths from the forest as $E(h(x))$. By using BST [14], the non-successful search length is computed, which is denoted as $c(n)$. An illustration of the Isolation Forest is shown in Figure 2. Outliers are defined based on shorter paths from the root node of the created Isolation Trees.

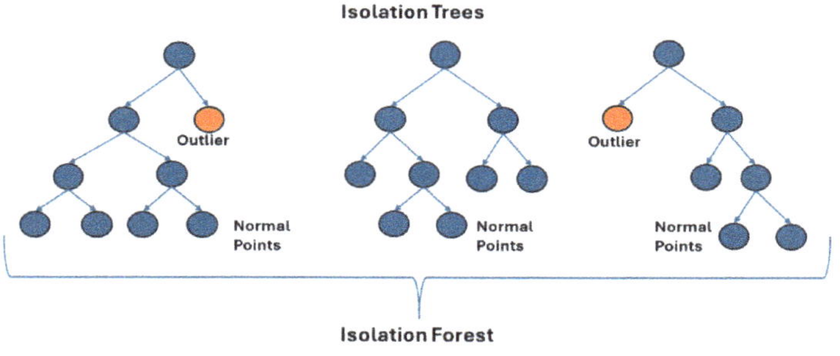

Figure 2. Isolation Forest illustration.

5.2. One-Class SVM

Support Vector Machines (SVMs) are a class of supervised machine learning methods that are used in a variety of applications for classification. The basic idea behind SVM is to find an optimal hyperplane to separate the data. Linear SVM is used to find an optimal 'maximum-margin' hyperplane to separate data. To deal with nonlinear classification tasks, kernel tricks are used. Kernels are used to map the data space into a higher-dimensional space, the feature space, where classification can be achieved. The functions that are used more as kernels are polynomial, radial basis function (rbf), and sigmoid.

One Class Support Vector Machines (One Class SVMs) have been proposed for novelty and outlier detection [15]. The goal is to find a function that separates the data from the origin through separating normal data in a specific region. Data points outside that region can be considered anomalies. Consider a dataset X with n samples and m features, Φ, a feature mapping from data space X to feature space F, and a kernel:

$$k(x,y) = (\Phi(x) \cdot \Phi(y)) \tag{11}$$

The quadratic problem to be solved in the case of One Class SVM is the following:

$$min\frac{1}{2}||w||^2 + \frac{1}{v \cdot m}\sum_i \xi_i - \rho \tag{12}$$

subject to:

$$w \cdot \Phi(x_i) \geq \rho - \xi_i,\ \xi_i \geq 0 \tag{13}$$

where w, ρ are the weight and bias terms of the hyperplane, ξ_i are the slack variables, and v is an adjustable parameter.

The dual-Lagrangian problem is derived as:

$$\min_a \frac{1}{2}\sum_{ij} a_i a_j k(x_i \cdot y_i) \tag{14}$$

Solving the above problem, the coefficients of the following decision function are derived:

$$f(x) = sign\left(\sum_i a_i k(x_i,\ x) - \rho\right) \tag{15}$$

In Equation (12), the parameter $v \in (0,1)$ is used in the second term to adjust the slack variables' effect on the quadratic problem. The parameter controls the trade-off between the number of outliers and the margin from the origin. If $v \to 0$, then the boundaries of the hyperplane are loose, and then it separates all data from the origin. If $v \to 1$, then the boundaries of the hyperplane become tighter, and it separates less data from the origin. An illustration of One Class SVM is shown in Figure 3.

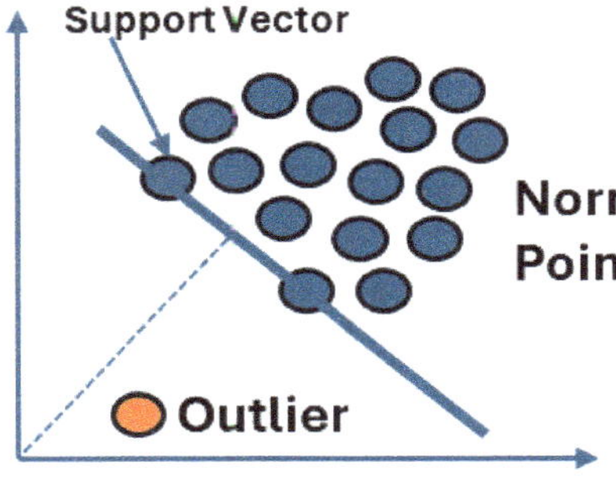

Figure 3. One Class SVM illustration.

5.3. Robust Covariance Ellipse

For a dataset X with n samples and m features, the Mahalanobis distance is given as:

$$MD(x) = \sqrt{(x - \overline{x})^T S^{-1}(x - \overline{x})} \tag{16}$$

where $\overline{x}$ is the mean, S is the covariance matrix of the samples, and T is the annotation of the transpose operation. The calculated distance of every data point can be used to define a region of normal data and outliers. In the calculation of mean and covariance, outliers can affect the calculated values, thus leading to the identification of outliers as normal data, an effect also known as the 'Masking Effect' [16]. To enhance robustness against this effect, Minimum Covariance Determinant (MCD) is used to estimate the mean and the covariance matrix. The MCD computes the mean and the covariance matrix for h observations. where $h > m$ and can be calculated as $h = \frac{(n+m+1)}{2}$. FAST-MCD [19] is employed for MCD implementation due to computation speed and efficiency. After the estimation of robust mean and covariance, the Mahalanobis distance for every sample is calculated. An illustration of the computed Robust Covariance Ellipse is shown in Figure 4.

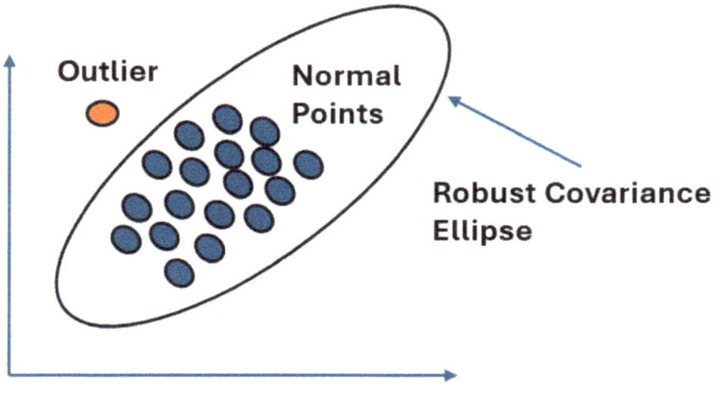

Figure 4. Robust Covariance Ellipse illustration.

5.4. Outlier Ensemble Approach

Outlier ensembles have been quite categorized in [16]. Depending on the structure of the ensemble, Sequential and Independent Ensemble or Model-Centered and Data-Centered Ensembles can be employed [20]. The approach used in this work falls into Independent and Model-Centered categories, where the whole dataset is used from all the outlier detection models, as shown in Figure 5. Each model generates anomaly scores for each sample point. The use of scores from multiple different models may require normalization as each model generates different arithmetic values. Averaging Ensemble can be used to extract averaging anomaly scores and can be carried out by simply averaging the values of each model. Weighted Averaging can also be employed, where values of each model are multiplied by a certain weight to place emphasis on certain values. Typical weights include statistical, adaptive, or case-specific ones [18]. Instead of Averaging, Max or RMS value of the overall normalized scores for each sample point can also be used. Like Scores Ensemble, Average, Max, and RMS value of predictions from each model can be employed. Majority Voting can also be used, where the majority of predictions are selected. Various approaches are assessed and compared through experimental tests in Section 6.

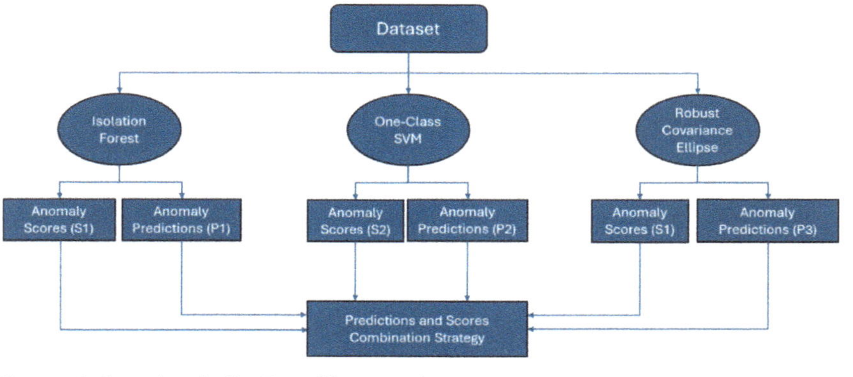

Figure 5. Independent Outlier Ensemble approach.

6. Experimental Procedure and Results

The test bench that was used for the experimental procedure is shown in Figure 6. The test rig consists of a Nidec's (Sycracuse, NY, USA) PMSM and DC Generator, a resistive load, an Inverter, a Current Measurement Unit, a Data Acquisition Unit, and a PC. The PMSM is coupled to the DC Generator through a flexible coupling. The DC Generator with the resistive load is used as a load for the PMSM. The PMSM parameters are shown in Table 3.

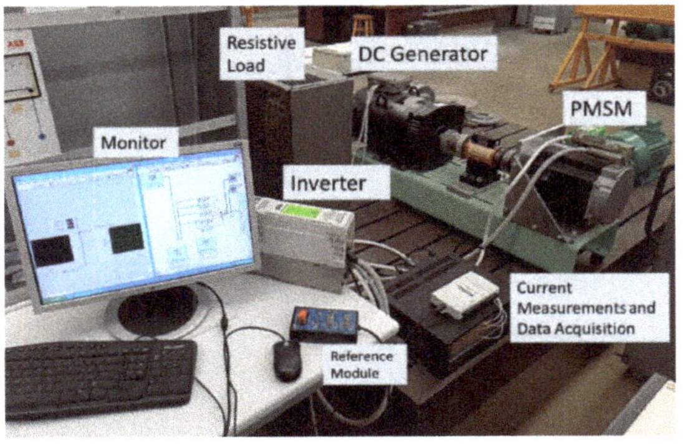

Figure 6. Configuration of the PMSM test bench for fault detection and severity estimation.

Table 3. PMSM parameters.

Parameters	Value
Rated Power	6.16 kW
Rated Voltage	380 V
Stall Current	14 A
Rated Speed	2000 rpm
Max Speed	2685 rpm
Number of Poles	8

The acquisition of current measurements is performed with a sampling frequency of 5 kHz using an NI Daq and LABVIEW. The fault considered in this case is a misalignment between PMSM and the DC Generator, causing eccentricity effects in the PMSM's shaft. It was achieved by placing metal shims in the support base of the motor. Two levels of fault severity were considered by different shim widths.

The collected data consist of three-phase current measurements for healthy and faulty operating conditions with increasing severity. In Figures 7 and 8, abc current time waveforms for healthy and faulty operating conditions are presented, respectively. The operating speed of the PMSM is 1400 rpm, with a load of 6 Nm. We can notice the deformation of the waveform's envelope in the faulty state due to the misalignment. To extract useful features of the faulty conditions, such as time and frequency domain features, d-q transformation and FFT analysis are employed. In Figures 9 and 10, the corresponding waveforms for q and d axis current in a rotating reference frame are presented. The d current waveform remains at zero, as Field Oriented Control (FOC) is used to drive the PMSM. We observe an increased ripple and distortion in q current signal over time, due to the additional harmonics caused by the misalignment condition.

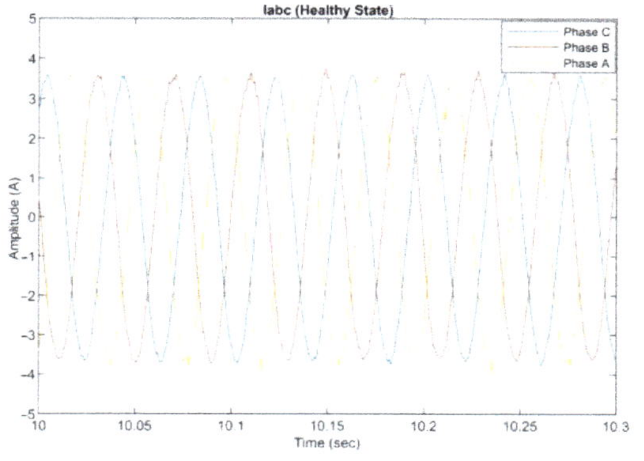

Figure 7. Motor ABC current waveforms in healthy state.

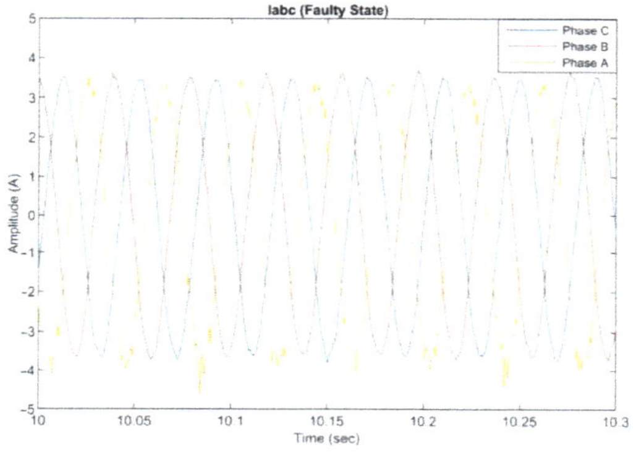

Figure 8. Motor ABC current waveforms in faulty state.

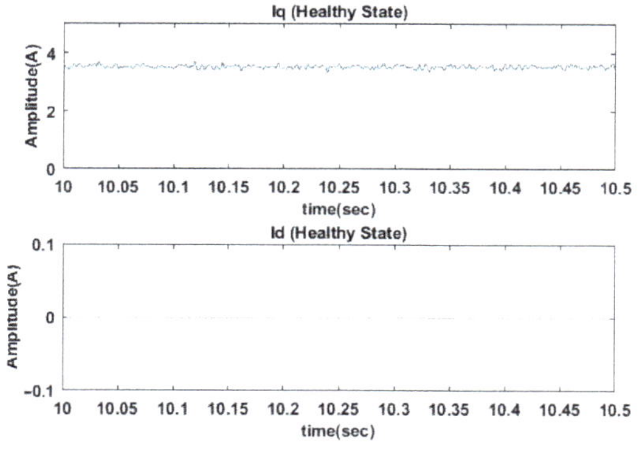

Figure 9. 3 Motor d-q current waveforms in healthy state.

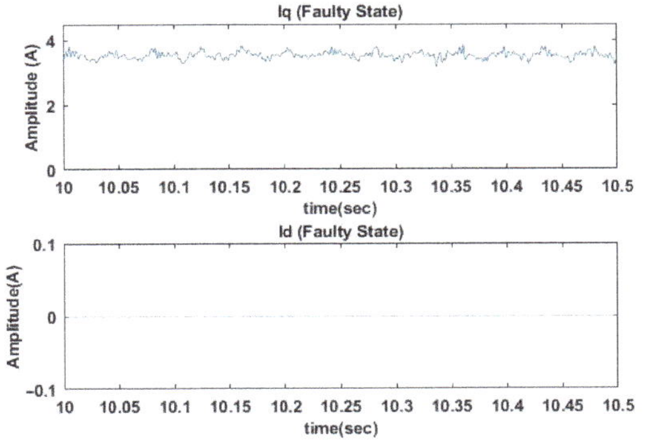

Figure 10. Motor d-q current waveforms in faulty state.

To derive the frequency characteristics of the fault, analysis of the current in the frequency domain is required. For this reason, in Figures 11 and 12, Power Spectral Density is calculated. In Figure 11, the spectrum of the healthy state (blue color) and faulty state (red color) are placed together. The specific eccentricity-related fault harmonics are indicated with red circles. In Figure 11, the q-axis Power Spectral Density is shown for the same conditions. Due to design and operation parameters of the PMSM test bench, some harmonics that follow the eccentricity-related pattern are also evident in the healthy state.

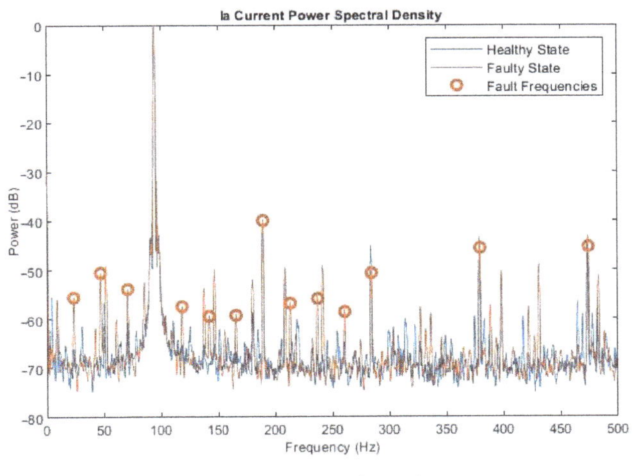

Figure 11. A-phase current Power Spectral Density.

From the waveform of the current over time, the statistical features of standard deviation, variance, skewness, and kurtosis are calculated, and from the signal's power spectrum, spectral density, spectral centroid, spectral spread, spectral skewness, and spectral kurtosis indices are calculated. The above characteristics are calculated for four different speeds and four different load levels under a healthy and faulty (misalignment) condition. The overall dataset consists of the above features for different operating modes and with an increasing severity of the fault.

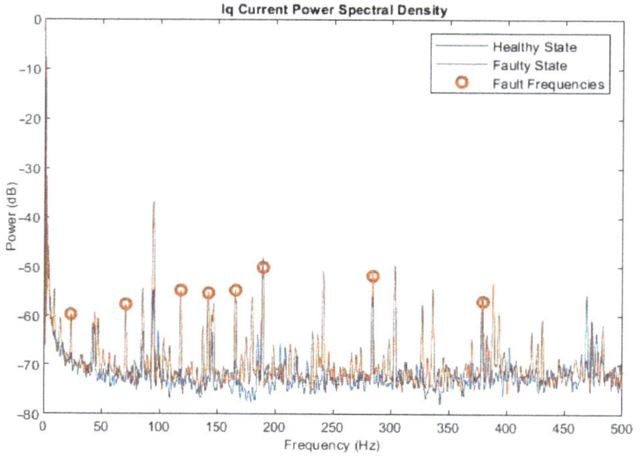

Figure 12. q-axis current Power Spectral Density.

Initially, we investigate the three different methods separately and compare them. The implementation of all methods is performed using Python and Scikit-learn. First, the Isolation Forest is investigated. The training of the model is performed with healthy data, while the test is performed with a dataset of healthy data and faulty data with increasing severity of the misalignment fault. The model's performance is investigated for different contamination parameter values and number of trees. As the number of samples and features is small in this dataset, the number of maximum features and maximum samples that can be adjusted as parameters are kept to the default values. However, it is important to note that in larger datasets and features, the above two parameters are important, as they can reduce computational complexity and indicate redundant features. Below, in Figure 13 are the evaluation metrics for different values of the contamination parameter and different number of trees. We observe that in all cases (50, 100, 150, 200 iTrees), better evaluation metric values appear in the range of 0.4–0.5 for the contamination parameter. The best evaluation metric values for 100 Isolation Trees and contamination parameter in the range of 0.4–0.5 are shown in Table 4.

Table 4. Isolation Forest evaluation metrics for 100 Isolation Trees and contamination parameter in the range of 0.4–0.5.

Evaluation Metrics	Value
Accuracy	0.82
Precision	0.81
Recall	0.95
F1-Score	0.88
ROC AUC	0.75

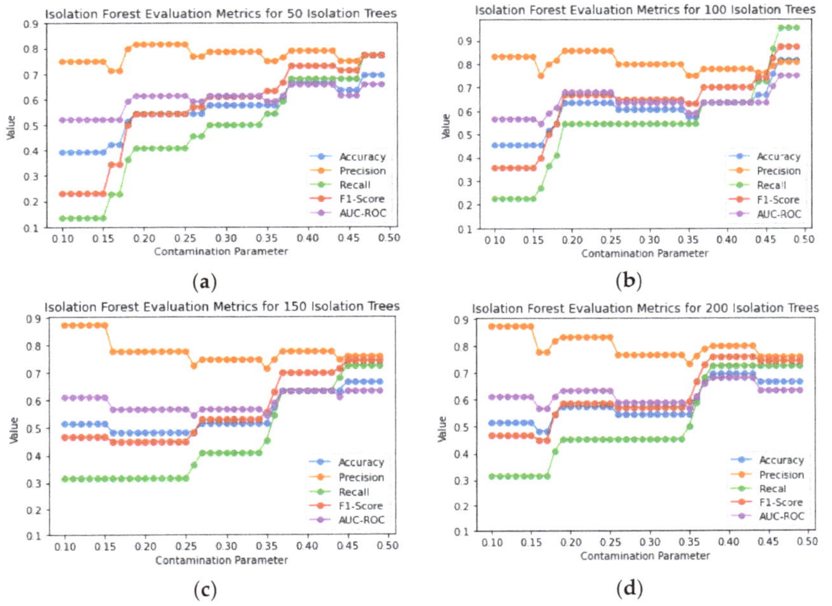

Figure 13. Isolation Forest evaluation metrics for 50 (**a**), 100 (**b**), 150 (**c**), 200 (**d**) Isolation Trees and contamination parameter in the range of 0.1–0.5.

Subsequently, One Class SVM is investigated. Like iForest, only healthy data were used to train the model and then tested with a dataset of healthy and faulty states with increasing severity. One Class SVM was tested for the Radial Basis Function kernel for different gamma and fraction parameter values. The gamma range of the test is between 0.005 and 0.04, as there are no further improvements in the performance of the model above 0.04. This can be seen in the corresponding figures for gamma = 0.02, 0.03, and 0.04, in Figure 14. The best values of evaluation metrics appear for gamma = 0.2 and fraction parameter = 0.6. For these cases, the evaluation metrics are shown in Table 5.

Table 5. One Class SVM evaluation metrics for gamma = 0.2 and fraction parameter = 0.6.

Evaluation Metrics	Value
Accuracy	0.97
Precision	0.96
Recall	1
F1-Score	0.98
ROC AUC	0.95

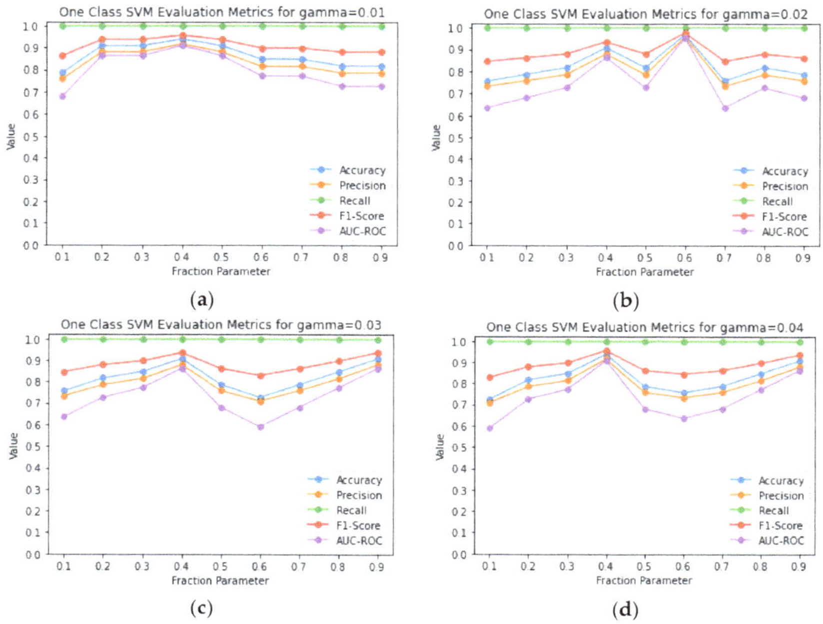

Figure 14. One Class SVM evaluation metrics for gamma equal to 0.01 (**a**), 0.02 (**b**), 0.03 (**c**), 0.04 (**d**) and fraction parameter in the range 0.1–0.9.

Lastly, Robust Covariance Elliptic Envelope was assessed for the same dataset. The influence of different contamination parameters is shown in Figure 15. We can observe that the best values of the evaluation metrics arise for a contamination parameter equal to 0.2 or 0.3. The evaluation metrics for the above parameters are shown in Table 6. The main difference lies in the decrease in precision and ROC AUC and the increase in Recall and F1-score for contamination parameter equal to 0.2 and 0.3, respectively.

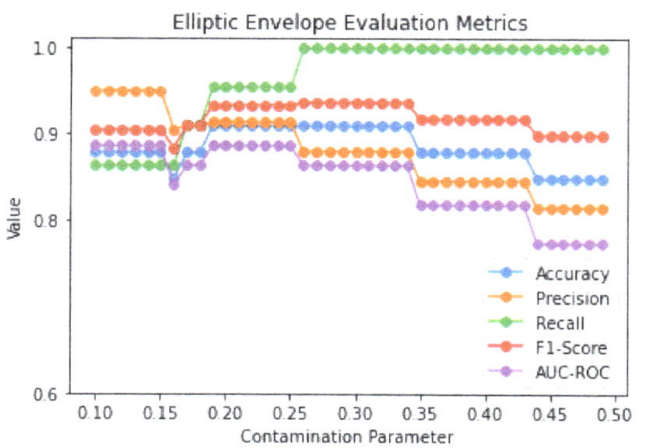

Figure 15. Robust Covariance Ellipse evaluation metrics for contamination parameter 0.1–0.5.

Table 6. Robust Covariance Ellipse evaluation metrics for gamma = 0.2 and contamination parameter = 0.2 and 0.3.

	Value	
Evaluation Metrics	**Contamination Parameter a = 0.2**	**Contamination Parameter a = 0.3**
Accuracy	0.91	0.91
Precision	0.91	0.88
Recall	0.95	1
F1-Score	0.93	0.94
ROC AUC	0.89	0.86

The above models can be combined through different ensemble techniques, as discussed in Section 5. More specifically, Majority Voting Ensemble can be used, where the majority of predictions are selected, and the Average Ensemble, where the average of the predicted values is calculated. Note that in the case of Averaging, rounding of the predicted values is required. The evaluation metrics for the above cases are shown in Table 7.

Table 7. Ensemble approaches' evaluation metrics.

	Ensemble Approach	
Evaluation Metrics	**Majority Voting**	**Averaging**
Accuracy	0.94	0.97
Precision	0.92	1
Recall	1	0.95
F1-Score	0.96	0.98
ROC AUC	0.91	0.98

To estimate the severity of the fault, the anomaly scores generated by each model are employed. Based on how anomaly scores are generated, each of the above methods generates a different range of values. For this reason, the values are normalized. Higher anomaly scores indicate faulty operating conditions in the PMSM. In the test dataset, the first 11 samples respond to a healthy working state, while the following 11 respond to a faulty state with a low fault severity, and then the last 11 respond to an increased fault severity. In the case of the One Class SVM, in Figure 16, we can see that healthy samples are distinguished from faulty samples, and especially the latest samples of increased severity. However, the difference between each sample for the latest samples and the increased severity is not clear. In the case of the Isolation Forest, in Figure 17, we see that we have increased anomaly scores for faulty cases that can be used as indicators of fault occurrence and increase in severity. In the case of the Elliptic Envelope, in Figure 18, we have the clearest picture, as we observe that anomaly scores are low for healthy samples while increasing with the appearance of the fault and especially at higher severity.

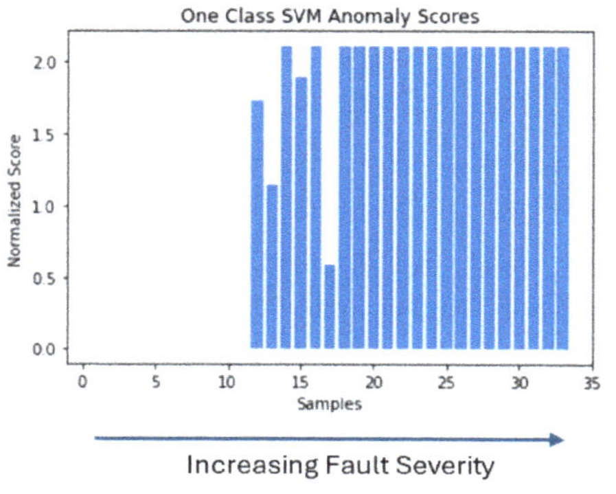

Figure 16. One Class SVM generated anomaly scores.

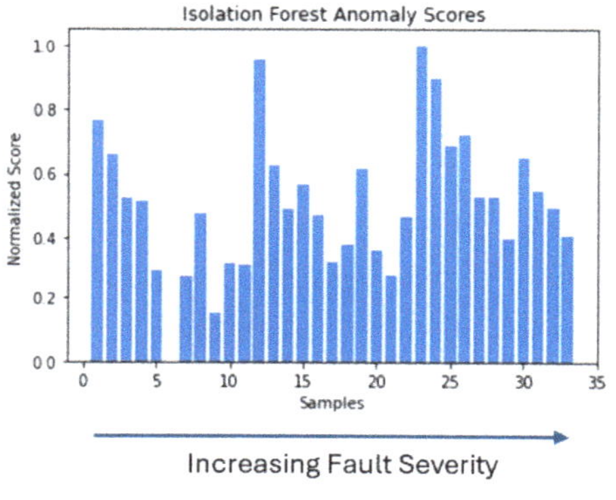

Figure 17. Isolation Forest generated anomaly scores.

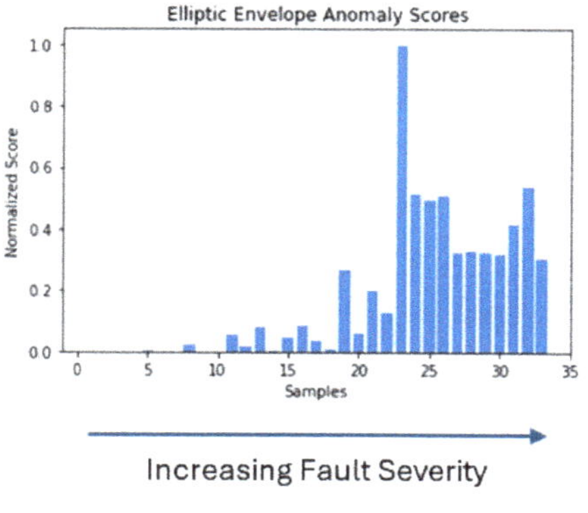

Figure 18. Robust Covariance Ellipse generated anomaly scores.

Ensemble techniques can be employed to utilize the scores generated from the models and improve severity estimation. For each sample of the three models, the average values are calculated. This results in the mean anomaly scores, shown in Figure 19. The evaluation metrics extracted in the previous section can be used to introduce weights to each anomaly score produced by each model, respectively. However, in this case, there was no notable change from the mean ensemble, so it was not examined further. Other than Mean

Ensemble, Max Ensemble can be employed, where the max values from each model are used. The corresponding anomaly scores are shown in Figure 20.

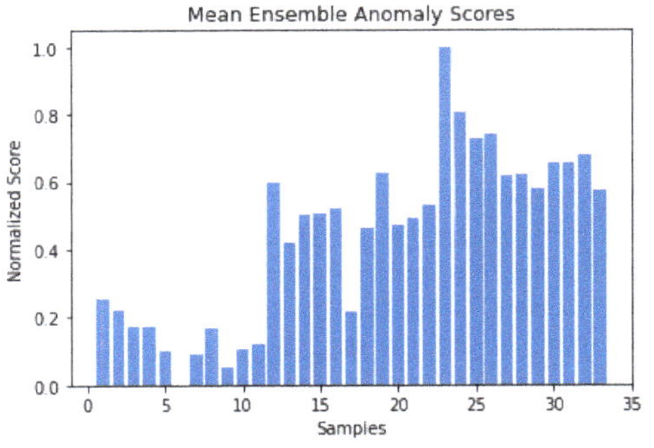

Figure 19. Mean Ensemble anomaly scores.

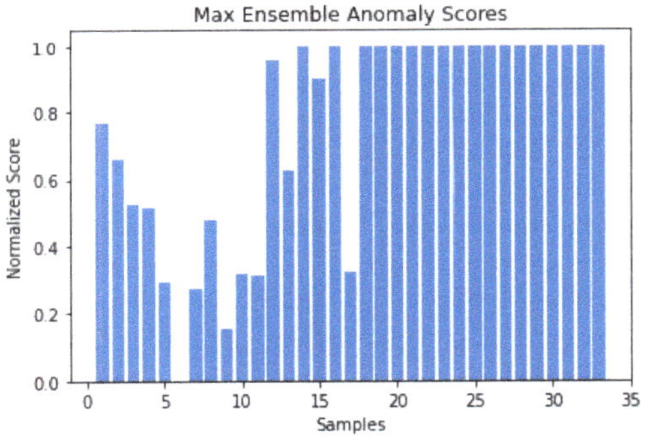

Figure 20. Max Ensemble generated anomaly scores.

An additional important piece of information from the generated anomaly scores is related to the detection of conditions where the fault is more intense or detectable. It is known [4,6] that the operating conditions of motor speed and load affect the occurrence and detectability of the fault. By using anomaly scores, we can see in which operating condition the highest anomaly score is displayed, as well as compare each operating condition with the corresponding one with a fault.

7. Conclusions

The proposed methodology utilizes PMSM's three-phase currents and speed measurements for online, non-invasive, and cost-effective condition monitoring of the PMSM. To extract fault-related features from the measurements, d-q transform was used. Distortions in time waveforms and several eccentricity-related frequencies in the power spectral density were observed for different speed and load conditions of the PMSM. Then, to extract useful indicators of fault conditions, statistical measures in time and frequency domain were used. The extracted statistical features were used for outlier detection by means of fault detection and severity estimation through Isolation Forest, One Class Support Vector Machine (SVM), and Robust Covariance Ellipse.

First, Isolation Forest was investigated for different isolation trees and contamination parameters. The best evaluation metrics were extracted for 100 Isolation Trees and a contamination parameter in the range of 0.4–0.5. The accuracy of Isolation Forest reached 0.82. One Class SVM was employed for the same task. Radial Basis Function was selected as the kernel and different gamma and fraction parameters were investigated. The best evaluation metrics were extracted for gamma equal to 0.2 and fraction parameter equal to 0.6. The accuracy reached 0.97. Lastly, Robust Covariance Ellipse fitting was tested. The highest accuracy achieved was 0.91 for gamma and contamination parameters equal to 0.2 and 0.3, respectively. One Class SVM was the best candidate in terms of Accuracy, Recall, Precision, F1-Score, and ROC AUC. For Severity Estimation, the extracted Outlier Anomaly Scores from the above methods were used. Comparing the three methods, increasing fault severity was better observed in Outlier Scores generated by Robust Covariance Ellipse fitting, then Isolation Forest, and lastly, One Class SVM. To combine the predictions and outlier scores, and so the advantages of each method, Independent Ensemble approaches are proposed. Majority Voting and Averaging Ensemble of the predictions led to Accuracy equal to 0,94 and 0,97, respectively. Max and Mean Ensemble of the Outlier Scores led to better observability of the increasing severity by each sample of the tested dataset.

Author Contributions: Conceptualization, K.K. and E.M.; methodology, K.K.; software, K.K.; validation, K.K., investigation, K.K.; resources, K.K.; data curation, K.K. and E.M.; writing—original draft preparation, K.K and E.M.; writing—review and editing, K.K and E.M.; supervision, E.M.; project administration, E.M. All authors have read and agreed to the published version of the manuscript.

Funding: This research received no external funding.

Institutional Review Board Statement: Not applicable.

Informed Consent Statement: Not applicable.

Data Availability Statement: The raw data supporting the conclusions of this article will be made available by the authors on request.

Acknowledgments: The present work was financially supported by the "Andreas Mentzelopoulos Foundation" (Corresponding Author: K.K.).

Conflicts of Interest: The authors declare no conflicts of interest.

References

1. Koutrakos, K.; Mitronikas, E. Detection of Shaft Misalignment of a PMSM using Zoom-FFT. In Proceedings of the 2023 IEEE 14th International Symposium on Diagnostics for Electrical Machines, Power Electronics and Drives (SDEMPED), Chania, Greece, 28–31 August 2023; pp. 96–102. [CrossRef]
2. Rosero, J.; Romeral, J.L.; Cusido, J.; Ortega, J.A.; Garcia, A. Fault detection of eccentricity and bearing damage in a PMSM by means of wavelet transforms decomposition of the stator current. In Proceedings of the 2008 Twenty-Third Annual IEEE Applied Power Electronics Conference and Exposition, Austin, TX, USA, 24–28 February 2008; pp. 111–116. [CrossRef]
3. Lamprokostopoulos, A.; Mitronikas, E.; Barmpatza, A. Detection of Demagnetization Faults in Axial Flux Permanent-Magnet Synchronous Wind Generators. *Energies* **2022**, *15*, 3220. [CrossRef]
4. Qi, Y.; Bostanci, E.; Gurusamy, V.; Akin, B. A Comprehensive Analysis of Short-Circuit Current Behavior in PMSM Interturn Short-Circuit Faults. *IEEE Trans. Power Electron.* **2018**, *33*, 10784–10793. [CrossRef]
5. Osmani, K.; Haddad, A.; Lemenand, T.; Castanier, B.; Alkhedher, M.; Ramadan, M. A critical review of PV systems' faults with the relevant detection methods. *Energy Nexus* **2023**, *12*, 100257. [CrossRef]
6. Niu, G.; Dong, X.; Chen, Y. Motor Fault Diagnostics Based on Current Signatures: A Review. *IEEE Trans. Instrum. Meas.* **2022**, *72*, 3520919. [CrossRef]
7. Henao, H.; Capolino, G.-A.; Fernandez-Cabanas, M.; Filippetti, F.; Bruzzese, C.; Strangas, E.; Pusca, R.; Estima, J.; Riera-Guasp, M.; Hedayati-Kia, S. Trends in Fault Diagnosis for Electrical Machines: A Review of Diagnostic Techniques. *IEEE Ind. Electron. Mag.* **2014**, *8*, 31–42. [CrossRef]
8. Orlowska-Kowalska, T.; Wolkiewicz, M.; Pietrzak, P.; Skowron, M.; Ewert, P.; Tarchala, G.; Krzysztofiak, M.; Kowalski, C.T. Fault Diagnosis and Fault-Tolerant Control of PMSM Drives–State of the Art and Future Challenges. *IEEE Access* **2022**, *10*, 59979–60024. [CrossRef]
9. Ebrahimi, B.M.; Roshtkhari, M.J.; Faiz, J.; Khatami, S.V. Advanced Eccentricity Fault Recognition in Permanent Magnet Synchronous Motors Using Stator Current Signature Analysis. *IEEE Trans. Ind. Electron.* **2014**, *61*, 2041–2052. [CrossRef]
10. Cai, B.; Zhao, Y.; Liu, H.; Xie, M. A Data-Driven Fault Diagnosis Methodology in Three-Phase Inverters for PMSM Drive Systems. *IEEE Trans. Power Electron.* **2017**, *32*, 5590–5600. [CrossRef]
11. Lee, H.; Jeong, H.; Koo, G.; Ban, J.; Kim, S.W. Attention Recurrent Neural Network-Based Severity Estimation Method for Interturn Short-Circuit Fault in Permanent Magnet Synchronous Machines. *IEEE Trans. Ind. Electron.* **2021**, *68*, 3445–3453. [CrossRef]

12. Skowron, M.; Orlowska-Kowalska, T.; Kowalski, C.T. Detection of Permanent Magnet Damage of PMSM Drive Based on Direct Analysis of the Stator Phase Currents Using Convolutional Neural Network. *IEEE Trans. Ind. Electron.* **2022**, *69*, 13665–13675. [CrossRef]
13. El-Dalahmeh, M.; Al-Greer, M.; Bashir, I.; El-Dalahmeh, M.; Demirel, A.; Keysan, O. Autonomous fault detection and diagnosis for permanent magnet synchronous motors using combined variational mode decomposition, the Hilbert-Huang transform, and a convolutional neural network. *Comput. Electr. Eng.* **2023**, *110*, 108894. [CrossRef]
14. Boukerche, A.; Zheng, L.; Alfandi, O. Outlier Detection: Methods, Models, and Classification. *ACM Comput. Surv.* **2020**, *53*, 55. [CrossRef]
15. Liu, F.T.; Ting, K.M.; Zhou, Z.-H. Isolation Forest. In Proceedings of the 2008 Eighth IEEE International Conference on Data Mining, Pisa, Italy, 15–19 December 2008; pp. 413–422. [CrossRef]
16. Shin, H.J.; Eom, D.-H.; Kim, S.-S. One-class support vector machines—An application in machine fault detection and classification. *Comput. Ind. Eng.* **2005**, *48*, 395–408. [CrossRef]
17. Márquez-Vera, M.; Ramos-Velasco, L.; López-Ortega, O.; Zúñiga-Peña, N.; Ramos-Fernández, J.; Ortega-Mendoza, R. Inverse fuzzy fault model for fault detection and isolation with least angle regression for variable selection. *Comput. Ind. Eng.* **2021**, *159*, 107499. [CrossRef]
18. Hubert, M.; Debruyne, M. Minimum covariance determinant. *WIREs Comput. Stat.* **2009**, *2*, 36–43. [CrossRef]
19. Aggarwal, C.C. *Outlier Analysis*; Springer Science and Business Media LLC: Dordrecht, The Netherlands, 2017. [CrossRef]
20. Mienye, I.D.; Sun, Y. A Survey of Ensemble Learning: Concepts, Algorithms, Applications, and Prospects. *IEEE Access* **2022**, *10*, 99129–99149. [CrossRef]

Disclaimer/Publisher's Note: The statements, opinions and data contained in all publications are solely those of the individual author(s) and contributor(s) and not of MDPI and/or the editor(s). MDPI and/or the editor(s) disclaim responsibility for any injury to people or property resulting from any ideas, methods, instructions or products referred to in the content.

applied sciences

MDPI

Article

A One-Class-Based Supervision System to Detect Unexpected Events in Wastewater Treatment Plants

Paula Arcano-Bea [1,†], Míriam Timiraos [1,2,†], Antonio Díaz-Longueira [1,3,†], Álvaro Michelena [1,3,*,†], Esteban Jove [1,3,†] and José Luis Calvo-Rolle [1,3,†]

1 CTC Research Group, University of A Coruña, Calle Mendizábal s/n, 15403 Ferrol, Spain; paula.arcano@udc.es (P.A.-B.); miriam.timiraos.diaz@udc.es (M.T.); a.diazl@udc.es (A.D.-L.); esteban.jove@udc.es (E.J.); jlcalvo@udc.es (J.L.C.-R.)

2 Fundación Instituto Tecnológico de Galicia, Department of Water Technologies, Calle Cantón Grande, 9, 15008 A Coruña, Spain

3 Centro de Investigación en Tecnologías de la Información y las Comunicaciones (CITIC), University of A Coruña, Campus de Elviña s/, 15071 A Coruña, Spain

* Correspondence: alvaro.michelena@udc.es

† These authors contributed equally to this work.

Abstract: The increasing importance of water quality has led to optimizing the operation of Wastewater Treatment Plants. This implies the monitoring of many parameters that measure aspects such as solid suspension, conductivity, or chemical components, among others. This paper proposes the use of one-class algorithms to learn the normal behavior of a Wastewater Treatment Plants and detect situations in which the crucial parameters of Chemical Oxygen Demand, Ammonia, and Kjeldahl Nitrogen present unexpected deviations. The classifiers are tested using different deviations, achieving successful results. The final supervision systems are capable of detecting critical situation, contributing to decision-making and maintenance effectiveness.

Keywords: WWTP; one class; faul detection; supervision system; kmeans; autoencoder; Gaussian model; NCBoP

1. Introduction

The current reality of water scarcity, influenced by factors like climate change, is widely acknowledged [1]. Simultaneously, the global population is steadily increasing [2], and with an improved standard of living, there is a noticeable rise in water consumption [3]. Predictably, this heightened water usage leads to an upsurge in wastewater production. The increased wastewater production poses a challenge for treatment plants, as they often struggle to keep up with the growing energy and resource demands needed to process these substantial water quantities [4].

In light of these challenges, optimizing the operation of Wastewater Treatment Plants (WWTPs) becomes crucial [5]. Various efforts have been made to enhance the efficiency of these facilities through different approaches. Depending on the sewer network type, WWTPs may receive either domestic and industrial wastewater in separative sewer networks or a combination of rainwater, domestic, and industrial wastewater in combined sewer networks. In the latter case, although the pollution level in wastewater is generally lower during rainfall events, the volume of water significantly increases, occasionally surpassing the treatment capacity of WWTPs [6,7].

The unique characteristics and location of each WWTP, along with temporal variations in the quantity and quality of the wastewater, necessitate thorough monitoring in various treatment processes to optimize their operation and comply with water purification regulations [8,9]. Monitoring in WWTPs is crucial for several reasons:

1. Real-time control of treatment processes by tracking key parameters.

2. Optimization of energy consumption based on treatment demands.
3. Early detection of anomalies and implementation of predictive maintenance strategies.
4. Optimization of sludge generation and management.
5. Adjustment of WWTP operation in response to changes in pollutant load due to climatic factors or temporal patterns in water consumption.
6. Ensuring the quality of treated water discharged into water bodies.

Despite the numerous benefits in efficiently managing WWTPs, the initial economic investment required for extensive monitoring in all treatment processes is often substantial. Therefore, minimizing the number of sensors, identifying crucial variables for measurement, or employing virtual sensors is vital in optimizing management and cost savings associated with WWTP operation [10–12]. For instance, monitoring parameters like total nitrogen in WWTPs, which quantifies the amount of organic matter, proteins, and amino acids in water, is crucial for understanding contamination levels and serves as an indicator to optimize various treatment processes. Additionally, measuring this variable extends beyond the use of a simple sensor, requiring focus, experience, and specific measurement methods for reliable results [13–15].

The investigation of approaches to enhance the energy efficiency of WWTPs is a crucial area of study, as evidenced by a study outlined in [16]. This research explores various options for optimizing the energy consumption of Italy's largest WWTP, acknowledging that energy usage constitutes a significant operational cost for these facilities and that efficiency improvements can result in substantial economic advantages.

In a complementary modeling approach, as discussed in [17], determining the optimal solid retention time is identified as a key aspect of effectively reducing operating expenses. The optimization of solid retention time contributes to the efficient removal of pollutants from wastewater, streamlining the treatment process and enhancing cost-effectiveness.

Additionally, the enhancement of ozonation processes by eliminating standard substances is addressed in [18]. This approach is crucial for improving the overall treatment efficiency of WWTPs by focusing on optimizing a specific treatment step, ultimately leading to more effective pollutant removal and reduction in operational costs.

Multiple studies support the idea that optimizing WWTPs brings tangible benefits [19]. For example, in [20], an index is presented that allows evaluating the monitoring and diagnostic performance of fault detection methods which takes into account several characteristics such as false alarms, false acceptances, and undesirable changes from correct detection to non-detection during a fault event.

Considering that many wastewater treatment facilities are publicly owned, the need for cost optimization becomes even more pronounced [21]. Publicly funded facilities must operate efficiently to ensure responsible resource allocation and effective waste management [22]. Moreover, the significance of treated water as a valuable resource is emphasized, particularly in regions facing frequent droughts [5]. This highlights the dual benefit of wastewater treatment optimization, not only in terms of cost reduction but also in the sustainable production of a valuable water resource, addressing challenges posed by water scarcity in drought-prone areas.

In [23], a novel technique is proposed for real-time monitoring of foam presence in WWTP tanks using texture segmentation models trained with centralized and federated approaches. The proposed methodology integrates into an image processing chain that involves capturing images with a professional camera, ensuring the absence of anomalies in the captured images, and implementing a real-time communication method for event notifications to plant operators.

While the studies mentioned earlier [4,16–22] have made valuable contributions to the field by investigating strategies like enhancing energy efficiency, determining the optimal retention time of solids, and refining the ozonation process, there remains a need for further research to address specific limitations. For instance, the study's exclusive focus on Italy's largest WWTP might restrict the applicability of the findings to diverse geographical and operational settings. Additionally, the optimization strategies proposed in [18] may not comprehensively consider the myriad challenges encountered by WWTPs, including the identification of anomalies in sensor measurements within the plant.

In light of these considerations, this work presents a novel method for optimizing WWTPs, offering a unique approach to address the complex and dynamic nature of wastewater treatment. Unlike the studies mentioned above, the proposed approach emphasizes the importance of continuous monitoring of a wide set of parameters, diagnosing and detecting possible faults in the sensors' measurements. This control not only minimizes the economic investment associated with the exhaustive review of measurements by an operator but also guarantees that the sensors measure correctly, a crucial point for regulatory compliance [24,25].

This paper presents an approach to optimize WWTP through fault detection. During plant operation, numerous parameters are monitored with the aim of applying different techniques to detect anomalies that may occur in certain plant areas. The approach is based on one-class classifiers trained with data registered during normal operation and tested with synthetic anomalies involving critical parameter deviations.

The document is structured as follows: after this introduction, the materials and methods are presented, then the experiments and results are described. Finally, conclusions are drawn, and future work is proposed.

2. Materials and Methods

2.1. Dataset Description

A WWTP constitutes a collection of facilities, typically situated within a population center, with its primary purpose being the reduction in wastewater pollution to acceptable levels before discharge into the aquatic environment. Depending on their size and the specific pollutants targeted for treatment, WWTPs consist of various treatment lines and processes. Generally, two main operational lines are recognized: the water line, which is dedicated to wastewater purification, and the sludge line, which focuses on managing solids (sludges) generated during the treatment processes. This study was conducted on a medium-sized wastewater treatment plant located in a Mediterranean climate setting, serving a population of approximately 15,000 individuals. Figure 1 illustrates the operational scheme of the WWTP utilized in this investigation, covering the primary wastewater treatments from the intake of raw wastewater into the WWTP to the release of the treated effluent into the aquatic environment. The water line of this WWTP primarily includes pretreatment, secondary treatment, and tertiary treatment stages.

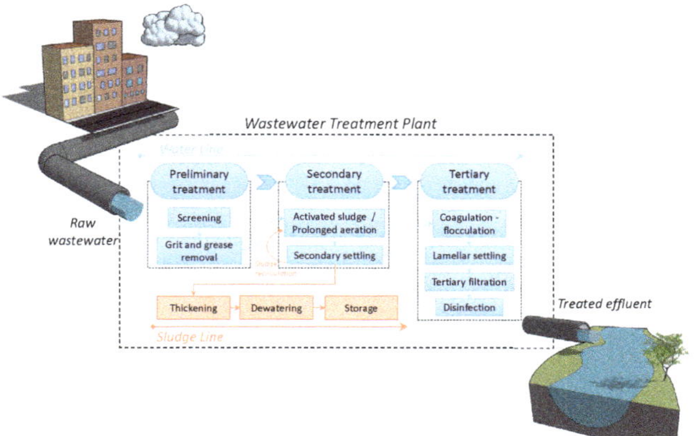

Figure 1. Scheme of the operation of the WWTP under study.

The initial phase of the water treatment process in the WWTP involves the introduction of raw wastewater into the water line, where pretreatment occurs. The primary objective of pretreatment is the elimination of coarse and fine solids through the screening stage, as well as the removal of greases and oils via the grit and grease removal system. Following pre-treatment, the resulting wastewater undergoes secondary treatment to eliminate dissolved and suspended organic matter that persists after the initial pretreatment. The WWTP in question employs activated sludge technology and prolonged aeration for this process, providing oxygen to aerobic microorganisms responsible for breaking down organic matter, as referenced in [26]. Subsequently, secondary settling takes place to separate solid waste or sludge formed during preceding biological processes.

The water treatment line concludes with tertiary treatment, aiming for a more extensive removal of specific contaminants, such as phosphorus, which may still be present in the water post secondary settling. In the examined WWTP, physicochemical treatments including coagulation–flocculation, lamellar settling, and filtration are implemented, as documented in [27]. The tertiary treatment process culminates with a disinfection step, involving the exposure of treated water to ultraviolet rays and the addition of sodium hypochlorite to eliminate pathogenic microorganisms. After traversing all the processes in the water line, the treated water is discharged into a water body.

Concerning the sludge line, a portion of the sludge generated during wastewater treatment is collected from the bottom of the secondary settling tanks of the WWTP, and subsequently recirculated to the biological reactor to become part of the biological concentrate. Simultaneously, any excess sludge is directed to the sludge line for treatment. This sludge often contains a significant water content, prompting its passage through a gravity-thickening process to increase the concentration of solids, facilitating handling and further processing. Following thickening, centrifugal dewatering is employed to further reduce the water content. The sludge is then temporarily stored at the WWTP until its final transfer for various applications, such as agricultural fertilizers.

In the scope of this research, a dataset comprising 23 monitored variables within the WWTP was examined. The samples constituting the dataset were collected over nine months, with a recording frequency of one value per day. Table 1 summarizes the description of each variable and the tags used for each one, indicating the key variables in bold font.

Table 1. Variables in the dataset.

Description of the Measured Variable	Variable Name
pH at the Entrance	PH_E
pH on Exit	PH_S
Conductivity at the Entrance	Conductivity_E
Conductivity at the Exit	Conductivity_S
V60 at the Entrance	V60_E
Solids in Suspension at the Entrance	SS_E
Solids in Suspension on Exit	SS_S
Biological Oxigen Demand on Exit	BOD_S
Chemical Oxygen Demand at the Entrance	COD_E
Chemical Oxygen Demand on Exit	**COD_S**
Total Nitrogen at the Entrance	NITROGEN_T_E
Total Phosphoro at the Entrance	PHOSPHORO_T_E
Total Phosphoro on Exit	PHOSPHORUS_T_S

Table 1. *Cont.*

Description of the Measured Variable	Variable Name
Ammonia at the Entrance	NH3_E
Total Kjeldahl Nitrogen at the Entrance	NTK_E
Nitrate at the Entrance	NO3_E
Nitrogen dioxide at the Entrance	N02_E
Ammonia on Exit	**NH3_S**
Total Kjeldahl Nitrogen on Exit	**NTK_S**
Nitrate on Exit	NO3_S
Nitrogen dioxide on Exit	N02_S
Thickener input	INPUT_ESP
Total Nitrogen on Exit	NITROGEN_T_S

2.2. Unexpected Events Description

The present document deals with the detection of unexpected events in WWTPs that may represent a potential risk for final water quality. This offers help in managing undesired working situations, contributing to enhancement of optimum decision-making. However, given the unfeasibility of registering datasets in that kind of situation, a synthetic dataset is required to emulate anomalies. Different experts in the field of chemistry and water treatment were consulted to determine which variables are especially significant in detecting dangerous situations in a WWTP. Following these criteria, the output variables of Chemical Oxygen Demand (COD), Ammonia (NH3), and Total Kjeldahl Nitrogen (NTK) were identified as the most critical due to their relevance in assessing water quality and environmental impact. Specific reasons why these variables are essential in environmental and water treatment studies are described below.

- Chemical Oxygen Demand:
  - Organic Pollution Indicator: COD measures the amount of oxygen needed to oxidize organic matter in a water sample. It is a direct indicator of the amount of organic contaminants present [28].
  - Environmental Impact: High levels of COD in water bodies can result in decreased dissolved oxygen, affecting aquatic life and ecosystems [29].
  - Process Control: In wastewater treatment plants, COD is essential to monitor and control the efficiency of treatment processes [30].
- Ammonia:
  - Direct Toxicity: Ammonia is toxic to many aquatic species. Elevated levels can cause adverse effects in fish and other aquatic organisms [31].
  - Eutrophication: Ammonia is a nutrient that, in excess, can contribute to the eutrophication of water bodies, promoting extensive growth of algae and aquatic plants, which can lead to water quality degradation and the death of aquatic fauna by anoxia [32].
  - Pollution Indicator: The presence of ammonia can indicate recent contamination by agricultural, industrial, or domestic waste [33].
- Total Kjeldahl Nitrogen:
  - Total Nutrient Component: TKN measures the total amount of nitrogen in the form of ammonia and organic matter, providing a more complete view of the nitrogen load in water [34].
  - Eutrophication and Water Quality: Like ammonia, TKN is related to eutrophication. Excess nitrogen can promote the growth of oxygen-consuming organisms, degrading water quality and negatively affecting aquatic ecosystems [35].

- Evaluation of Sources of Pollution: TKN helps to identify and quantify the sources of nitrogen pollution, whether agricultural, industrial, or urban [36].

In summary, these parameters are critical because they provide a comprehensive view of the pollutant load, potential toxicity, and environmental impact of water bodies. Their monitoring and control are essential for sustainable water resource management and protecting aquatic ecosystems. Hence, the generation of unexpected events followed the next steps:

1. From the initial 237 samples, 16% were randomly selected to be converted to anomalies.
2. For each instance selected, one of the three variables (COD_S, NH3_S, and NTK_S) was randomly selected.
3. Once the variable to be modified was selected, its value was deviated by a given percentage.

Figure 2 represents the process of selecting random instances for anomaly generation. The original set is represented as a blue-colored square, and then, four random instances, represented in grey color, are converted to anomalies. The final set is divided into normal samples (target samples) and anomalous ones. Once the anomalous set is selected, Figure 3 represents the process to generate anomalies: one of the three key variables (COD_S, NH3_S, and NTK_S) is randomly selected, and its value is modified by a given percentage.

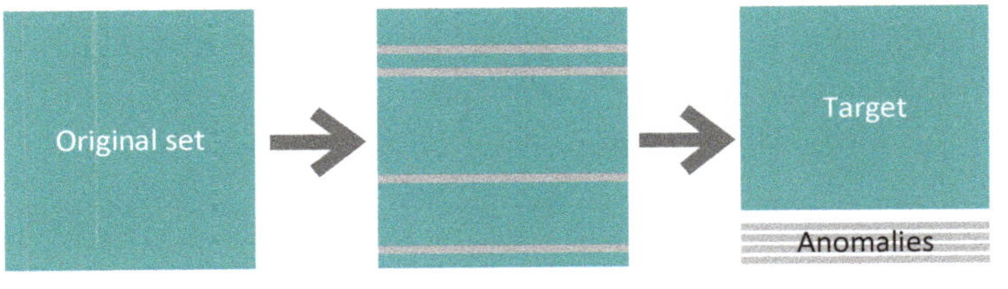

Figure 2. Random selection of instances for anomaly generation.

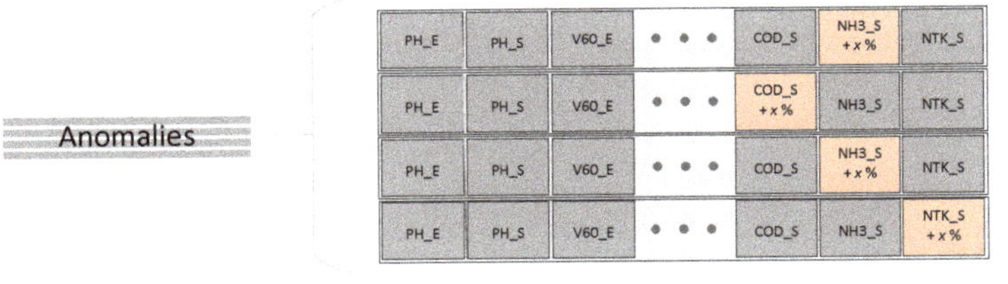

Figure 3. Procedure of generation of anomalies.

The final goal of this synthetic dataset is to determine situations that represent an emergency situation derived from abnormal values of critical variables such as COD_S, NH3_S, and NTK_S. Furthermore, the possibility of modifying the percentage x tested can provide an idea of classifier sensitivity.

2.3. One-Class Techniques

This paper proposes an intelligent classifier to detect risky situations in a WWTP. The proposal is based on learning the patterns of data registered only during normal operation. Once the model is trained, unexpected events can be determined if data differ from the learned patterns. Four different one-class techniques are proposed and described below to achieve this objective.

2.3.1. Autoencoder

This method utilizes an Autoencoder which relies on an Artificial Neural Network (ANN) to detect anomalies by determining their characteristics. Typically, the ANN setup includes an input layer, one or more hidden layers, and an output layer interconnected by weighted links [37]. The fundamental concept behind this approach involves reconstructing the input data (d) into output data (d_{rec}) through nonlinear dimensional reduction within the hidden layer, as depicted in Figure 4. Consequently, the number of neurons in the hidden layer is smaller than the input dimensions.

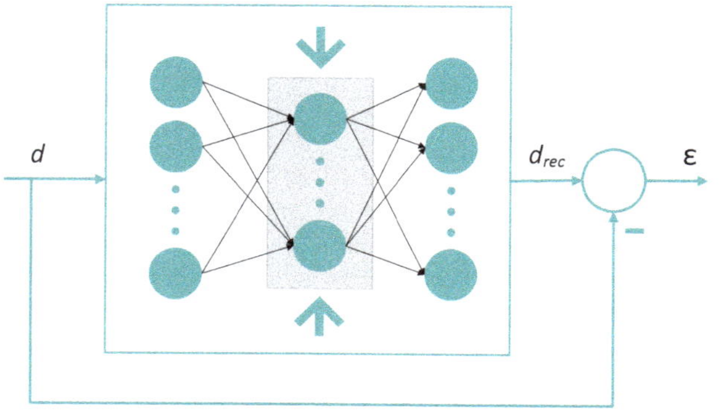

Figure 4. Cluster division depending on the value of *K*.

After the network undergoes training exclusively on the training dataset, data exhibiting distinct behaviors are expected to exhibit notable distinctions within the hidden layer space. Consequently, a test data point (q) is anticipated to demonstrate a substantial reconstruction error, quantified as $|d - d_{rec}|$. This metric serves as the criterion for categorizing the data as anomalous [38].

2.3.2. Gaussian Model

An alternative method for tackling anomaly detection, particularly in facial recognition, involves employing one-class techniques that rely on density functions. One straightforward approach within this paradigm is to apply a normal or Gaussian distribution function to the target dataset [38]. This function is derived from the same dataset used for training purposes.

After establishing the mean vector and covariance matrix, the method for identifying the anomalous status of a test sample relies on its evaluation within the Gaussian function. This straightforward approach offers a low computational overhead, making it advantageous, particularly for large datasets with a normal distribution shape.

In Figure 5, a simplified representation of a Gaussian function (depicted by the blue line) with a one-dimensional set is illustrated. The determination of whether a test instance is anomalous or not is based on comparing it to a predefined threshold (indicated by the red lines), which is set during the training phase.

2.3.3. The K-means Algorithm

The K-means algorithm, an unsupervised technique widely employed in various domains like machine learning, image processing, and pattern recognition [39], aims to partition the dataset into clusters containing data points with similar characteristics [39].

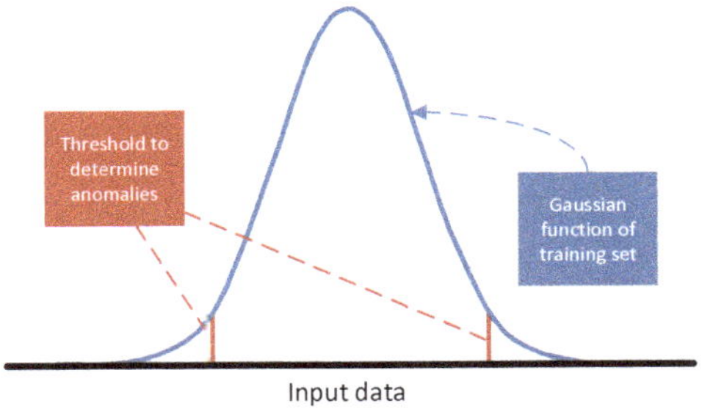

Figure 5. Example of Gaussian Model for one-class task in $\mathbb{R}^1$.

Given a set $X = x_1, x_2, \ldots, x_N$, where $x_i \in \mathbb{R}^n$, the K-means algorithm partitions the data into K subsets $G_1, G_2, \ldots, G_K$, each associated with centroids $C = c_1, c_2, \ldots, c_K$, where $c_j \in \mathbb{R}^n$. This partitioning is performed based on a clustering error criterion [40], typically computed as the sum of Euclidean distances between each point $x_i \in \mathbb{R}^n$ and its centroid c_j.

The application of this algorithm using a one-class approach is based on the distance from test instances to their nearest centroid. It is considered anomalous if this value is greater than the maximum distance registered for that cluster during the training stage. Figure 6 shows an example in $\mathbb{R}^2$ where the training set is grouped into two clusters. Then, the green dot represents a test instance that is labeled as normal because the distance to its nearest centroid is below the maximum distance of that cluster. On the contrary, the long distance from the red dot to its centroid reveals the anomalous nature of the test instance.

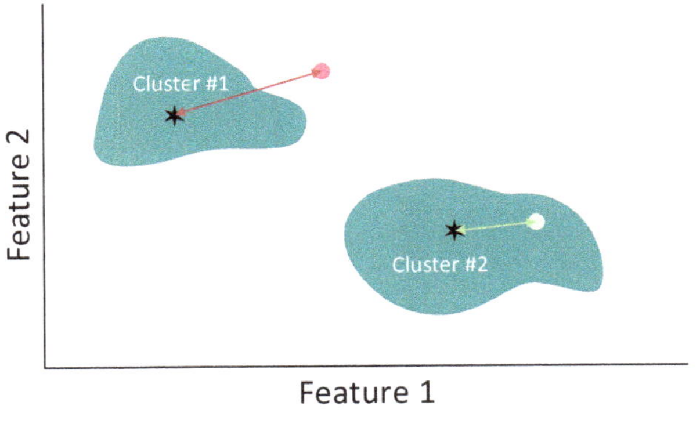

Figure 6. An example of a K-means algorithm applied to anomaly detection in 2D.

2.3.4. Non-Convex Boundary over Projections

The Non-Convex Boundary over Projections (NCBoP) algorithm innovatively utilizes non-convex hull calculations to model the shape of the target dataset [41]. This approach addresses limitations observed in the conventional convex hull method applied to random projections [42]. The fundamental principle of NCBoP involves approximating dataset boundaries in $\mathbb{R}^n$ by employing non-convex hulls across π random projections onto 2D planes. Subsequently, non-convex limits are determined on these planes, thereby reducing the computational complexity associated with calculating non-convex limits across $\mathbb{R}^n$.

The NCBoP algorithm computes a non-convex polygon for each π 2D random projection by initially selecting a starting point, determined as the lowest y coordinate among all points projected onto the p_i plane. Subsequently, it identifies the K-nearest points and arranges them based on the polar angle, retaining only the furthest point from the starting point. This process is repeated for each 2D random projection.

After computing the initial points, a stack structure is established to accommodate the next third point. Subsequently, it evaluates whether the subsequent point in the list turns left (added to the stack) or right (removes the top point from the stack). This iterative process continues until it returns to the starting point. Once the training process concludes, all points are situated within the non-convex polygon generated by the algorithm.

Upon completion of the training phase, the criterion for identifying whether a new test point is anomalous is as follows: if the data fall outside of at least one of the t projected hulls, it is deemed anomalous. This determination is illustrated in Figure 7.

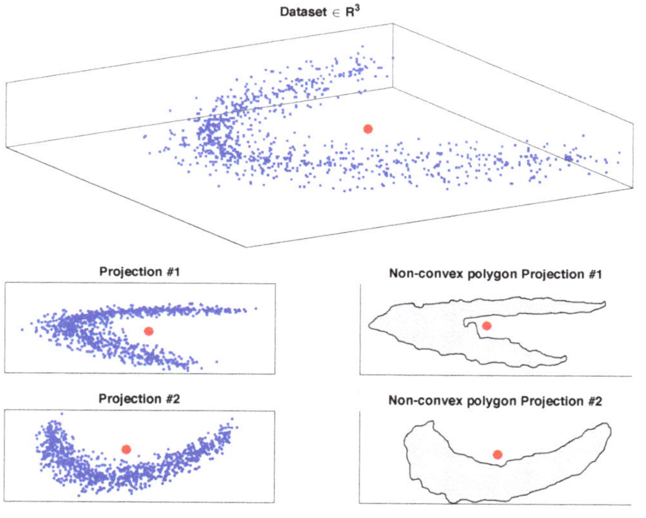

Figure 7. Example of NCBoP for one-class techniques in $\mathbb{R}^3$.

3. Experiments and Results

This section describes the different configurations considered for each one-class technique and the results derived from the proposed experiments.

3.1. Experiment Setup

As stated in previous sections, an initial dataset corresponding to normal working operations is used to train one-class classifiers. Then, unexpected and undesired data are introduced into those classifiers to determine whether they are capable of detecting hazardous situations. The severity of the anomalous event is checked with the aim of evaluating the classifier's sensitivity. The percentage deviation is swept from 5 % to 95 % with a 10 % step. As no information about those anomalies is available during the training process, it is not feasible to apply feature selection techniques to evaluate the importance of each input variable in the classification process, so all variables are considered. In this sense, using an Autoencoder can contribute to reducing data dimension, compressing the data, and identifying sub-spaces with better performance.

To implement the best classifier for each type of anomaly, the experiments are configured according to the following parameters:

- Autoencoder: the input data of the neural network are comprised of the 23 variables monitored in the WWTP. The system is trained to learn the patterns from normal operation and replicate, at the output, the same value as the input. Once the model is trained, anomalous instances replicate the data at the output with significant error.

This error, known as the reconstruction error, is the criteria to determine the anomaly detection. The number of layers is swept from 1 to 22, which is the number of variables minus one. The possibility of considering a percentage of anomalous points (OP) in the training set is checked: 0%, 5%, 10%, and 15%.

- Gaussian Model: the regularization parameter is swept from 0 to 0.005 with a step of 0.001. The possibility of considering a percentage of anomalous points (OPs) in the training set is checked: 0%, 5%, 10%, and 15%.
- K-means: the number of clusters is swept from 1 to 30. The possibility of considering a percentage of anomalous points (OPs) in the training set is checked: 0%, 5%, 10%, and 15%.
- NCBoP: the projections π tested are 10, 50, 100, 500 and 1000. Furthermore, the λ parameter in charge of reducing and expanding the boundaries is set to 0.6, 0.8, 1, 1.2, and 1.4.

All these configurations are tested through a five K-fold cross-validation process with three different data preparation methods: no preparation, Z-score, and 0–1 normalization. The mean values of the confusion matrix components during the five folds are registered. Furthermore, the F1-score parameter (%) is also included since it is an underscored metric in the field of anomaly detection, particularly in evaluating classifier ability to detect irregularities in complex systems [43,44]. To ensure classifier robustness, three aspects are taken into consideration:

- The instances converted to anomalies are randomly selected from the initial dataset.
- From each selected instance, the variable modified is also randomly selected.
- The process is repeated following K-fold cross-validation, ensuring that all data are subjected to both training and test phases.

3.2. Results

Autoencoder. The results achieved using Autoencoder classifiers are shown in Table 2. This technique shows a remarkable improvement in performance when the deviation percentage increases. In most cases, the best result corresponds to a configuration that considers 5% of the training set as anomalous. A common trend in all experiments consists of the use of a high number of neurons in the hidden layer, meaning that almost all variables are important to achieve the best classifier.

Table 2. Results and configuration for the best Autoencoder classifier depending on anomaly percentage deviation.

Dev (%)	Preproc	Hidden Layer	OP (%)	TP	TN	FP	FN	F1 (%)
5	Zscore	19	0	38.3	8.5	28.5	1.7	71.7
15	Zscore	20	5	34.2	20.9	16.1	5.8	75.7
25	Zscore	20	0	39.5	18.8	18.2	0.5	80.9
35	Zscore	20	5	34.3	33.4	3.6	5.7	88.1
45	Zscore	20	5	34	34.4	2.6	6	88.8
55	Zscore	18	5	34.1	36.4	0.6	5.9	91.3
65	Zscore	18	5	34.9	36.2	0.8	5.1	92.2
75	Zscore	20	5	34	35.8	1.2	6	90.4
85	Zscore	20	5	34.6	36.1	0.9	5.4	91.7
95	Zscore	18	5	35.8	37	0	4.2	94.5

Gaussian Model. The Gaussian model results are shown in Table 3. This statistical technique leads to successful results with more than 95% of the F1-score in 9 out of 10 anomaly sets. The lowest value is achieved when the anomalies are generated by deviating only by 5% from the original value. In all cases, the anomalies are correctly classified (True Negative), with a really low number of situations in which a false alarm is set. In contrast to Autoencoder classifiers, an outlier percentage of 0% in the training set is the best possible configuration.

Table 3. Results and configuration for best Gaussian classifier depending on anomaly percentage deviation.

Dev (%)	Preproc	RP	OP (%)	TP	TN	FP	FN	F1 (%)
5	Norm	0	5	35.9	37.0	0.0	4.1	94.6
15	Norm	0	0	38.7	37.0	0.0	1.3	98.3
25	Zscore	0	0	38.7	37.0	0.0	1.3	98.3
35	Norm	0	0	38.8	37.0	0.0	1.2	98.5
45	Zscore	0.001	0	38.8	37.0	0.0	1.2	98.5
55	Norm	0.001	0	38.9	37.0	0.0	1.1	98.6
65	Norm	0.001	0	39.0	37.0	0.0	1.0	98.7
75	Zscore	0.003	0	39.1	37.0	0.0	0.9	98.9
85	Zscore	0.003	0	38.9	37.0	0.0	1.1	98.6
95	Zscore	0.001	0	39.0	37.0	0.0	1.0	98.7

K-means. The results achieved using K-means classifiers are shown in Table 4. Although the performance improves as the deviation increases, the F1 performance is below 80% in all cases, with consequent worse results than those of Autoencoder and Gaussian classifiers. The proper number of clusters to divide the dataset does not follow a common trend, varying from 1 to 29. The most critical aspect consists of the false positives that start with 37 (100% of the anomalies labeled as normal instances) and finish with 12, which is an improvable value.

Table 4. Results and configuration for the best K-means classifier depending on anomaly percentage deviation.

Dev (%)	Preproc	OP (%)	k	TP	TN	FP	FN	F1 (%)
5	NoNorm	0	1	39.8	0.0	37.0	0.2	68.2
15	NoNorm	0	5	39.9	0.0	37.0	0.1	68.3
25	Zscore	0	3	40.0	0.0	37.0	0.0	68.4
35	Norm	0	29	39.4	1.0	36.0	0.6	68.3
45	Norm	0	12	39.7	0.7	36.3	0.3	68.4
55	Zscore	15	9	32.3	18.8	18.2	7.7	71.4
65	Zscore	10	9	34.3	17.3	19.7	5.7	73.0
75	Zscore	10	13	33.9	20.2	16.8	6.1	74.8
85	Zscore	10	17	33.0	24.1	12.9	7.0	76.8
95	Zscore	10	13	33.3	25.1	11.9	6.7	78.2

NCBoP. Table 5 summarizes the performance of NCBoP classifiers in the different experiments. As in the case of K-means, the F1-score does not exceed 80% in any configuration with a great number of anomalous instances classified as normal. The value of λ is 1.4 in 8 out of 10 cases, which means that an increment in the decision boundary leads to better results. However, when the deviation used to generate the anomaly is 5%, the decision boundary should be reduced. The number of projections does not follow a recognizable pattern, varying depending on the experiment.

To understand the improvement of each algorithm with the percentage variation of COD_S, NH3_S, and NTK_S, Figure 8 represents the values of the F1-score for each technique. The Gaussian model approach is the best algorithm to detect test parameter deviations.

Table 5. Results and configuration for best NCBoP classifier depending on anomaly percentage deviation.

Dev (%)	Preproc	λ	π	TP	TN	FP	FN	F1 (%)
5	NoNorm	0.8	500	37.1	0.5	36.5	2.9	65.3
15	Norm	1.4	10	36.5	3.5	33.5	3.5	66.4
25	Norm	1.4	10	37.7	2.9	34.1	2.3	67.4
35	Zscore	1.4	10	37.5	3.6	33.4	2.5	67.6
45	Zscore	1.4	100	30.5	20.3	16.7	9.5	70.0
55	Zscore	1.4	50	32.8	18.8	18.2	7.2	72.1
65	Zscore	1.2	10	33.2	21.4	15.6	6.8	74.8
75	Zscore	1.4	50	33.3	22.8	14.2	6.7	76.1
85	Zscore	1.4	100	31.3	27.4	9.6	8.7	77.4
95	Zscore	1.4	50	32.0	27.7	9.3	8.0	78.7

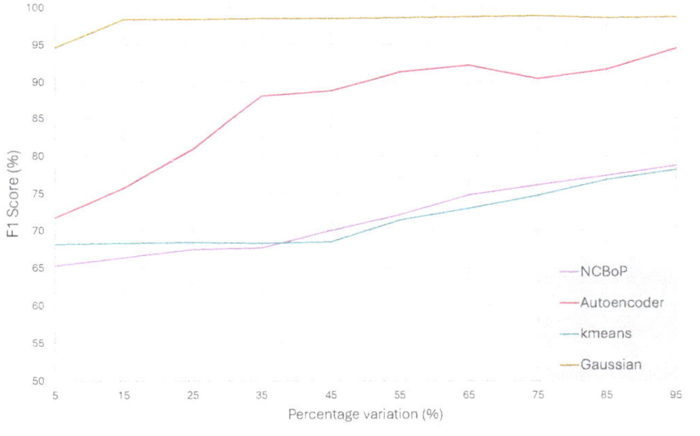

Figure 8. F1-score for each classifer.

4. Discussion

The presented work proposed a semi-supervised methodology to detect alarm situation in a WWTP. The importance of the proposal lies on the fact of implementing classifiers that take into consideration all system variables registered from normal operation with no need of human expertise to determine unexpected and anomalous events. In the current context of digitalization and quality standard increment, this tool takes advantage of data availability to enhance the supervision and diagnostics in critical infrastructure. The proposal can complement traditional diagnosis and supervision systems, offering an alarm in case of early anomaly detection. This contributes to maintenance decision making, easing effectiveness, and economic and energy efficiency. Furthermore, the results show how classifier performance improves significantly when the deviation is greater, resulting in a more dangerous situation. Despite the clear advantages of the proposal, there are several factors to which a one-class approach cannot contribute. As this is a semi-supervised methodology, it is necessary to have an initial set of normal operation instances, so human expertise is needed at least during the first data registration stage to ensure data quality. Once this limitation is overcome, the combination of the proposal with other supervised techniques will be suitable for detecting sensor misreadings to complement the one-class classifier.

5. Conclusions and Future Works

This paper presents the use of different one-class techniques to determine the appearance of unexpected events in a WWTP facility. Those events are synthetically generated to simulate deviations in COD_S, NH3_S, and NTK_S, representing critical measures to determine potential risks. With the aim of evaluating how effective each technique is, the tested deviations varied from 5% to 95%. In general terms, an increase in that deviation results in better performance in four techniques. Although NCBoP, K-means, and Autoencoder present a remarkable improvement with each deviation step, Gaussian classifiers achieve the best results. The experiments and results for the technique showed that, after a five K-fold validation, all negative instances (unexpected and hence undesired events) were correctly labeled. The false alarm cases were significantly low, with a mean value below 1.5 samples in all experiments above 15% deviation.

The results indicated that the proposed system represents a valuable contribution to the WWTP monitoring process. Taking into consideration the 23 variables along the facility, it is possible to detect when three critical parameters present unexpected values, according to the patterns learned from normal operation sets.

In future works, there are three alternative ways to supplement the contribution. The first would consider individual classifiers for each tested parameter: COD_S, NH3_S, and NTK_S. Although the classified proposed detects all unexpected events with the Gaussian model, this approach could improve the rest of the classifiers. Furthermore, instead of semi-supervised algorithms in which only information from normal operation is available, the use of supervised classifiers could be tested. This approach would implement a classifier for normal operation and another one for unexpected deviation. In the case of good results, these classifiers would indicate the specific parameter with unexpected performance. Finally, it could be possible to consider the modification of other variable instead of COD_S, NH3_S, and NTK_S to detect an unexpected situation that correspond to less severe scenarios.

Author Contributions: Conceptualization, M.T. and A.D.-L.; methodology, M.T. and P.A.-B.; software, A.D.-L. and Á.M.; validation, P.A.-B., Á.M. and J.L.C.-R.; formal analysis, J.L.C.-R. and E.J.; investigation, E.J.; resources, M.T.; data curation, A.D.-L.; writing—original draft preparation, P.A.-B. and E.J.; supervision, E.J.; project administration, J.L.C.-R. All authors have read and agreed to the published version of the manuscript.

Funding: Míriam Timiraos's research was supported by the Xunta de Galicia (Regional Government of Galicia) through grants to industrial Ph.D. (http://gain.xunta.gal (accessed on 12 June 2024)), under the Doutoramento Industrial 2022 grant with reference 04_IN606D_2022_2692965. Álvaro Michelena's research was supported by the Spanish Ministry of Universities (https://www.universidades.gob.es/ (accessed on 12 June 2024)), under the "Formación de Profesorado Universitario" grant with reference FPU21/00932. Antonio Díaz-Longueira's research was supported by the Xunta de Galicia (Regional Government of Galicia) through grants to Ph.D. (http://gain.xunta.gal (accessed on 12 June 2024)), under the "Axudas á etapa predoutoral" grant with reference ED481A-2023-072. This work was supported by Xunta de Galicia through Axencia Galega de Innovación (GAIN) by grant IN853C 2022/01, Centro Mixto de Investigación UDC-NAVANTIA "O estaleiro do futuro", which is ongoing until the end of September 2025. The support was inherited from both the starting and consolidation stages of the same project throughout 2015–2018 and 2018–2021, respectively. This stage was also co-funded by ERDF funds from the EU in the framework of program FEDER Galicia 2021–2027. CITIC, as a center accredited for excellence within the Galician University System and a member of the CIGUS Network, receives subsidies from the Department of Education, Science, Universities, and Vocational Training of the Xunta de Galicia. Additionally, it is co-financed by the EU through the FEDER Galicia 2021-27 operational program (Ref. ED431G 2023/01). This research is the result of the Strategic Project "Critical infrastructures cybersecure through intelligent modeling of attacks, vulnerabilities and increased security of their IoT devices for the water supply sector" (C061/23), as a result of the collaboration agreement signed between the National Institute of Cybersecurity (INCIBE) and the University of A Coruña. This initiative is carried out within the framework of the funds of the Recovery Plan, Transformation and Resilience Plan funds, financed by the European Union (Next Generation).

Institutional Review Board Statement: Not applicable.

Informed Consent Statement: Not applicable.

Data Availability Statement: The data presented in this study are available on request from the corresponding author. The data are not publicly available due to privacy.

Conflicts of Interest: The authors declare no conflicts of interest.

References

1. Şenol, R.; Salman, O.; Kaya, Z. Potable water production from ambient moisture. *Appl. Water Sci.* **2023**, *13*, 10. [CrossRef]
2. Brown, T.C.; Mahat, V.; Ramirez, J.A. Adaptation to future water shortages in the United States caused by population growth and climate change. *Earth's Future* **2019**, *7*, 219–234. [CrossRef]
3. Boretti, A.; Rosa, L. Reassessing the projections of the world water development report. *NPJ Clean Water* **2019**, *2*, 15. [CrossRef]
4. Safarpour, H.; Tabesh, M.; Shahangian, S.A. Environmental Assessment of a Wastewater System under Water demand management policies. *Water Resour. Manag.* **2022**, *36*, 2061–2077. [CrossRef]
5. Spellman, F.R. *Handbook of Water and Wastewater Treatment Plant Operations*; CRC Press: Boca Raton, FL, USA, 2013.
6. Mascher, F.; Mascher, W.; Pichler-Semmelrock, F.; Reinthaler, F.F.; Zarfel, G.E.; Kittinger, C. Impact of Combined Sewer Overflow on Wastewater Treatment and Microbiological Quality of Rivers for Recreation. *Water* **2017**, *9*, 906. [CrossRef]
7. Ianes, J.; Cantoni, B.; Remigi, E.U.; Polesel, F.; Vezzaro, L.; Antonelli, M. A stochastic approach for assessing the chronic environmental risk generated by wet-weather events from integrated urban wastewater systems. *Environ. Sci. Water Res. Technol.* **2023**, *9*, 3174–3190. [CrossRef]
8. Lu, J.Y.; Wang, X.M.; Liu, H.Q.; Yu, H.Q.; Li, W.W. Optimizing operation of municipal wastewater treatment plants in China: The remaining barriers and future implications. *Environ. Int.* **2019**, *129*, 273–278. [CrossRef] [PubMed]
9. Bertanza, G.; Boiocchi, R.; Pedrazzani, R. Improving the quality of wastewater treatment plant monitoring by adopting proper sampling strategies and data processing criteria. *Sci. Total Environ.* **2022**, *806*, 150724. [CrossRef] [PubMed]
10. Longo, S.; d'Antoni, B.M.; Bongards, M.; Chaparro, A.; Cronrath, A.; Fatone, F.; Lema, J.M.; Mauricio-Iglesias, M.; Soares, A.; Hospido, A. Monitoring and diagnosis of energy consumption in wastewater treatment plants. A state of the art and proposals for improvement. *Appl. Energy* **2016**, *179*, 1251–1268. [CrossRef]
11. Martínez, R.; Vela, N.; el Aatik, A.; Murray, E.; Roche, P.; Navarro, J.M. On the Use of an IoT Integrated System for Water Quality Monitoring and Management in Wastewater Treatment Plants. *Water* **2020**, *12*, 1096. [CrossRef]
12. Kizgin, A.; Schmidt, D.; Joss, A.; Hollender, J.; Morgenroth, E.; Kienle, C.; Langer, M. Application of biological early warning systems in wastewater treatment plants: Introducing a promising approach to monitor changing wastewater composition. *J. Environ. Manag.* **2023**, *347*, 119001. [CrossRef] [PubMed]
13. Bagherzadeh, F.; Mehrani, M.J.; Basirifard, M.; Roostaei, J. Comparative study on total nitrogen prediction in wastewater treatment plant and effect of various feature selection methods on machine learning algorithms performance. *J. Water Process Eng.* **2021**, *41*, 102033. [CrossRef]
14. Ye, G.; Wan, J.; Deng, Z.; Wang, Y.; Chen, J.; Zhu, B.; Ji, S. Prediction of effluent total nitrogen and energy consumption in wastewater treatment plants: Bayesian optimization machine learning methods. *Bioresour. Technol.* **2024**, *395*, 130361. [CrossRef] [PubMed]
15. Murei, A.; Kamika, I.; Momba, M.N.B. Selection of a diagnostic tool for microbial water quality monitoring and management of faecal contamination of water sources in rural communities. *Sci. Total Environ.* **2024**, *906*, 167484. [CrossRef]
16. Borzooei, S.; Campo, G.; Cerutti, A.; Meucci, L.; Panepinto, D.; Ravina, M.; Riggio, V.; Ruffino, B.; Scibilia, G.; Zanetti, M. Optimization of the wastewater treatment plant: From energy saving to environmental impact mitigation. *Sci. Total Environ.* **2019**, *691*, 1182–1189. [CrossRef]
17. Muoio, R.; Palli, L.; Ducci, I.; Coppini, E.; Bettazzi, E.; Daddi, D.; Fibbi, D.; Gori, R. Optimization of a large industrial wastewater treatment plant using a modeling approach: A case study. *J. Environ. Manag.* **2019**, *249*, 109436. [CrossRef]
18. Cunha, D.L.; da Silva, A.S.; Coutinho, R.; Marques, M. Optimization of ozonation process to remove psychoactive drugs from two municipal wastewater treatment plants. *Water Air Soil Pollut.* **2022**, *233*, 67. [CrossRef]
19. Garcia-Alvarez, D.; Fuente, M.; Vega, P.; Sainz, G. Fault Detection and Diagnosis using Multivariate Statistical Techniques in a Wastewater Treatment Plant.* *This work was supported in part by the national research agency of Spain (CICYT) through the project DPI2006-15716-C02-02 and the regional government of Castilla y Leon through the project VA052A07. *IFAC Proc. Vol.* **2009**, *42*, 952–957. [CrossRef]
20. Corominas, L.; Villez, K.; Aguado, D.; Rieger, L.; Rosén, C.; Vanrolleghem, P.A. Performance evaluation of fault detection methods for wastewater treatment processes. *Biotechnol. Bioeng.* **2011**, *108*, 333–344. [CrossRef]
21. Schraa, O.; Tole, B.; Copp, J.B. Fault detection for control of wastewater treatment plants. *Water Sci. Technol.* **2006**, *53*, 375–382. [CrossRef]
22. Ruiz, M.; Sin, G.; Berjaga, X.; Colprim, J.; Puig, S.; Colomer, J. Multivariate Principal Component Analysis and Case-Based Reasoning for monitoring, fault detection and diagnosis in a WWTP. *Water Sci. Technol.* **2011**, *64*, 1661–1667. [CrossRef] [PubMed]
23. Carballo Mato, J.; González Vázquez, S.; Fernández Águila, J.; Delgado Rodríguez, A.; Lin, X.; Garabato Gándara, L.; Sobreira Seoane, J.; Silva Castro, J. Foam Segmentation in Wastewater Treatment Plants. *Water* **2024**, *16*, 390. [CrossRef]
24. Lin, H.; Lin, C.; Xie, D.; Acuna, P.; Liu, W. A Counter-Based Open-Circuit Switch Fault Diagnostic Method for a Single-Phase Cascaded H-Bridge Multilevel Converter. *IEEE Trans. Power Electron.* **2023**, *39*, 814–825. [CrossRef]

25. Lin, H.; Cai, C.; Chen, J.; Gao, Y.; Vazquez, S.; Li, Y. Modulation and control independent dead-zone compensation for H-bridge converters: A simplified digital logic scheme. *IEEE Trans. Ind. Electron.* **2024**, 1–6. [CrossRef]
26. Orhon, D. Evolution of the activated sludge process: The first 50 years. *J. Chem. Technol. Biotechnol.* **2014**, *90*, 608–640. [CrossRef]
27. Matamoros, V.; Salvadó, V. Evaluation of a coagulation/flocculation-lamellar clarifier and filtration-UV-chlorination reactor for removing emerging contaminants at full-scale wastewater treatment plants in Spain. *J. Environ. Manag.* **2013**, *117*, 96–102. [CrossRef] [PubMed]
28. Eddy, M.; Abu-Orf, M.; Bowden, G.; Burton, F.L.; Pfrang, W.; Stensel, H.D.; Tchobanoglous, G.; Tsuchihashi, R.; Firm, A. *Wastewater Engineering: Treatment and Resource Recovery*; McGraw Hill Education: New York, NY, USA, 2014.
29. Sawyer, C.N.; McCarty, P.L.; Parkin, G.F. *Chemistry for Environmental Engineering and Science*; McGraw-Hill: New York, NY, USA, 2003.
30. Tchobanoglus, G.; Burton, F.; Stensel, H.D. Wastewater engineering: Treatment and reuse. *Am. Water Work. Assoc. J.* **2003**, *95*, 201.
31. Huff, L.; Delos, C.; Gallagher, K.; Beaman, J. *Aquatic Life Ambient Water Quality Criteria for Ammonia-Freshwater*; US Environmental Protection Agency: Washington, DC, USA, 2013.
32. Carpenter, S.R.; Caraco, N.F.; Correll, D.L.; Howarth, R.W.; Sharpley, A.N.; Smith, V.H. Nonpoint pollution of surface waters with phosphorus and nitrogen. *Ecol. Appl.* **1998**, *8*, 559–568. [CrossRef]
33. World Health Organization. Ammonia in Drinking-Water: Background Document for Development of WHO Guidelines for Drinking-Water Quality. WHO. 2003. Available online: https://cdn.who.int/media/docs/default-source/wash-documents/wash-chemicals/ammonia.pdf?sfvrsn=3080badd_6 (accessed on 12 June 2024)
34. Holmes, D.E.; Dang, Y.; Smith, J.A. Nitrogen cycling during wastewater treatment. *Adv. Appl. Microbiol.* **2019**, *106*, 113–192.
35. Smith, V.H.; Tilman, G.D.; Nekola, J.C. Eutrophication: Impacts of excess nutrient inputs on freshwater, marine, and terrestrial ecosystems. *Environ. Pollut.* **1999**, *100*, 179–196. [CrossRef]
36. Camargo, J.A.; Alonso, Á. Ecological and toxicological effects of inorganic nitrogen pollution in aquatic ecosystems: A global assessment. *Environ. Int.* **2006**, *32*, 831–849. [CrossRef] [PubMed]
37. Sakurada, M.; Yairi, T. Anomaly detection using autoencoders with nonlinear dimensionality reduction. In Proceedings of the MLSDA 2014 2nd Workshop on Machine Learning for Sensory Data Analysis, New York, NY, USA, 2 December 2014; p. 4.
38. Tax, D.M.J. One-Class Classification: Concept-Learning in the Absence of Counter-Examples. Ph.D. Thesis, Delft University of Technology, Delft, The Netherlands, 2001.
39. Ahmed, M.; Seraj, R.; Islam, S.M.S. The k-means algorithm: A comprehensive survey and performance evaluation. *Electronics* **2020**, *9*, 1295. [CrossRef]
40. Chong, B. K-means clustering algorithm: A brief review. *Acad. J. Comput. Inf. Sci.* **2021**, *4*, 37–40.
41. Jove, E.; Casteleiro-Roca, J.L.; Quintián, H.; Méndez-Pérez, J.A.; Calvo-Rolle, J.L. A new method for anomaly detection based on non-convex boundaries with random two-dimensional projections. *Inf. Fusion* **2021**, *65*, 50–57. [CrossRef]
42. Sartipizadeh, H.; Vincent, T.L. Computing the approximate convex hull in high dimensions. *arXiv* **2016**, arXiv:1603.04422.
43. Zakariah, M.; Almazyad, A.S. Anomaly Detection for IOT Systems Using Active Learning. *Appl. Sci.* **2023**, *13*, 12029. [CrossRef]
44. Almotairi, A.; Atawneh, S.; Khashan, O.A.; Khafajah, N.M. Enhancing intrusion detection in IoT networks using machine learning-based feature selection and ensemble models. *Syst. Sci. Control Eng.* **2024**, *12*, 2321381. [CrossRef]

Disclaimer/Publisher's Note: The statements, opinions and data contained in all publications are solely those of the individual author(s) and contributor(s) and not of MDPI and/or the editor(s). MDPI and/or the editor(s) disclaim responsibility for any injury to people or property resulting from any ideas, methods, instructions or products referred to in the content.

applied sciences

MDPI

Article

Genetic Multi-Objective Optimization of Sensor Placement for SHM of Composite Structures

Tomasz Rogala [1], Mateusz Ścieszka [1], Andrzej Katunin [1,*] and Sandris Ručevskis [2]

[1] Department of Fundamentals of Machinery Design, Faculty of Mechanical Engineering, Silesian University of Technology, Konarskiego 18A, 44-100 Gliwice, Poland; tomasz.rogala@polsl.pl (T.R.)

[2] Institute of Materials and Structures, Riga Technical University, Kipsalas iela 6A, LV-1048 Riga, Latvia; sandris.rucevskis@rtu.lv

* Correspondence: andrzej.katunin@polsl.pl; Tel.: +48-32-237-1069

Abstract: Increasingly often, due to the high sensitivity level of diagnostic systems, they are also sensitive to the occurrence of a significant number of false alarms. In particular, in structural health monitoring (SHM), the problem of optimal sensor placement (OSP) is appearing due to the need to reach a balance between performance and cost of the diagnostic system. The applied approach of considering nondominated solutions allows for adaption of the system parameters to the user's expectations, treating this optimization problem as multi-objective. For this purpose, the NSGA-II algorithm was selected for the determination of an optimal set of parameters in the OSP problem for the detection of delamination in composite structures. The objectives comprise minimization of type-I and type-II errors, and number of sensors to be placed. The advantage of the proposed approach is that it is based on experimental data from the healthy structure, whereas all cases with a presence of delamination were acquired from numerical experiments. This makes it possible to develop a customized SHM system for the arbitrary location of damage.

Keywords: optimal sensor placement; structural health monitoring; multi-objective optimization; delamination; NSGA-II algorithm; numerical experiment; type-I and type-II errors

1. Introduction

Structural health monitoring (SHM) became a paradigm in the operation of numerous structures and transport means, especially in civil and mechanical engineering, due to the ability to continuously evaluate structural conditions based on measured response and building prognostic models based on this. Currently, SHM is incorporated into the latest design and operation concepts, such as condition-based maintenance (CBM), the philosophy used in aviation, where maintenance is scheduled according to the current condition of the structure only [1,2]. SHM is also an integral part of the newest solutions in civil engineering, such as building information modelling (BIM) [3,4] and the concept of digital twins [5], where it plays a crucial role in the acquisition and management, being the main source of data for the evaluation of structural condition. Regardless of the advancement of the mentioned concepts and approaches based on them, there are still unsolved problems influencing their performance. One of these problems is an appropriate sensor placement that determines the effectiveness of the condition evaluation of a tested structure. It is especially important in the case of SHM systems based on modal analysis, one of the oldest and most widely used approaches in both civil and mechanical engineering [6]. Recent studies of the authors [7,8] demonstrated the need for the application of an advanced machine-learning approach for this task to find a balance between the performance and cost of an SHM system.

There are a plenty of different optimal-sensor-placement (OSP) strategies for SHM. A systematic review on different approaches can be found in [9–11]. From these optimization strategies, one group is of sequential-based sensor placement strategies where subsequent steps are based on sequentially eliminated sensor locations that contribute less to the objective functions [10,12] or otherwise by adding sensor locations that enhance the value of the evaluation function [10,13]. Another group are simple or metaheuristic optimization strategies such as greedy algorithms [14], and bioinspired e.g., evolutionary, ant colony, particle swarm, or physically inspired e.g., simulated annealing algorithms [10]. Among the above meta heuristics, evolutionary methods have become evident [10]. These algorithms can be easily customized to the needs of the optimization problem because the objective function can be independently constructed without the need for reconfiguration of the optimization routine, similarly to other heuristic approaches. The objective functions can be built based on classical objectives, e.g., Fisher information matrix (FIT), mutual information (MI), etc., [10] as well as based on outcomes of the SHM system [9] which allows for building different optimization strategies. Increasingly often, different cost functions are taken in order define objective of SHM with an OSP problem. This requires defining of the multiterm weighted objective function or the necessity to build an appropriate aggregation operator [7]. Because in OSP various evaluation functions can be considered and for the SHM systems the performance and costs are counterparts [9], the multi-objective optimization strategies have been recognized as useful strategies. In [15], a decision-making multi-objective optimization based on a particle-swarm algorithm was used. The authors have used a minimum dependency based on Modal Assurance Criteria (MAC), minimum of redundancy based on similarity measure, and the maximization of signal-quality strength as a multi-objective. The obtained Pareto front solution was filtered using the decision-making membership faction to select a subset of obtained solutions. Another study, shown in [16], was based on multi-objective genetic optimization with the application of two objectives using modal shapes. One of the objective was used to compare the modes of the structure amongst themselves, and the other was used to compare the modes of the same structure in two different configurations. The authors also showed comparison of the multi-objective approaches with a single-objective approach and sequential-based sensor-placement strategies, indicating robustness of the sensor networks selected using the multi-objective approach for damage detection of civil construction. Another case study with multi-objective optimization of sensor placement on a composite rotor blade is presented in [17]. The objective function is based on the minimalization of the sensor number and, simultaneously, the minimalization of the error between the reconstructed real mode and the mode based on the information from the candidate sensor networks. Another study [18], presents multi-objective optimization with the combination of maximization of the nominal performance that could be defined using the mean and maximum of MAC, the conditional number, or the effective independence and minimalization of the data uncertainty.

Many of the above-mentioned multi-objective-optimization problems use different algorithms; most of them are dominance-based methods based on popular NSGA-II [17–20]. This algorithm is often used as long as no more than three objectives are taken into account [21,22]. There are also decomposition-based methods which decompose multi-objective problems into a set of single-objective problems. For instance, Downey et al. [23] presented a bi-optimization objective function based on the accuracy of the strain-map reconstruction based on the optimal-sensor-placement network. They used the concept of minimalization of type-I and type-II errors of the SHM system assuming that these errors can be reduced by minimizing the mean absolute error between the real system and estimated strain map response and by minimalizing the maximum difference between them. It should be noted that the use of multi-objective optimization is an important direction in developing OSP strategies in SHM systems.

In this study, the SHM system is based on the classification-of-information system, which allows to estimate the sensitivity and specificity of damage recognition. These characteristic features, together with the number of used sensors, are taken as individual objectives in the multi-objective optimization of sensor placement. The outcomes are presented as Pareto solutions which, linked to design variables, allow the user to obtain required performance and costs of an SHM system. This is an ongoing study within a series of previously performed studies [7,8] with a general focus on development of a customizable SHM system with a trade-off between performance and cost effectiveness.

2. Tested Structure and Experiments

2.1. Parameters of the Tested Structure

The structure considered in this study to test the developed OSP approach is a carbon-fiber-reinforced polymeric (CFRP) composite plate with the dimensions of 490 × 240 mm, thickness of 1.84 mm, and the following stacking sequence: $(0/90)_{5s}$ (see Figure 1a). In the study, the healthy plate was considered as a reference for building the numerical models described in the next subsection. The detailed description of the manufacturing process and material properties of this plate are presented in [7].

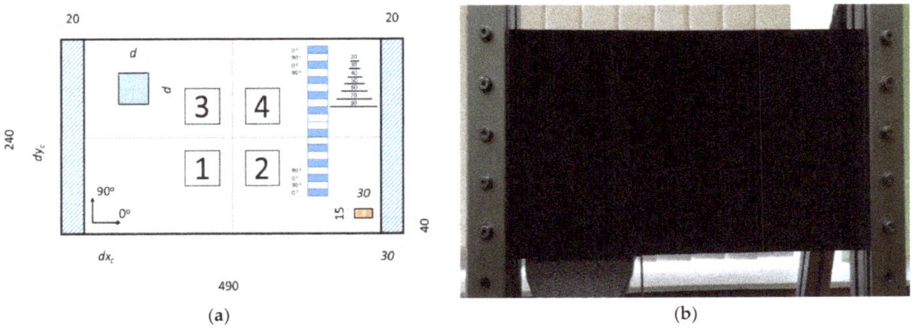

(a) (b)

Figure 1. The scheme of the tested CFRP plate (**a**), and the experimental setup (**b**).

The plate was fixed on the shorter side to the aluminum frame at a distance of 20 mm from the edge using screws located with a distance of 50 mm on the clamped sides and a tightening torque of 20 Nm. The plate was then subjected to mechanical excitation using the Penton CAD10T/EN loudspeaker (Worthing, UK) using the periodic chirp signal in the bandwidth of 0 ÷ 625 Hz with a resolution of 0.195 Hz (see Figure 1b).

The measurements of a structural response were performed on a defined grid of 25 × 13 equidistantly spaced measurement points using a Polytec® PSV-500-3D scanning laser vibrometer (Waldbronn, Germany) with three averaging cycles to improve the accuracy of the measurements. During measurements, the vibration velocity in the defined measurement points was acquired. Details on the experimental setup are provided in [7]. As a result of the performed measurements, the structural response of a plate was acquired in the form of a frequency-response function (FRF). The natural frequencies determined from the modal analysis are presented in Table 1.

Table 1. Natural frequencies of the tested plate.

Mode No.	Frequency, Hz	Vibration Velocity, 10^{-4} m/s
1	58.20	0.2
2	67.58	0.17
3	159.76	0.31
4	174.22	0.48
5	232.81	1.1
6	294.92	0.66
7	299.61	1.37
8	339.45	0.41
9	434.76	0.53
10	543.36	0.47
11	562.11	0.18

To determine the strain values in both directions for the tested plate, Sx and Sy, respectively, the transverse displacements w were considered. Further, from transverse displacements smoothed using the least squares linear regression, the strain fields were determined. Each smoothed value is determined using the adjacent data points. The strain value maps were normalized in such a way that the largest value was assigned to a unit and the remaining values were scaled relative to the maximum value. The deformation values on the plate surface in appropriate directions are calculated using the following formula:

$$Sx_{(i,j)} = \left(\frac{\partial^2 w}{\partial x^2}\right)_{(i,j)} \cdot \frac{1}{2}h;\ Sy_{(i,j)} = \left(\frac{\partial^2 w}{\partial y^2}\right)_{(i,j)} \cdot \frac{1}{2}h, \quad (1)$$

where h is the thickness of the plate.

2.2. Numerical Experiments

Data for the intact plate were collected from the laboratory experiment. Additionally, numerical experiments have been performed to collect data from various conditions of structural damage. In order to develop a system for recognizing other health conditions (the presence of delamination in this case), it was necessary to supplement the set of learning examples with data coming from damaged conditions. For this purpose, it was decided to supplement these data using an appropriately tuned numerical model. The model was tuned based on data for the intact condition, and then, after an appropriate level of compliance between the model and the experimental data was reached, the condition with the presence of delamination was simulated. The model developed in this way made it possible to obtain a set of learning examples covering damage conditions. The detailed description of the definition of the numerical model and performing numerical simulations as well as further processing of data was presented in the previous work [7].

The parameters and constraints of the model were assigned to be consistent with the physical tested plate. Each layer of the composite plate is represented by a single element throughout the thickness. Numerical modal analysis was used to determine the first seven dominated eigenvalues and eigenvectors. To obtain the strain values, the absolute values of the transverse displacements were normalized in descending order. The strain values were calculated according to Formula (1). To obtain a denser mesh, the strain fields were smoothed using regression analysis. An example comparison of experimental and numerical transverse displacements shapes and adequate strain shapes is shown in Figure 2.

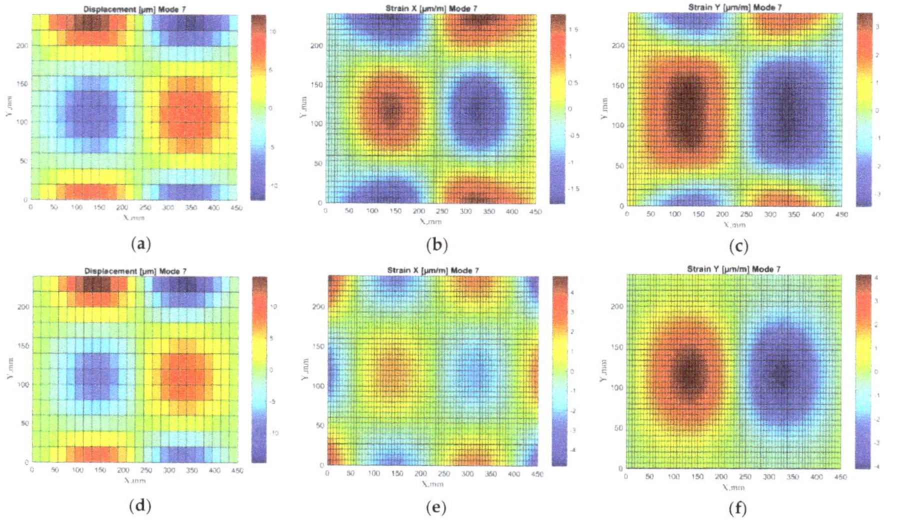

Figure 2. Transverse displacement shapes and strain fields in X and Y directions for 7th mode: experimental (**a**–**c**) and numerical (**d**–**f**).

3. Multi-Objective Optimization

3.1. OSP Problem

The primary goal of OSP is to localize the positions of the sensors in order to collect relevant information about the analyzed structure. This information, together with the available knowledge about the observed structure, allows for estimating its health condition which is, in turn, one of the main objectives of an SHM system. Usually, the process of determining sensor placement can be a part of the design of an SHM system [20]. This is often preceded by the evaluation of the structural characteristic and dynamics of the structure and also the selection of appropriate sensor types. The selection of the sensor type and the evaluation of the dynamic characteristic of the composite plate was a part of the research in [7], and is based on the application of strain gauges oriented horizontally or vertically to monitor modal strain of the composite plates. Usually, the next step comprises the necessity of sensor placement. This optimal placement strongly depends on the objective function of the unconstrained or constrained optimization process. There are many objective functions widely used for that purpose, e.g., a function based on Fisher matrix indices [24,25], MAC [20,24], condition number [24], modal kinetic-energy-based index [26], and many others, e.g., [9]. Most of them are mainly based on the assumption of widely understood criteria of independency, which allow one to select these locations of the sensors with the minimum redundancy of the captured data. In addition to that, from the perspective of the effectiveness of the whole SHM system, including the OSP problem, information on the health condition of the analyzed structure is the primary outcome of any SHM system.

Taking the above into account, the spatial placement of the sensor network can be based on the final evaluation of the SHM system, such as the effectiveness of a properly recognized heath condition of the structure. This allows one to establish adequate input to design parameters of the optimization process such as number of sensors, number of analyzed modes, etc., and to link them directly to the final effectiveness of the SHM system instead of using intermediate criteria such as statistical independency criteria, mutual information criteria, and any others [20,24–26].

In this study, the minimalization of the type-I and type-II errors were taken separately as the main criteria for the evaluation of the effectiveness of the SHM system [10,23,27]. The minimalization of type-I error allows for increasing the specificity of the system leading to a lower number of false alarms. The minimalization of the type-II error allows one to enhance the sensitivity of the system in accurate detection of the health condition minimizing any misses in the condition recognition. Keeping both of these criteria in balance requires a trade-off and is crucial for designing SHM systems. For instance, too-high sensitivity may lead to low specificity with a high number of false alarms, which may lead to information overload of the human operator of the SHM system. On the other hand, maintaining high specificity may lead to poor performance in the recognition of all faulty states of the structure. In the paper [23], it was shown that both sensitivity and specificity are often opposite counterparts in classification problems. Both types of errors are also naturally opposite to the minimalization criteria of the number of sensors used. Usually more relevantly selected and placed sensors lead to minimization of the type-I and type-II errors, but are more costly and require more data to analyze.

In many cases, the designer of the SHM system can be interested in solutions with enhanced sensitivity rather than false-alarm avoidance. In this case, a weighed objective function in multi-objective optimization allows for analysis of a specific region of the Pareto front, biasing the search towards a solution that optimizes weighed objective function. In this study, for the purpose of not losing the wider diversity of Pareto front solutions, no weights have been considered, leaving the designer to select results from the wider range of possible solutions and its design parameters.

Taking all of these multiple conflicting objectives into account, the multi-objective optimal-sensor-placement problem can be defined as follows, where the objective functions can be represented as

$$f_1(x_1, x_2, x_3) = \frac{FP(x_1, x_2, x_3)}{TN(x_1, x_2, x_3) + FP(x_1, x_2, x_3)}, \tag{2}$$

$$f_2(x_1, x_2, x_3) = \frac{FN(x_1, x_2, x_3)}{TP(x_1, x_2, x_3) + FN(x_1, x_2, x_3)}, \tag{3}$$

$$f_3(x_1) = n_x(x_1) + n_y(x_1), \tag{4}$$

where, $f_1,\ f_2,\ f_3$ are objective functions associated with the minimalization of type-I error, type-II error, and the minimalization of the total sensor used, respectively. The vector of $\boldsymbol{x} = [x_1.x_2,\ x_3]$ is a vector of design variables, where x_1 represents the total number of sensors, x_2 is number of used shape modes, and x_3 indicates the required minimal distance between the sensors measuring the strain displacements in the same direction, because the modal modes, especially for the healthy condition, are symmetrical. The minimum required distance allows for maintaining the spread of the sensors at the surface of the plate avoiding placing them in the same quadrant. *FP*, *TN*, *FN*, and *TP* correspond to false-positive, true-negative, false-negative, and true-positive cases, respectively [25]. Because the proposed SHM is a classification-based information system, these parameters are evaluated using classification statistics obtained using *k*-fold validation based on classification-error statistics [9,25]. The n_x, n_y corresponds to the number of strain sensors used for measuring along the x and y direction, respectively. Finally, multi-objective optimization is defined to minimize

$$F(\boldsymbol{x}) = \min([f_1,\ f_2,\ f_3]), \tag{5}$$

subject to

$$3 \leq x_1 \leq 10\ ;\ 2 \leq x_2 \leq 7\ ;\ 120\ \leq x_3 \leq 200,$$

$$d(sx_i(x,y), sx_j(x,y)) \geq x_3;\ d\left(sy_i(x,y), sy_j(x,y)\right) \geq x_3;\ for\ i \neq j,$$

where $sx_i(x,y)$, $sy_i(x,y)$ are the positions of the *i*-th and *j*-th strain gauges oriented horizontally, $sx_{i,}$ and vertically, sy_i, at the coordinates (x,y), and d is the Euclidean distance between the considered sensors. The range from 2 up to 10 sensors was used to maintain the consistency of this paper with previous studies [7]. The lower limit of 2 sensors requires the use of at least one sensor in every X or Y direction. The upper boundary has been selected based on the limited number of channels in our measurement path presented in [8]. The number of modes was also selected according to the number of modes used in previous research [7]. The upper boundary of the number of modes was limited to seven most-dominated modes in the frequency-response function; therefore, the 7th, 5th, 6th, 9th, 4th, 10th, and 8th mode from Table 1. were taken into account. The last mode from this set, the 8th mode, has about 10% of the energy of the highest mode. Other modes in the considered band rate have less than or equal to 5% of the energy in comparison to the most dominated mode. The value of minimum required distance was established based on the occurrence of the symmetric modes' shapes of which the strain values are the same (for the healthy condition of the plate) in four imaginary quadrants of the plate visible in Figure 1a. To force the placing sensor in separate imaginary quadrants, this value of 120 mm was selected as the low boundary. A similar approach was used in the study [7]. On the other hand, the upper limit of the required distance between the sensors was also limited to avoid considering unfeasible solutions where a large number of sensors with the same orientation could not be placed in the limited plate size.

To limit the number of possible solutions, it has been assumed that the position of the sensors are assigned to the closest node of the virtually superimposed grid on the composite plate, where nodes are vertically and horizontally spaced with the distance of 5 mm:

$$x, y \in N \mid\ x_i - x_j = 5k,\ y_i - y_j = 5m\ \ k, m \in N \tag{6}$$

Due to the dimensions of the side clamps of the considered composite plate and the size of strain gauges, the coordinate values are limited to the following limits: $20 \leq x \leq 470$ and $5 \leq y \leq 235$. The number of possible combinations of sensor placement, excluding constraint of the required distance between sensors is still huge and comprises, e.g., for three sensors 1.18×10^{10}, candidate solutions.

The multi-objective approach allows one to obtain decision variables of an SHM system to be independent of any weighed multi-objective function approach, which usually is difficult to determine at the early stage of the designing process. For SHM design purposes, the proposed multi-objective approach should return nondominated solutions using the approximation of the Pareto front for the considered individual objectives. Nondominated solutions obtained in this way give an opportunity for the designer to select appropriate parameters of the SHM system, adjusting it to the appropriate sensitivity level of fault recognition and at the same time keeping the appropriate level of false alarms and considering the appropriate cost of the system with the number of sensors used.

3.2. General Optimization Procedure

To estimate individual objectives of $F(\boldsymbol{x})$, the SHM effectiveness through the estimation of classification performance statistics need to be completed. For this purpose, a classifier should be prepared that takes information on the strain values for the given coordinates according to the sensor's network placement, and returns the information about the health condition of the composite plate. Separately, to avoid using the coordinates of the location of the sensors as design variables, the appropriate method of sensor placement was used internally in this proposed optimization procedure. This allows one to take advantage of the available knowledge on the effectiveness of the selected sensor-placement algorithm used for the composite plate based on the results obtained in [7].

The general optimization procedure is based on the application of a heuristic approach with elitist modification of the NSGA-II algorithm [18,20,28–31]. The algorithm allows approximation of the Pareto front which represents the equivalent trade-off between considered objectives f_1, f_2, f_3. In this paper, the optimization framework launch execution of other processes such as OSP and validation of classification. The general scheme of the optimization process is shown in Figure 3.

The NSGA-II executes the process of multi-objective optimization by defining an initial and any after generation, where individuals represent candidate sensor networks. These sensor networks are generated using the sensor-placement method based on selected design parameters and available strain maps for different shape modes obtained experimentally, according to [7], for the heathy condition of the composite plate. During the process of optimization, some of the parameters are treated as constant parameters, among others, they include the clamping area of the composite plate, where placing sensors is not possible. Another constant parameter is the relative additive noise level used in data augmentation for the purpose of obtaining balanced dataset for learning of the classifiers.

In this research, one of the algorithms for sensor placement shown and evaluated in the paper [7] was used. This algorithm is based on the selection of the position of the sensor candidates on the rank defined on the basis of the sum of values of the mode shapes for individual coordinates. Initially, the mode shapes are ordered according to their higher dominance in frequency-response function shown in Table 1. Next, individual strain maps of selected modes are normalized from 0 to 1, assigning 1 to the highest strain value. Then, the sum of the strains of selected modes is performed, and the strain values in the x and y direction are organized separately as vectors of values in descending order [7,8]. Afterwards, based on this order, the sensors are sequentially placed in the coordinates, which corresponds to the highest value in the mentioned vector, taking into account other requirements such as maintaining the required distance between sensors, etc. The effectiveness of this algorithm was compared with other algorithms, among others, with a modified Effective Independence Method [12]. In this comparison, we also took into account the influence of the noise level on the accuracy of damage detection. More details can be found in [7].

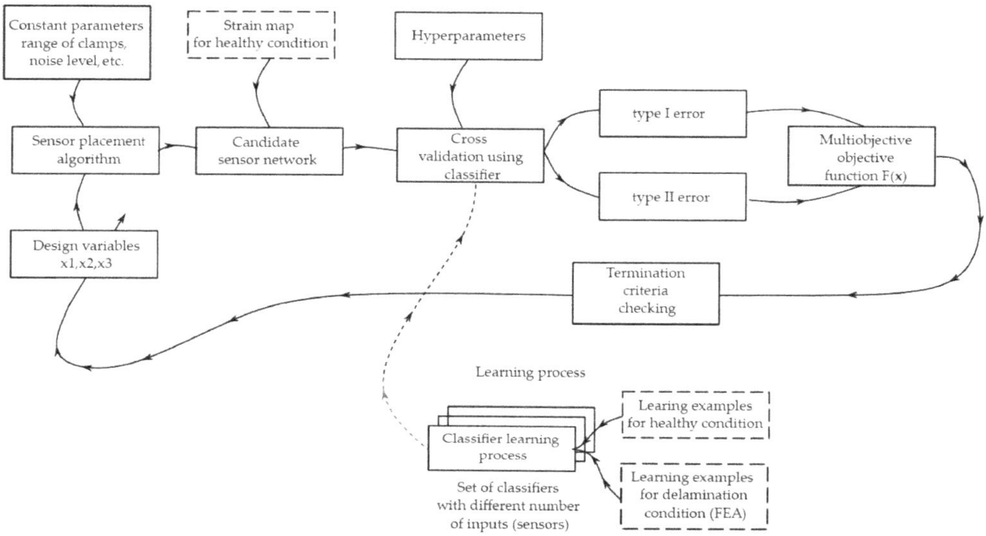

Figure 3. The general scheme of the multi-objective optimization process for optimal-sensor-placement problem.

In this study, the sensor network obtained using the above algorithm is mainly evaluated using the prepared classifier, which is validated using k-fold cross-validation. The results of this validation make it possible to compute the objective function f_1, f_2, which represents the number of type-I errors and type-II errors. This allows one to compute a multi-objective function for the individuals in the considered generation. The selection process of individuals for the next generation is performed using the tournament selection [23] with a fraction value of 0.8. During the creation of offspring generations, the mutation operator that allows individuals to generate feasible values that meet the bound constraints shown in Equations (5) and (6) was used [32]. The one-point crossover operation on the basis of an intermediate combination [32] was used.

In the process of the optimization evaluation, the following termination criteria were used. If the spread distance of the solutions across the Pareto front changes less in subsequent iterations than an established tolerance, the optimization process should stop. Otherwise, if the limited number of generations will be computed, the process may stop as well. Due to the heuristic characteristics of the conducted optimization, e.g., making the process of optimization independent of the process of the initial generation of the design variables, the process was repeated several times. In addition to the optimization process, the obtained Pareto front was evaluated using the standard deviation of the Pareto distance [28]—a standard deviation of distances between two closest neighbors for non-dominated solutions. In these calculations, the dominated solutions at the ends of the Pareto fronts are omitted. The second evaluation parameter was defined as the spread distance, which takes the standard deviation of the Pareto distance as an internal parameter. Apart from the Pareto distance [28], the spread distance [28] allows one to estimate the change of the Pareto fronts in a subsequent iteration of the optimization process. The lower the value, the more equally can distributed solutions on the Pareto front be reached.

The evaluation of the current health condition of the analyzed plate is based on the observed relevant sensor network, and based on the appropriate information obtained from the acquired knowledge about the dynamic modal behavior of the composite plate for healthy and damage conditions. This approach requires an appropriate representation of knowledge, which, in this case, takes the form of an intrinsic mathematical representation of the trained k nearest-neighbor classifier (k-NN).

For the purpose of the estimation of hyperparameters, a separated validation dataset was prepared. This validation set consisted of 54 sensor networks obtained individually for all combinations of the number of sensors from 2 to 10 and the number of modes from 2 to 7. These sensors positions were established based on the criteria shown in paper 7. These solutions are dominated solutions used only for the purpose of hyperparameters' selection. Every dataset for every sensor network contains a set of 2560 validation examples. The Bayesian optimization in the space of the considered hyperparameters was performed using Matlab© fitcknn routine. The objective function used for this purpose was the loss function estimated as 1-classification accuracy. In order to maintain good generalization properties of the classifier and to avoid overlearning, the estimated nearest classifier numbers were selected based on the criteria that this hyperparameter does not lead to a higher difference than 1% of the loss function between the best selection indicated by the routine, usually with a high number of neighbors and one that is selected manually with a very limited number of nearest neighbors. The analysis indicated that squared Euclidean, Minkowski, Euclidean, Chebyshev, Cityblock, and Mahalonobis do not indicate any significant differences between classification accuracy (more than 5%).

The selection of the hyperparameters mentioned above allows us to estimate the hyperparameters separately for the set of classifiers with different input numbers (number of sensors). Finally, the squared Euclidean distance with four nearest neighbors and inverse distance weighing were efficient to obtained high-accuracy classifiers for the set of used classifiers. For data sets that contain information from more than five modes, even three or two nearest neighbors can be applied without loss of classification accuracy.

The general optimization procedure described in this section was implemented in the Matlab© 2023a release (MathWorks Inc., Natick, MA, USA) using k-NN and NSGA-II Matlab built-in algorithms.

3.3. Data Preparation

The process of multi-objective optimization requires preparing a high-accuracy trained classifier which allows the evaluation of the value of the multi-objective function for all individuals. In this study, the classifiers are learned and validated during every iteration of the optimization procedure for the known number of inputs (sensor numbers) and selected strain modes. Due to the stochastic nature of the selection of the subset of the data sets for learning and validation for the same set of design variables, small differences can be observed for the same value of the input variable vector as a result of this stochastic nature. To decrease the variability of the results of the type-I and type-II errors due to the mentioned stochastic nature, a ten-fold cross-validation procedure was used to evaluate the classification statistics.

The proper evaluation of the objective function using type-I and type-II errors requires also a balanced data set preparation. For this purpose, the learning data set was augmented using replication of the learning examples with additive noise with normal distribution. The additive noise of the percentage k is defined as of 95% probability with the value ε falling within the range of $\varepsilon \in [-a, +a]$, which can be written as follows:

$$\Phi_{\mu,\sigma^2}(a) - \Phi_{\mu,\sigma^2}(-a) = 95\%, \tag{7}$$

where Φ_{μ,σ^2} is a cumulative normal distribution with the mean μ and the standard deviation σ^2. This requires the estimation of the standard deviation σ^2 of the zero-mean normal distribution based on the following equality equation:

$$\Phi^{-1}\left(97.5\%;\ \sigma^2\right) - \Phi^{-1}\left(2.25;\ \sigma^2\right) = 2a \tag{8}$$

where Φ^{-1} is the inverse cumulative normal distribution of the zero mean estimated at the boundaries of the 95% probability interval, and σ^2 is seeking standard deviation. Solving Equation (8) provides the required value for the standard deviation. In this case, for normalized values of strain maps, a 5% normal additive noise has a standard deviation of 2.551×10^{-2}.

The noise level was established on the basis of the initial evaluation of an accuracy of the damage recognition on the basis of a random subset of input variables with a lower number of sensors and a lower number of modes. The goal of the selection of this noise level was to select the highest value of the noise level to obtain the highest possible generalization properties of the classifier, which still guarantee obtaining a satisfactory level of damage recognition. For this purpose, a five-fold cross-validation procedure was used for the selected subset of input variables with different numbers of sensors, modal shape numbers, and different noise levels from 2 up to 5%. As a satisfactory level of the classifier validation accuracy, at least 95% of classification accuracy [25] was used. It has been observed that for the first three mode shapes and two sensors with a level of 5% of additive random noise, the validation accuracy leads to 95.6%. For other cases and lower noise levels, the validation outcome resulted in higher accuracy. Due to this, the noise level of 5% was attributed as a noise level in the augmentation process of a leaning dataset. The idea of using a high possible noise level results from using numerous numerical cases in the learning examples to maintain high generalization properties of the used classifiers.

3.4. Optimization Results

The results of the optimization process allows for estimating the approximation of the Pareto front for the multi-objective function shown in Equation (5). Finally, the front is represented by 38 solutions. The optimization process was terminated, since the average change in the spread of the Pareto solutions was less then 10^{-4}. The standard deviation of an average distance for the solution equals 10.41×10^{-2} and the spread distance equals 29.51×10^{-2}. The solutions have been obtained after 51 generations. To avoid the influence of the randomly generated initial population and also to avoid the influence of any other stochastic processes (e.g., a 10-fold validation process), the process was repeated several times. No significant changes have been observed in the repetitions of the optimization process.

Pareto solutions in the multi-objective function domain are shown in Figure 4a. The parameter space can be observed in Figure 4b. Additionally, some projections of the Pareto solutions for type-I and type-II errors are presented separately in Figures 4c and 4d, respectively. In Figure 4c,d, dominated solutions from the results of paper [7] have been added as well. These results have been obtained using the same OSP algorithm for single-objective optimization based on the aggregated t-norm evaluation function presented in [7] and with the constant required minimum distance of 127.5 mm.

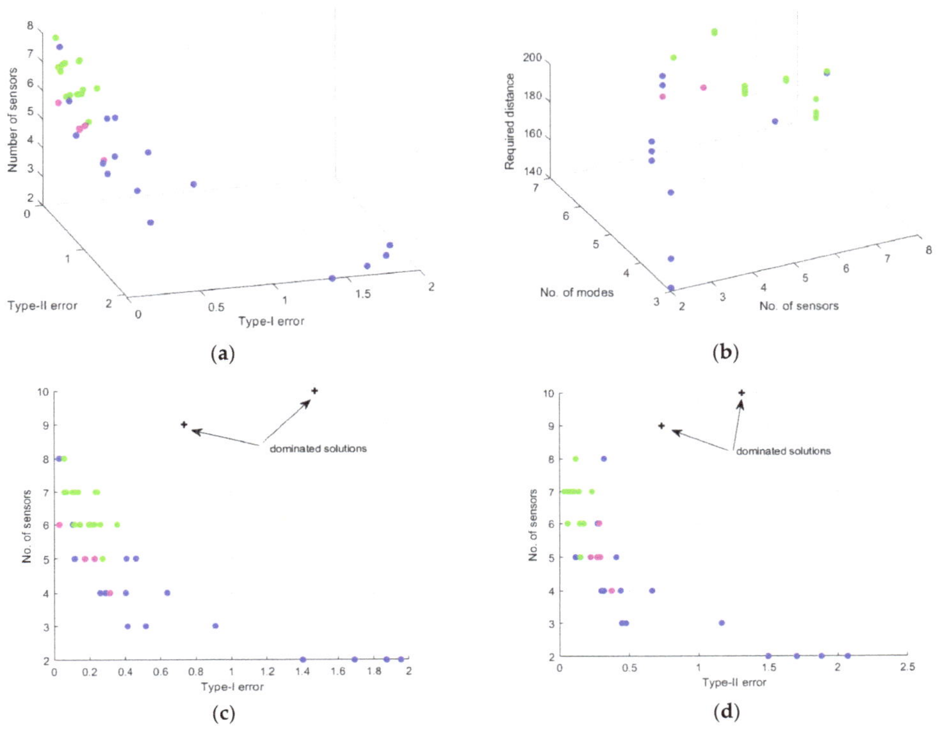

Figure 4. Results of the multi-objective optimization process: (**a**) scatter plot of the results representing approximation of three-dimensional Pareto front in the multi-objective function domain, (**b**) scatter plot of parameter space of design variables, (**c**) a side view Pareto front presenting perspective of type-I error value versus number of sensors, (**d**) a side view Pareto front presenting perspective of type-II error value versus number of sensors.

4. Analysis and Discussion

The analysis of the obtained results indicates that the process of multi-objective optimization based on Pareto front leads to many nondominated solutions, which allows for binding effectiveness of the SHM system with design variables. Based on this information, the designer of the SHM system may decide on the number of sensors in the individual direction, their placement, and the number of modes, simultaneously, while keeping an appropriate level of sensitivity in the system in damage detection as well as an appropriate level of possible false alarms.

To facilitate the SHM user to make decisions about the characteristic sensitivity, specificity, and the utilized sensor number, the obtained results of optimization can be further postprocessed to support this selection. A common-sense approach that can be used for that purpose is a Pareto principle that 80% of the effects come from 20% causes. Using this principle, the Pareto front solutions have been divided into the following subset:

$$\frac{\sum_{z\in A} g(z)}{\sum_{z\in Z} g(z)} \geq 0.8\ for\ g(z) = -\log(z+1)\ and\ g(z_1) \leq g(z_2) \leq \cdots \leq g(z_n) \tag{9}$$

where

$$\{z \in A \mid z = f_1 + f_2 + \alpha \cdot f_3\} \text{ or } \{z \in B \mid z = \beta \cdot f_1 + f_2\}.$$

Annotation z_i represents individual solutions of a Pareto front, $g(z)$ is the negation transformation function, and $z \in A$ represents the subset of Pareto solutions with a cumulative contribution of the top 20% solutions, when all objective functions f_1, f_2, f_3 are taken into account, but with limited influence of the number of sensors represented by weight α. Similarly, $z \in B$ represents the subset of Pareto solutions with cumulative contribution of the top 20% solutions when the sum of type-I and type-II errors are taken into account, with the limited influence of the type-I error represented by weight β. In the present study, the following exemplary weights were assumed, $\alpha = 0.2$ and $\beta = 0.5$. These weights can be appropriately estimated for the known costs of sensors and required unbalanced performance toward type-I or type-II errors. The Pareto solutions shown in Figure 4 represented by the magenta-filled circles show the top 20% of solutions for set A, and the green-filled circles show solutions for the weighed sum of the observed type-I and type-II errors represented by set B.

The analysis of Figure 4b together with Figure 4c,d of the obtained results indicates that optimal solutions, which belong to the set A set, require the use of four to six sensors and the first six dominated modes (7th, 5th, 6th, 9th, 4th, 10th). An exemplary sensor network for this subset of possible sensor networks is presented in Figure 5.

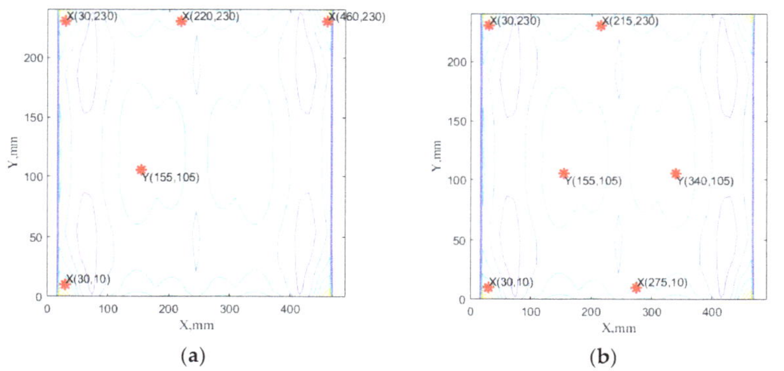

Figure 5. The example of nondominated solutions from Pareto solutions belonging to subset A: (**a**) the case with 5 sensors, 4 sensors in x direction, and 1 sensor in y direction for the required distance of 188 mm and the first six modes; (**b**) the case with 6 sensors, 4 sensors in x direction, and 2 sensors in y direction for the required distance of 184 mm. In both figures, sum of first six dominated mode shapes are presented (7th, 5th, 6th, 9th, 4th, 10th) as a background contour plot.

In that case, when set B is considered, five to eight sensors are necessary to consider with the use of at least first five to seven dominated modes. This can be observed in Figure 4c,d, and also in Figure 6, where the sum of the errors versus the number of sensors are presented.

The analysis of the design variables of the solutions belonging to set B, shown in Figure 4b. indicates that for six to seven sensors it requires consideration of at least six modes to effectively recognize the healthy conditions of the composite plate. The solution with the minimal error type-II is presented in Figure 7a. This sensor networks consist of seven sensors, five in the x direction and two in the y direction for the required distance of 177 mm and the first five dominated modes (7th, 5th, 6th, 9th, 4th). The exemplary sensor network for solutions of set B is presented in Figure 7.

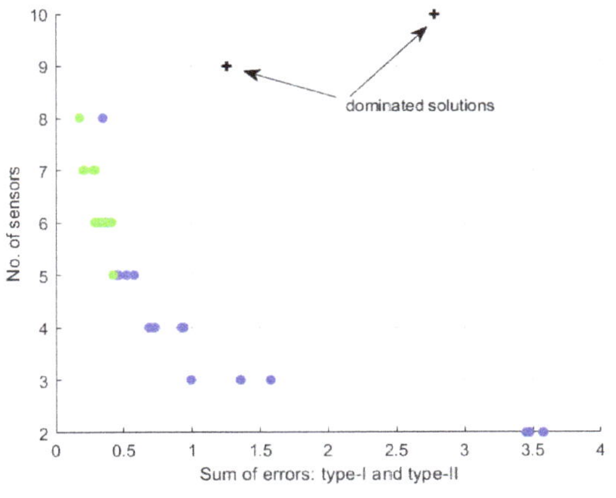

Figure 6. The sum of type-I and type-II errors versus total number of sensors.

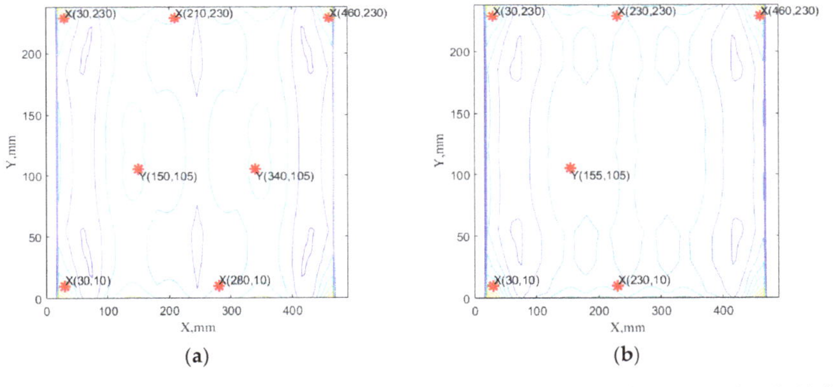

Figure 7. The example of nondominated solutions from Pareto solutions belonging to subset B: (**a**) the case with 7 sensors, 5 in x direction and 2 in y direction for the required distance of 177 mm and first five dominated modes (7th, 5th, 6th, 9th, 4th). (**b**) The case with 6 sensors, five in x and 1 in y directions for the required distance of 194 mm and the first seven dominated modes (7th, 5th, 6th, 9th, 4th, 10th, 8th).

This case study is based on the experience gathered in the ongoing studies presented in [7,8]. In paper [7], specific optimal-sensor-placement algorithms and different ways of evaluating sensor networks using different aggregated criteria were selected and compared. Additionally, the selected classification models that allow the mapping of signals from sensor networks to the evaluation of the condition of the composite plate were selected and validated. In [8], validation of optimal-sensor-placement networks based on the experimental data was shown. It should be emphasized that the ongoing studies presented in [7,8] were aimed at finding the optimal solutions based on the accuracy of composite-plate state recognition, but the candidate sensor-placement networks were selected based on the intermediate criteria such as Fisher matrix indices, etc. These results are also presented in Figures 4c,d and 6. In the present study, the selection of the optimal sensor network is driven by the evaluation of the SHM system using minimization of type-I and type-II errors and the number of sensors. Comparison of the results shows the advantage of the optimization framework presented in this study over the previous results where the optimal-sensor-network selection was not dependent on the final outcome of the SHM system.

5. Conclusions

This study presents the results of multi-objective optimization of the sensor-placement network evaluated using the effectiveness of the SHM system. This effectiveness was achieved by evaluating the sensitivity to recognize the damage condition of the composite plate and also by minimizing false alarms. Additionally, this evaluation can be completed using the minimization of the sensor number, which allows for finding the trade-off between the cost and performance of an SHM system. A specific genetic approach using NSGA-II allowed for recognizing Pareto front solutions and allowed the designer of an SHM system the opportunity to select appropriate values of design variables. The examples with five, six, and seven sensors in the sensor network were demonstrated as the solutions of the performed optimization in this study. This decision can be taken by directly linking the selected results with the design variables. In this study, a selection of design variables based on the Pareto principle is also proposed.

Author Contributions: Conceptualization, T.R., A.K. and S.R.; methodology, T.R., S.R., A.K. and M.Ś.; software, T.R., M.Ś. and S.R.; validation, T.R. and M.Ś.; formal analysis, T.R. and M.Ś.; investigation, S.R.; resources, S.R.; data curation, S.R., T.R. and M.Ś.; writing—original draft preparation, T.R., A.K. and S.R.; writing—review and editing, T.R., A.K. and S.R.; visualization, T.R., A.K. and S.R.; project administration, S.R.; funding acquisition, S.R. and A.K. All authors have read and agreed to the published version of the manuscript.

Funding: This research was partially funded by the Latvian Council of Science project "Smart Materials, Photonics, Technologies and Engineering Ecosystem" (project No. VPP-EM-FOTONIKA-2022/1-0001).

Data Availability Statement: The data that support the findings of this study are available upon reasonable request. The data are not publicly available due to ongoing research process.

Conflicts of Interest: The authors declare no conflicts of interest. The funders had no role in the design of the study; in the collection, analyses, or interpretation of data; in the writing of the manuscript; or in the decision to publish the results.

References

1. Tinga, T.; Loendersloot, R. Aligning PHM, SHM and CBM by understanding the physical system failure behaviour. In Proceedings of the PHM Society European Conference, Nantes, France, 8–10 July 2014.
2. Broer, A.A.R.; Benedictus, R.; Zarouchas, D. The need for multi-sensor data fusion in structural health monitoring of composite aircraft structures. *Aerospace* **2022**, *9*, 183. [CrossRef]
3. Sadhu, A.; Peplinski, J.E.; Mohammadkhorasani, A.; Moreu, F. A review of data management and visualization techniques for structural health monitoring using BIM and virtual or augmented reality. *J. Stuct. Eng.* **2022**, *149*, 03122006. [CrossRef]
4. Fawad, M.; Salamak, M.; Poprawa, G.; Koris, K.; Jasinski, M.; Lazinski, P.; Piotrowski, D.; Hasnain, M.; Gerges, M. Automation of structural health monitoring (SHM) system of a bridge using BIMification approach and BIM-based finite element model development. *Sci. Rep.* **2023**, *13*, 13215. [CrossRef] [PubMed]
5. Bado, F.M.; Tonelli, D.; Poli, F.; Zonta, D.; Casas, J.R. Digital twin for civil engineering systems: An exploratory review for distributed sensing updating. *Sensors* **2022**, *22*, 3168. [CrossRef]
6. Janeliukstis, R.; Mironovs, D.; Safronovs, A. Statistical structural integrity control of composite structures based on an automatic operational modal analysis—A review. *Mech. Compos. Mater.* **2022**, *58*, 181–208. [CrossRef]
7. Ručevskis, S.; Rogala, T.; Katunin, A. Optimal sensor placement for modal-based health monitoring of a composite structure. *Sensors* **2022**, *22*, 3867. [CrossRef]
8. Ručevskis, S.; Rogala, T.; Katunin, A. Monitoring of damage in composite structures using an optimized sensor network: A data-driven experimental approach. *Sensors* **2023**, *23*, 2290. [CrossRef] [PubMed]
9. Barthorpe, R.J.; Worden, K. Emerging Trends in Optimal Structural Health Monitoring System Design: From Sensor Placement to System Evaluation. *J. Sens. Actuator Netw.* **2020**, *9*, 31. [CrossRef]
10. Hassani, S.; Dackermann, U. A Systematic Review of Optimization Algorithms for Structural Health Monitoring and Optimal Sensor Placement. *Sensors* **2023**, *23*, 3293. [CrossRef]
11. Wang, Y.; Chen, Y.; Yao, Y.; Ou, J. Advancements in Optimal Sensor Placement for Enhanced Structural Health Monitoring: Current Insights and Future Prospects. *Buildings* **2023**, *13*, 3129. [CrossRef]
12. Kammer, D.C. Sensor placement for on-orbit modal identification and correlation of large space structures. *J. Guid. Control Dyn.* **1991**, *14*, 251–259. [CrossRef]

13. Azin Mehrjoo, B.; Moaveni, B.; Papadimitriou, C.; Khan, U.; Rife, J. Iterative optimal sensor placement for adaptive structural identification using mobile sensors: Numerical application to a footbridge. *Mech. Syst. Signal Process.* **2023**, *200*, 110556.
14. Zhang, B.-Y.; Ni, Y.-Q. A data-driven sensor placement strategy for reconstruction of mode shapes by using recurrent Gaussian process regression. *Eng. Struct.* **2023**, *284*, 115998. [CrossRef]
15. Lin, T.-Y.; Tao, J.; Huang, H.-H. A Multiobjective Perspective to Optimal Sensor Placement by Using a Decomposition-Based Evolutionary Algorithm in Structural Health Monitoring. *Appl. Sci.* **2020**, *10*, 7710. [CrossRef]
16. Civera, M.; Pecoreelli, M.L.; Ceravolo, R.; Surace, C.; Fragonara, L.Z. A multi-objective genetic algorithm strategy for robust optimal sensor placement. *Comput.-Aided Civ. Infrastruct. Eng.* **2021**, *36*, 1185–1202. [CrossRef]
17. de Souza Mello, F.M.; Pereira, J.L.J.; Gomes, G.F. Multi-objective sensor placement optimization in SHM systems with Kriging-based mode shape interpolation. *J. Sound Vib.* **2024**, *568*, 118050. [CrossRef]
18. Yang, C.; Xia, Y. Interval Pareto front-based multi-objective robust optimization for sensor placement in structural modal identification. *Reliab. Eng. Syst. Saf.* **2024**, *242*, 109703. [CrossRef]
19. Deb, K.; Jain, H. An Evolutionary Many-Objective Optimization Algorithm Using Reference-Point-Based Nondominated Sorting Approach, Part I: Solving Problems with Box Constraints. *IEEE Trans. Evol. Comput.* **2014**, *18*, 577–601. [CrossRef]
20. Deb, K. *Multi-Objective Optimization Using Evolutionary Algorithms*; John Wiley & Sons, Ltd.: Chichester, UK, 2001.
21. Zhang, Z.; Peng, C.; Wang, G.; Ju, Z.; Ma, L. A new optimal sensor placement method for virtual sensing of composite laminate. *Mech. Syst. Signal Process.* **2023**, *195*, 110319. [CrossRef]
22. Guratzsch, R.F.; Mahadevan, S. Structural health monitoring sensor placement optimization under uncertainty. *AIAA J.* **2010**, *48*, 1281. [CrossRef]
23. Downey, A.; Hu, C.; Laflamme, S. Optimal sensor placement within a hybrid dense sensor network using an adaptive genetic algorithm with learning gene pool. *Struct. Health Monit.* **2018**, *17*, 450–460. [CrossRef]
24. Gomes, G.F.; de Almeida, F.A.; da Silva Lopes Alexandrino, P.; da Cunha, S.S.; de Sousa, B.S.; Ancelotti, A.C. A multi-objective sensor placement optimization for SHM systems considering Fisher information matrix and mode shape interpolation. *Eng. Comput.* **2019**, *35*, 519–535. [CrossRef]
25. Tharwat, A. Classification assessment methods. *Appl. Comput. Inform.* **2021**, *17*, 168–192. [CrossRef]
26. Nong, S.-X.; Yang, D.-H.; Yi, T.-H. Pareto-Based Bi-Objective Optimization Method of Sensor Placement in Structural Health Monitoring. *Buildings* **2021**, *11*, 549. [CrossRef]
27. Yang, Y.; Chadha, M.; Hu, Z.; Todd, M.D. An optimal sensor design framework accounting for sensor reliability over the structural life cycle. *Mech. Syst. Signal Process.* **2023**, *202*, 110673. [CrossRef]
28. Deb, K.; Pratap, A.; Agarwal, S.; Meyarivan, T. A fast and elitist multi-objective genetic algorithm: NSGA-II. *IEEE Trans. Evol. Comput.* **2002**, *6*, 182–197. [CrossRef]
29. An, H.; Youn, B.D.; Kim, H.S. A methodology for sensor number and placement optimization for vibration-based damage detection of composite structures under model uncertainty. *Compos. Struct.* **2022**, *279*, 114863. [CrossRef]
30. An, H.; Youn, B.D.; Kim, H.S. Optimal placement of non-redundant sensors for structural health monitoring under model uncertainty and measurement noise. *Measurement* **2022**, *204*, 112102. [CrossRef]
31. Mendler, A.; Döhler, M.; Ventura, C.E. Sensor placement with optimal damage detectability for statistical damage detection. *Mech. Syst. Signal Process.* **2022**, *170*, 108767. [CrossRef]
32. Deep, K.; Singh, K.P.; Kansal, M.L.; Mohan, C. A real coded genetic algorithm for solving integer and mixed integer optimization problems. *Appl. Math. Comput.* **2009**, *212*, 505–518. [CrossRef]

Disclaimer/Publisher's Note: The statements, opinions and data contained in all publications are solely those of the individual author(s) and contributor(s) and not of MDPI and/or the editor(s). MDPI and/or the editor(s) disclaim responsibility for any injury to people or property resulting from any ideas, methods, instructions or products referred to in the content.

applied sciences

MDPI

Article

Label-Free Fault Detection Scheme for Inverters of PV Systems: Deep Reinforcement Learning-Based Dynamic Threshold

Giup Seo, Seungwook Yoon, Junyoung Song, Ekta Srivastava and Euiseok Hwang *

Gwangju Institute of Science and Technology (GIST), 123, Cheomdangwagi-ro, Buk-gu, Gwangju 61005, Republic of Korea
* Correspondence: euiseokh@gist.ac.kr; Tel.: +82-62-715-3223

Abstract: Generally, photovoltaic (PV) fault detection approaches can be divided into two groups: end-to-end and threshold methods. The end-to-end method typically uses a deep neural network (DNN) to learn fault patterns from labeled datasets, which directly detect whether faults occur or not. The threshold method first estimates power generation and uses thresholds to detect atypical deviations of measured values from estimated ones. The former method heavily relies on fault-labeled data and, therefore, requires the collection of abnormal event records, which is usually difficult, due to the sparseness of these events. The latter method typically uses statistical approaches, such as 3-sigma, to find thresholds, and it can be practically utilized without fault labels. However, setting a threshold with a proper confidence interval is still challenging, as PV power generation is sensitive to variations in environmental conditions, such as irradiance, ambient temperature, wind speed and humidity. In this paper, we propose a novel deep reinforcement learning (DRL)-based label-free fault detection scheme in which thresholds are dynamically assigned with suitable confidence intervals under varying environmental conditions. Various weather properties were used as input features (i.e., states) to a DRL agent, and proper thresholds were estimated in real time from the actions of the DRL agent. To this end, a reward function was designed for learning proper thresholds without fault labels under different weather conditions. To evaluate the performance of the proposed scheme, the PV dataset of the National Institute of Standards and Technology (NIST) was used, as it includes paired records of local weather and PV generations. The DRL-based scheme was compared with static and conventional dynamic threshold methods, based on statistical approaches. The results revealed that the proposed scheme outperformed the existing methods, providing a 5.67% higher F_1-score in the NIST dataset.

Keywords: label-free fault detection; photovoltaic systems; deep reinforcement learning; dynamic threshold

1. Introduction

The capacity expansion of renewable energy sources is among the sustainable methods that can mitigate global warming issues [1]. Photovoltaic (PV)-based energy generation systems are broadly deployed due to their affordability and the growing concerns with regard to the environment [2]. The PV system, which has a scalability characteristic, can be installed in various areas, such as households, buildings, and utilities. Various approaches have been investigated to effectively accept such PV systems, including studies on energy forecasting [3,4], optimization [5,6], and monitoring [7] approaches. In particular, developing PV health monitoring systems is essential for the stable management of PV systems, as PV faults may cause energy losses or fatal accidents. A site in the UK experienced energy losses of 18.9% in a year due to PV system faults [8]. The faults often cause fires and can cause fatal problems in power grids [9,10].

Various fault detection approaches for PV systems have been studied based on the following approaches: modeling [11–15], electrical signal analysis [16–20], and electrical circuit simulation [21–24]. Each approach has its pros and cons according to user perspectives. In particular, model-based approaches are often employed because their simplicity, compared with electrical signal analysis or circuit analysis approaches. The model-based approaches can be typically divided into end-to-end and threshold methods. First, the end-to-end methods directly decide whether faults occur or not, which requires a huge amount of fault-labeled data to train fault detection models, such as deep neural network (DNN). Recently, to mitigate this issue, semi-supervised learning schemes were studied for PV fault detection and showed comparable performance with those of fully supervised learning methods [25,26]. Second, the threshold methods estimate the power generation of PV systems, and compare the estimations with actually measured power generation. By analyzing the differences, users detect different abnormalities, such as shading effects [27], snow accumulation [28], maximum power point tracking error [29], and faulty conditions of DC–AC converters [15,30,31]. The threshold methods have the advantage of working well even without fault-labeled data. However, the thresholds are typically predetermined [31], or determined by monitoring data profiles [30], which has room for improvement.

The accurate detection of faults and the reduction of false alarms are crucial for the effective operation of PV systems. Therefore, properly setting threshold confidence intervals is among the essential processes of the model-based approaches. However, renewable energy sources, such as wind or PV systems, are significantly affected by local weather conditions. Thus, users experience difficulties in manually setting thresholds under various weather conditions. To deal with this problem, in this paper, we propose a deep reinforcement learning (DRL)-based dynamic threshold scheme, in which an AI agent continuously updates thresholds depending on weather changes. We employed a proximal policy optimization (PPO) agent and used various fields of weather data, including irradiance, ambient temperature, humidity, and wind speed. The agent consists of continuous state and action spaces. A reward function was designed to learn proper thresholds without fault labels, which is useful, since fault-labeled data is very limited. To evaluate the proposed scheme, we used the PV data of the National Institute of Standards and Technology (NIST) [32], which is publicly accessible. The proposed model was compared to the existing static and dynamic threshold schemes. To the best of our knowledge, this is the first study in which a DRL agent with continuous action space was used as a label-free fault detection scheme to detect inverter faults in PV systems. The contributions of this paper are summarized as follows:

- A DRL-based dynamic threshold scheme is proposed for fault detection in PV inverters. The scheme automatically sets proper thresholds for the accurate detection of faults and the reduction of false alarms under various weather conditions. The scheme can mitigate the burdens of manually assigning confidence intervals for thresholds.
- A customized reward function was designed for learning proper thresholds for the detection of inverter faults under various weather conditions without fault labels, which can be helpful under the scarcity of fault-labeled data.

The rest of this paper is structured as follows: Section 2 describes precedent research. Section 3 conducts data analysis, and introduces the proposed scheme. Section 4 discusses the experimental results. Section 5 provides the conclusion of the study.

2. Related Works

2.1. Threshold-Based Fault Detection in PV Systems

Fault detection in PV systems has been investigated from various perspectives to avoid energy losses and accidents [33]. Image-based fault detection schemes were proposed with the rapid development of image sensors and AI approaches, such as convolution neural networks (CNNs) [34–40]. Despite the accurate diagnostic ability of such schemes, their infrastructural installations are very expensive [41]. On the other hand, model-based detection schemes using thresholds have the advantages of low cost and rapid detection, and they are regarded as leading methods in fault detection, due to their simplicity and decent performance [42].

However, the model-based approaches using thresholds face many challenges, because the performance of fault detection significantly depends on threshold settings [43]. Properly setting thresholds is essential to ensure accurate detection. Thus, many threshold methods have been proposed to improve accuracy [15,42,44–47]. Despite the efforts, there are still some issues to cover. First, fixed threshold methods show difficulties in practical situations, in which volatility and uncertainty exist. For example, Bressan et al. proposed a scheme for setting thresholds and detecting shading faults, where faults were detected if there were drops greater than 10% [45]. Rouani et al. introduced a shading fault detection method in a grid-connected PV system, in which vertices principal component analysis (VPCA) was compared with a standard PCA [48]. However, a trade-off relation existed between less missed detection and more false alarms, which is a critical problem of fixed threshold methods.

To deal with this problem, Platon et al. proposed an online fault detection method, in which different dynamic thresholds were used according to solar irradiance intervals [15]. Wang et al. improved setting thresholds, based on solar irradiance, by considering more features, such as current and voltage [49]. Pan et al. introduced a fault diagnosis threshold method based on non-parametric kernel density estimation (NKDE), which obtained thresholds by assigning the confidence values of models [42]. The above approaches require assigning confidence intervals for setting thresholds. Therefore, there is still a limitation in regard to finding proper confidence intervals for the PV system and situations where volatility and uncertainty exist.

In this paper, we propose a DRL-based dynamic threshold scheme in which a DRL agent automatically finds proper thresholds without assigning confidence intervals (e.g., 3-sigma) for various situations, while considering the various weather conditions, such as ambient temperature, irradiance, humidity, and wind speed. Using the proposed scheme, users have no burdens for manually setting thresholds.

2.2. Deep Reinforcement Learning-Based Fault Detection

The Markov decision process (MDP) is a general framework of reinforcement learning (RL), which comprises an agent and environment where the agent receives reward signals, such as positive or negative feedback, caused by its action at each state [50]. The agent continuously learns what an optimal action is for each state through interactions with the environment. As decision makers, the advantage of RL agents is consideration of long-term planning while deciding on an optimal action. Therefore, they choose the actions that can maximize cumulative rewards.

Huang et al. first introduced a value-based DRL anomaly detector, in which a Deep Q-Function Network (DQN) algorithm was utilized, and two actions for classifying abnormal and normal states were given [51]. Yu and Sun proposed a policy-based anomaly detector, in which Asynchronous Advantage Actor-Critic (A3C) algorithm was used, and the action space had two actions [52]. The policy-based anomaly detector outperformed the value-based approach. Khanzaeli applied a valued-based anomaly detector with state-space models to structural health monitoring, where Q-learning was used, and artificial anomalies were generated for test purposes [53].

Although the existing RL-based approaches had decent performances, they only used discrete action spaces with fault labels, which was the same problem as classification problems. Thus, it is difficult to apply these approaches under the scarcity of fault-labeled data. Moreover, these approaches do not utilize the continuous action space of DRL agents. In this study, we aimed to mitigate the difficulty of assigning threshold confidence by using a DRL-based dynamic threshold scheme. This can be applied under the scarcity of fault-labeled data because this scheme operates in an unsupervised manner (without fault labels). In the proposed scheme, a setting of continuous state and action spaces is applied, and the threshold values of each state are outputs of the DRL agent.

3. Methodology

Model-based approaches for inverter faults distinguish fault conditions by comparing residuals with thresholds representing normal states [15,54]. In this study, we designed an estimation model with an input of dominant weather data and proposed a DRL-based dynamic threshold model that can learn proper thresholds for detecting inverter faults without fault-labels under varying weather conditions. Figure 1 shows the configuration of the proposed threshold scheme, and Figure 2 compares the proposed scheme with conventional approaches. In the proposed scheme, first we performed data preprocessing to deal with missing data with scaling. For handling the missing data, we used linear interpolation, while deleting the periods that could not be recovered because of insufficient information. For data scaling, we utilized the minimum–maximum normalization. Then, we selected the weather data that affected power generation by analyzing the correlation coefficients between weather data and power generation. Once weather data was selected, we trained a DNN estimator to estimate power generation values from weather data. Afterwards, we trained a DRL agent to set dynamic thresholds for detecting the inverter faults of PV systems.

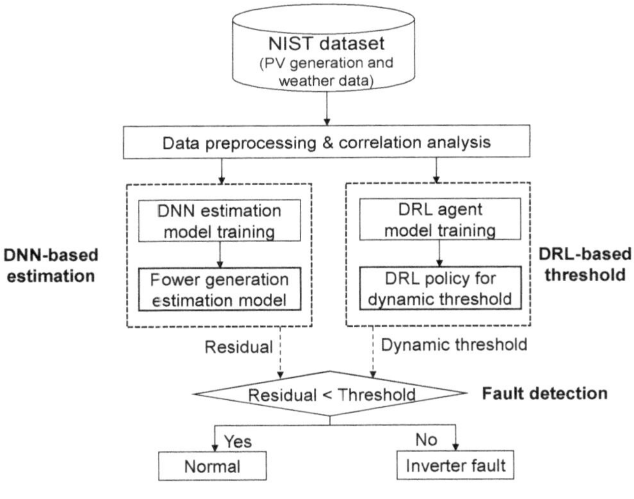

Figure 1. The proposed DRL-based threshold scheme for detecting the inverter faults of PV systems.

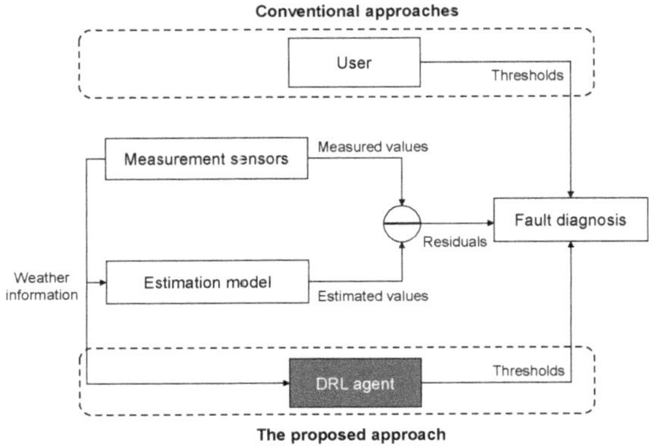

Figure 2. Structure comparison of model-based approaches for detecting inverter faults of PV systems.

3.1. Dataset

3.1.1. Data Description

The PV datasets of the NIST are publicly accessible, and they provide various measurement data, such as PV power generation and weather data. The measurement data were collected for four years, from January 2015, to December 2018, obtained at one-minute intervals. There are three types of PV arrays in the NIST dataset: canopy, ground, and roof. The specifications of PV arrays are explained in detail in [32].

3.1.2. Data Analysis

First, we analyzed the linear relationship between the PV generation data and weather data to identify which weather feature dominantly affected the power generation of PV systems in the NIST dataset, by investigating Pearson's correlations as follows:

$$r_{XY} = \frac{\sum_{i=1}^{N}(X_i - \overline{X})(Y_i - \overline{Y})}{\sqrt{\sum_{i=1}^{N}(X_i - \overline{X})^2}\sqrt{\sum_{i=1}^{N}(Y_i - \overline{Y})^2}} \tag{1}$$

where r_{XY} denotes a correlation coefficient between different random variables (irradiance, humidity, and etc.), X denotes PV generation, Y denotes environment variables, and i denotes a sample index, where N is the number of samples for each random variable. $\overline{X}$ and $\overline{Y}$ denote sample mean values, respectively.

Irradiance, ambient temperature, relative humidity, wind speed, and snow depth were considered to be environment variables. Canopy was selected among the three PV types. The power generation data was measured at the inverter of the PV system. The heatmap of the correlation matrix for PV generation and the local weather data fields during 2015 in the NIST dataset are illustrated in Figure 3. We could identify that irradiances mainly affected the power generation of the PV system, while ambient temperature and humidity showed weaker correlations with power generation. Even though the ambient temperature showed positive correlations with the power generation, because irradiance effects were related to temperature, it was reported that the temperature increase actually reduced power generation output [55]. To observe this, we empirically divided the data into different irradiance intervals (e.g., 0–100, 100–200, ..., 900–1000). We could observe that power generation decreased, and variation increased, when ambient temperature increased, as shown in Figure 4. This meant that the ambient temperature had negative correlations with power generation when it was decoupled from irradiance. Figure 5 also shows that the negative correlation relationship became stronger in high irradiance intervals. Especially, wind speed and snow depth seemed to have no correlations with power generation. Nevertheless, we observed that snow depth affected power generation in some periods, which is illustrated in Figure 6. The weather station, which measured the snow depth, was somewhat far from the PV systems. We had relatively fewer snow fall event periods compared to other weather conditions. This might have caused the very weak correlation between power generation and snow depth. For the implementation of our proposed scheme and benchmark models, we excluded the few periods when snow fell. Regarding wind speed, we could not directly observe an impact on power generation. However, it was reported that wind speed had an extreme effect the temperature of PV cells, equivalent to a cooling effect of 15–20 °C at wind speeds of 10 m/s [56,57], closely related to the efficiency of PV panels [55]. Therefore, we considered four environmental variables: irradiance, ambient temperature, humidity, and wind speed.

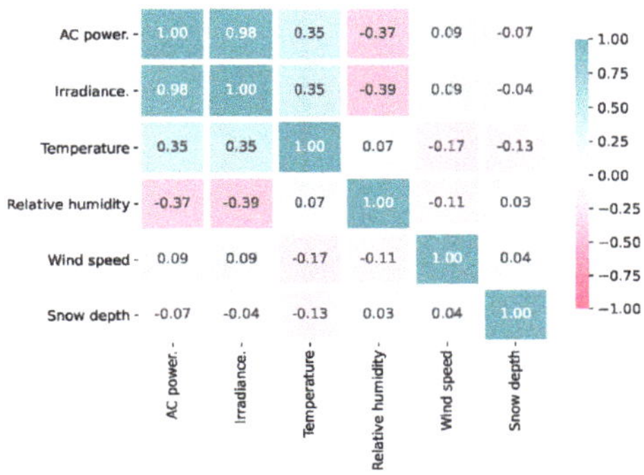

Figure 3. Heatmap of a correlation matrix for PV generation and local weather data fields in the NIST dataset.

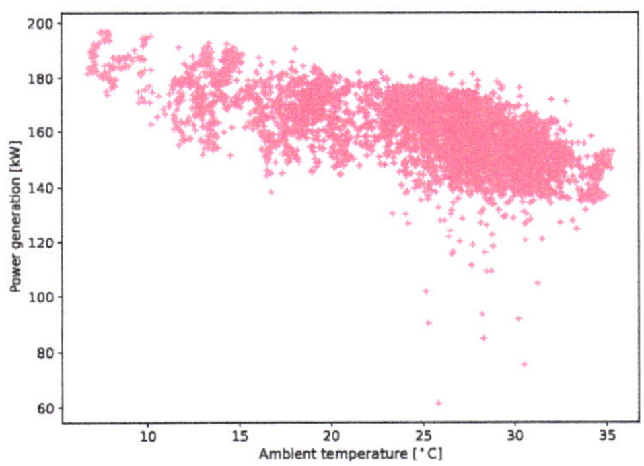

Figure 4. Scatter plot of power generation and ambient temperature within the irradiance interval (600∼700 [W/m^2]).

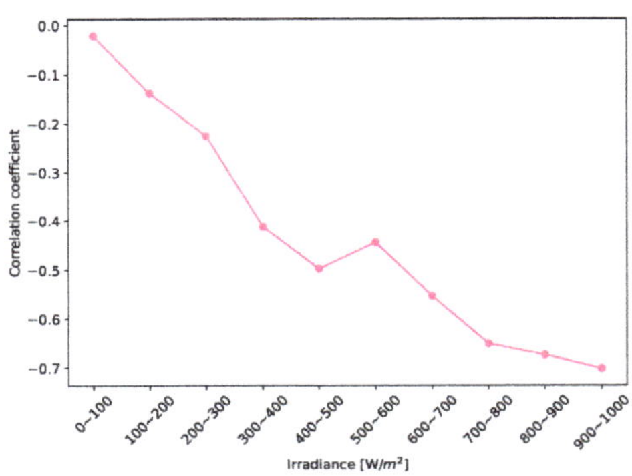

Figure 5. Correlation coefficients between ambient temperature and power generation in different irradiance intervals.

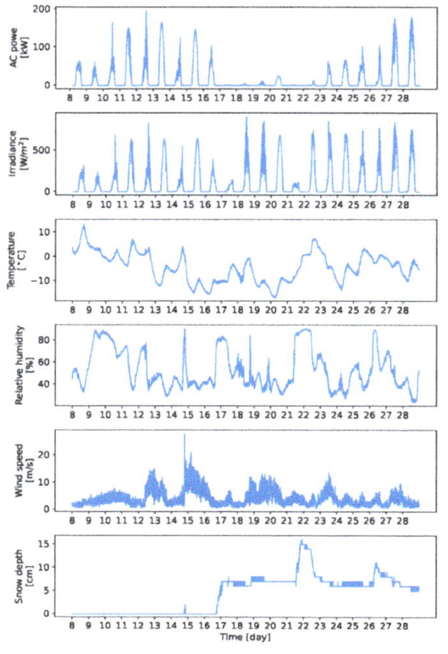

Figure 6. Profiles of power generation and local weather for 21 days in February 2015.

In addition, we observed some inverter faults in the dataset, and they were labeled by an expert who was managing the PV systems. It was also observed that AC power, current and voltage sharply decreased and increased in the PV system within a short time period, while irradiance showed a normal pattern, as illustrated in Figure 7.

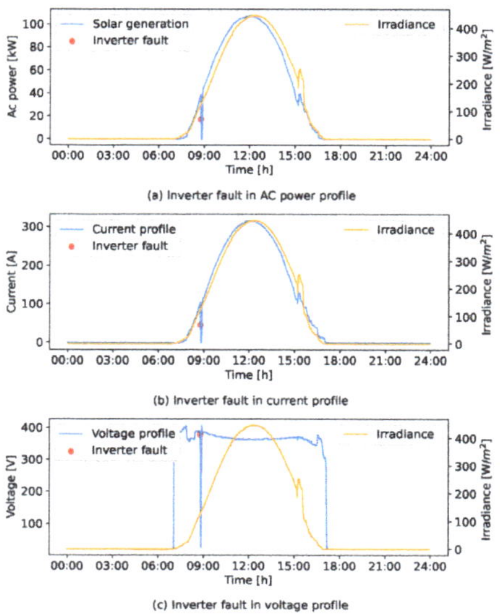

Figure 7. Typical inverter fault patterns observed in the NIST dataset.

3.2. Proposed Scheme

3.2.1. Estimation Model

We used a DNN as a estimation model to acquire residuals between the estimated values and measured values. Moreover, we used three separate models for AC power, current, and voltage. The weather information (irradiance D_t, temperature T_t, wind speed W_t, humidity H_t) was input to the estimation models, and the outputs were estimated values, such as $\hat{P}_t$, $\hat{I}_t$ and $\hat{V}_t$. Each model consisted of 4 layers, where the input layer had 4 nodes, two hidden layers had 128 nodes, and the output layer had 1 node, and the rectified linear unit (ReLU) function was used as an activation function for the hidden layers. We used Adam optimizer as a learning algorithm and we set a fixed learning rate at 0.0001.

$$\hat{P}_t = f_P(D_t, T_t, W_t, H_t), \tag{2}$$

$$\hat{I}_t = f_I(D_t, T_t, W_t, H_t), \tag{3}$$

$$\hat{V}_t = f_C(D_t, T_t, W_t, H_t) \tag{4}$$

We split the NIST PV canopy dataset into training, validation, and test datasets for model learning. The canopy PV dataset of 2015 was used for training and validation, where we selected the last day of every week as a validation set, (i.e., training and validation ratio was 6:1). The dataset of 2016 was used for testing. When we trained and evaluated the estimation model, the time period when irradiance somewhat existed was considered, from 9:00 a.m. to 16:00 p.m.

We utilized mean absolute error (MAE) and symmetric mean absolute percentage error (SMAPE) as metrics for the estimation performance evaluation.

$$\text{MAE} = \frac{1}{N}\sum_{t=1}^{N}|\hat{X}_t - X_t| \tag{5}$$

$$\text{SMAPE} = \frac{100}{N}\sum_{t=1}^{N}\frac{|\hat{X}_t - X_t|}{(|X_t|+|\hat{X}_t|)/2} \tag{6}$$

3.2.2. Deep Reinforcement Learning-Based Dynamic Threshold

PPO is one of the DRL agent algorithms that can be utilized for both discrete action and continuous action spaces. PPO was utilized because it outperformed other policy gradient methods, such as Advantage Actor–Critic (A2C) and Actor–Critic with Experience (ACER) [58]. The objective function of the PPO was the following:

$$L^{CLIP}(\theta) = \hat{\mathbb{E}}_t[\min(z_t(\theta)\hat{A}_t, \text{clip}(z_t(\theta), 1-\epsilon, 1+\epsilon)\hat{A}_t)] \tag{7}$$

In the clip term, ϵ was a hyperparameter for clipping the probability ratio $z_t(\theta)$. The default value of ϵ was 0.2. The clipping effect prevents an excessively large policy update. The details of the algorithms are described in [58]. In our proposed scheme, we used the PPO as a DRL agent for setting dynamic thresholds. For ϵ, the default value was used. We utilized each agent for each estimation model and denoted our DRL agent as DRL-DT.

In RL, an agent chooses an action for maximizing a cumulative reward after observing information on each state at a given weather condition. Therefore, designing a proper action space, state space, and reward function for a problem is essential. We designed a DRL scheme for detecting the inverter faults in the NIST PV dataset. Figure 8 illustrates the structure of the designed DRL scheme.

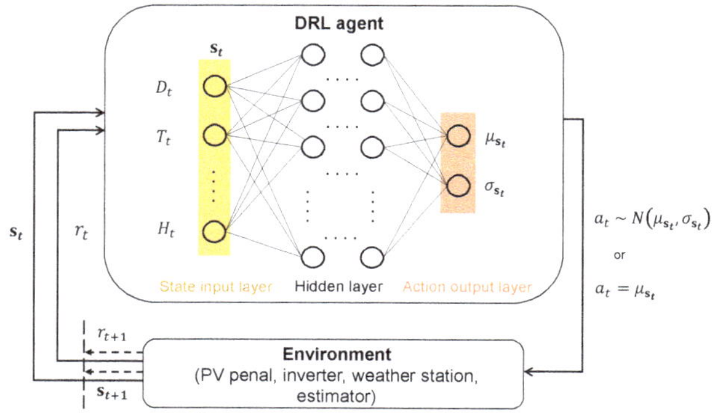

Figure 8. DRL structure for setting dynamic thresholds.

Action

For action, we used a continuous action space setting, in which the Gaussian policy was utilized. There were two output nodes in the last layer of the policy function of the DRL-DT: one was μ_{s_t}, and the other was σ_{s_t}. Sigmoid function was utilized to constrain a range of μ_{s_t}, and Softplus function was utilized for σ_{s_t}. During training, dynamic threshold values were sampled in the Gaussian distribution with μ_{s_t} and σ_{s_t}.

$$a_t \sim N(\mu_{s_t}, \sigma_{s_t}) \tag{8}$$

During the test, we used μ_{s_t} as threshold values because sampling thresholds from distribution caused performance uncertainties.

$$a_t = \mu_{s_t} \tag{9}$$

State

The same weather information (irradiance D_t, ambient temperature T_t, wind speed W_t, humidity H_t) was used as the input of the estimation model. This observation provided the agent information necessary for properly setting dynamic thresholds under varying weather conditions. A state vector, representing an observation at a time step, is as follows:

$$\mathbf{s}_t = (D_t, T_t, W_t, H_t) \tag{10}$$

Reward

We designed a reward function for setting dynamic thresholds to detect inverter faults under the condition that there were no fault labels. Feedback signals from the reward function can play a key role in anomaly or fault detection tasks in systems in which fault events are sparse. Once an estimation model is trained, residual values can be obtained by using an estimated value as follows:

$$e_t = |X_t - \hat{X}_t| \tag{11}$$

In the fault detection task, if e_t exceeded a threshold value $a_t(a_t \geq 0, \forall a_t)$, we considered that an inverter fault occurred at the used time step. Therefore, if $\frac{e_t}{a_t}$ was larger than, or equal to, 1, it was considered that a fault had occurred. Otherwise, it was considered that the PV systems were operating in a normal way.

Our proposed method was for dealing with cases in which fault labels were either sparse or not. For this reason, we trained our agent with only the normal data of the NIST canopy dataset. By utilizing the $\frac{e_t}{a_t}$, we could provide feedback to our agent on whether it chose proper thresholds without fault labels or not. Our designed reward function was as follows:

$$r_{t+1} = \begin{cases} \beta(\frac{e_t+\epsilon}{a_t+\epsilon}), & 0 \le \frac{e_t+\epsilon}{a_t+\epsilon} < 1 \\ \gamma(-\frac{e_t+\epsilon}{a_t+\epsilon}), & \frac{e_t+\epsilon}{a_t+\epsilon} \ge 1 \end{cases} \quad (12)$$

We assumed that the training data for the agent was normal. If the agent chose a threshold value, a_t, larger than a residual value, e_t, the agent received a positive reward, such as $\beta(\frac{e_t+\epsilon}{a_t+\epsilon})$. Otherwise, the agent received a negative reward, such as $\gamma(-\frac{e_t+\epsilon}{a_t+\epsilon})$. Here ϵ was used to prevent zero division, where ϵ was set as 1. As a result, after agent training, the agent tended to choose a tight, but robust, threshold at each state under varying weather conditions. By setting the β and γ as below, we could provide more weights on what users paid attention to more.

$$\beta + \gamma = 1 \quad (13)$$

The reward function can be illustrated as in Figure 9.

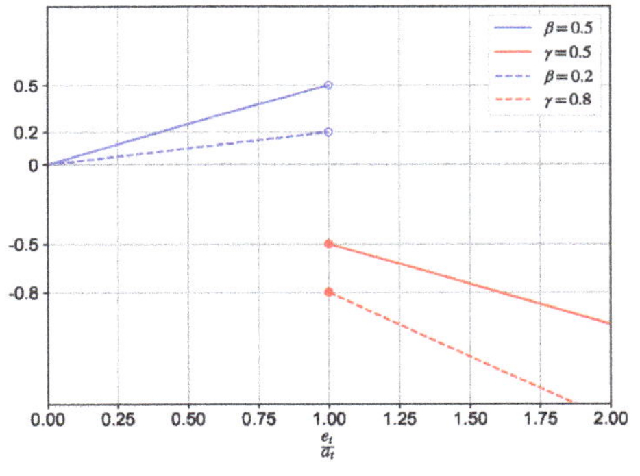

Figure 9. Reward function of the proposed DRL-DT with different β and γ.

For the training process of the DRL agent,

1. We randomly selected a day as an episode among the same training dataset that was used for the DNN estimator.
   (a) The same weather information (irradiance D_t, ambient temperature T_t, wind speed W_t, and humidity H_t) was used as a state to the DRL agent.
   (b) The DRL agent generated an action that was used as a threshold value.
   (c) The threshold value was compared with the difference between a measured value and an estimated value.
   (d) The DRL agent received a reward value, based on the designed reward function.
2. The DRL agent updated its policy to get higher reward values.

The (a)–(d) steps were repeated until an episode was done. The 1–2 steps were repeated until the DRL agent's policy converged to an optimal policy.

Even though the agent maximized the cumulative rewards, it was not guaranteed that the performance of the agent was practically getting better or not. Therefore, we used metrics, such as recall, precision, and F_1 score, to validate the performance of the proposed dynamic threshold scheme. The formulae of the metrics were as follows:

$$\text{Recall} = \frac{\text{TP}}{\text{TP + FN}} \quad (14)$$

$$\text{Precision} = \frac{\text{TP}}{\text{TP + FP}} \quad (15)$$

$$F_1\text{score} = \frac{2 \times \text{Precision} \times \text{Recall}}{\text{Precision+Recall}} \quad (16)$$

True positive, false positive, and false positive are denoted as TP, FP, and FN, respectively.

4. Results and Discussion

4.1. Performance Evaluation

4.1.1. Estimation Performance

We used the NIST canopy PV dataset as a training and validation set, in 2015. For the evaluation metrics, we used MAE and SMAPE. Figure 10 shows the training and validation errors of estimating power generation. We saved an estimation model at each epoch, and utilized the model that showed the best performance in the validation set. We tested the model using the NIST canopy PV dataset, in 2016. For the power generation estimation, the training, validation, and test errors were 4.99, 5.48 and 8.26, in terms of MAE, and they were 4.34, 6.32 and 8.30, in terms of SMAPE, respectively. Figure 11 shows a comparison between the measured and estimated profiles, while (a), (b) and (c) show the results of the AC power, current, and voltage estimations, respectively. Table 1 describes the performance details of each model.

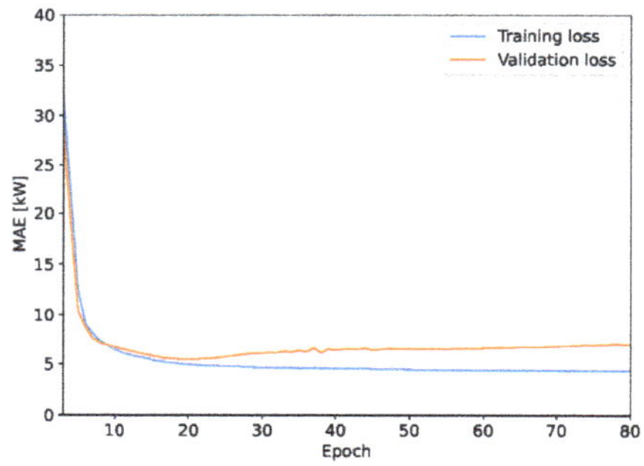

Figure 10. Loss curve in power generation estimation.

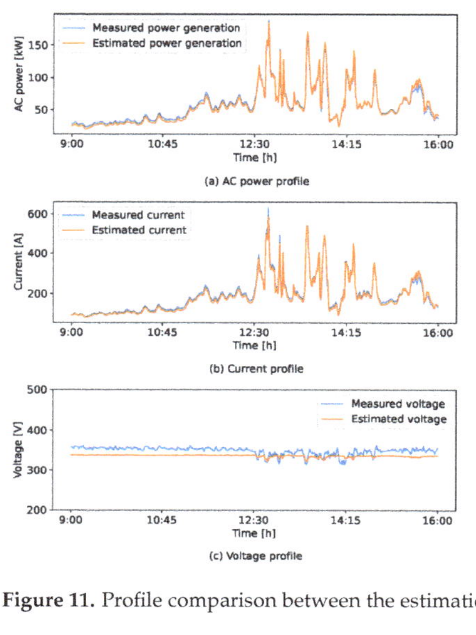

Figure 11. Profile comparison between the estimation and measurement values.

Table 1. Power generation estimation performance of the estimation model based on DNNs.

	Training Error		Validation Error		Test Error	
	MAE	SMAPE	MAE	SMAPE	MAE	SMAPE
AC power [kW]	4.99	4.34	5.48	6.32	8.26	8.30
Current [A]	15.75	5.16	16.02	6.88	25.18	7.87
Voltage [V]	6.61	1.95	6.80	2.33	11.38	3.17

4.1.2. Threshold Performance

After training the estimation model, we utilized the model and calculated the residuals between the measured values and estimated values. By using the same dataset as the estimation model, we trained the DRL-DT agent. Therefore, we used the dataset of 2015 for training and the dataset of 2016 for testing. To set β and γ, we searched 0.01 and 0.05 and, then, formed from 0.1 up to 0.9 by increasing 0.1 of β. We found that the F_1 score was best when β was 0.1. Figure 12 shows the performance of DRL-DT in terms of recall, precision, and F_1 score.

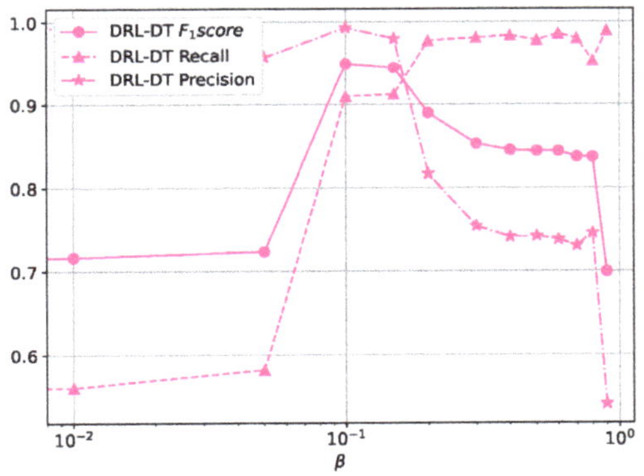

Figure 12. PV inverter fault detection performance of the proposed DRL-DT as a function of β.

To compare our proposed method with benchmark methods, we implemented existing statistical threshold approaches [15,49], a DNN classifier, and a DRL classifier. The classifiers had 7 input nodes for features (irradiance D_t, ambient temperature T_t, wind speed W_t, humidity H_t, the error of an estimated power e_t^P, the error of an estimated current e_t^I, the error of an estimated voltage e_t^V). Two output nodes were used to represent normal operation and inverter fault. Softmax function was used as an activation function in the output layer. In the case of the statistical approaches, fault labels were not required for the threshold. However, DNN classifier and DRL classifier required fault labels. Thus, we categorized the label-free groups according to whether fault labels were used or not. The used statistical approaches, and our proposed scheme, were label-free. We denoted the statistical approaches using dynamic thresholds of DT-1 [49] and DT-2 [15], and these approaches only used residuals between the estimated and measured values. In the case of the DNN and DRL classifiers, fault data and labels were used when they were trained. Figure 13 shows a comparison between the proposed scheme and benchmark approaches. The inverter fault data in the NIST canopy dataset had a relatively simple pattern that sharply decreased in a specific time period. Thus, the supervised manner (with fault labels) could almost completely classify inverter faults and normal data, showing an F_1 score of 0.99. Among the label-free approaches, we observed that our proposed method showed the best performance, which was closest to the performance of the supervised manner, showing an F_1 score of 0.94. As a result, the proposed scheme provided a 5.67% higher F_1-score, compared with the DT-2. The summary of the model performances are described in Table 2. The recall and precision curves were compared in Figure 14. We observed that DRL-DT had the best performance and that its performance was closest to that of the supervised manner at two recall standards (0.91, 0.95).

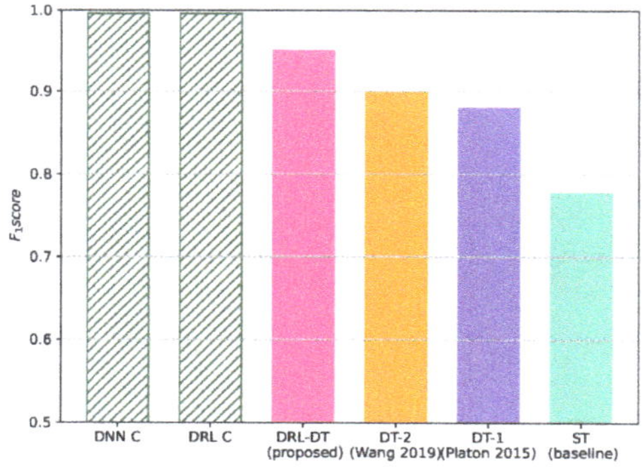

Figure 13. F_1 score comparison between model-based fault detection schemes [15,49].

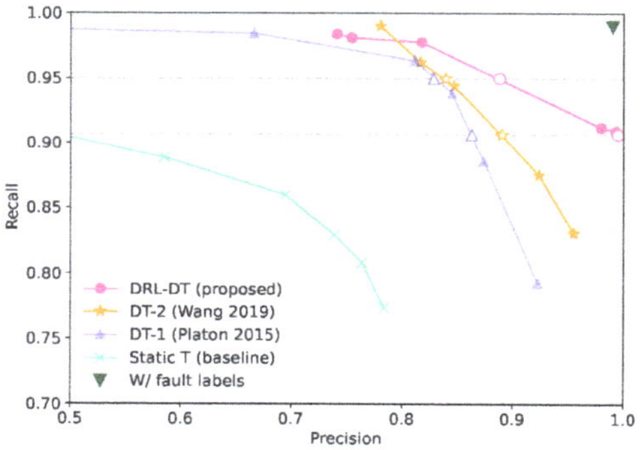

Figure 14. Recall and precision curves for proposed and conventional fault detection schemes [15,49].

Table 2. Performance comparison for inverter fault detection schemes.

	Recall	Precision	F_1 Score	Label-Free
DNN classifier	1.000	0.991	0.995	×
DRL classifier	1.000	0.991	0.995	×
DRL-DT (proposed)	0.911	0.990	**0.949**	◯
DT-2 [49]	0.875	0.923	0.898	◯
DT-1 [15]	0.886	0.874	0.880	◯
Static T	0.773	0.783	0.778	◯

DT-2 showed the second best performance. Both DT-2 and DT-1 only used irradiance for setting the thresholds. We shifted the confidence intervals by 1 sigma to search for the best threshold values at each irradiance interval, and used the best performances observed by both DT-2 and DT-1. The differences between them were as follows. DT-2 set thresholds for power, current, and voltage, while DT-1 set thresholds for only power. Our scheme was also applied to power, current, and voltage when comparing the performances (the number of threshold types was the same as DT-2). The irradiance interval of DT-2 (0–50, 50–100, . . ., 900–950, 950) was more granular than that of DT-1 (0–50, 50–250, 250–500, 500).

Figure 15 shows the threshold patterns of DT-2 and DRL-DT for e^P. We identified that DT-2 was more sensitive to irradiance interval, and DRL-DT showed more smoothed threshold values. e^P was almost the same as e^I. DT-2 and DT-1 showed decent performances by only considering irradiance intervals at cool temperature conditions (lower than 25 degrees). However, when the ambient temperature was higher than 25 degrees, the irradiance-based approaches clearly showed their weakness, as shown in Figure 16. Figure 17 shows false positive alarm occurrences with ambient temperatures for each label-free approach.

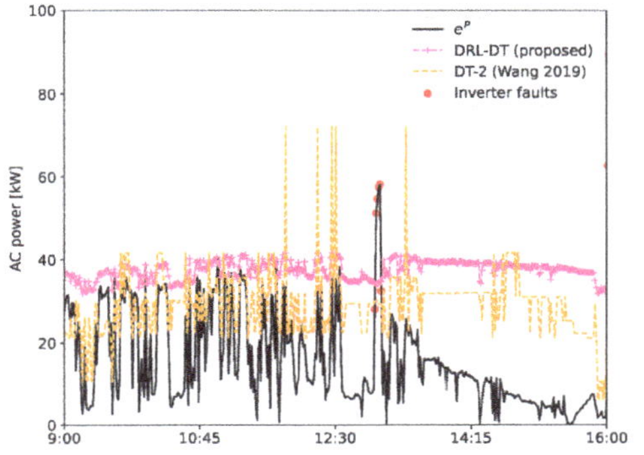

Figure 15. Power residuals and dynamic threshold profiles in a test day recorded at high ambient temperatures [49].

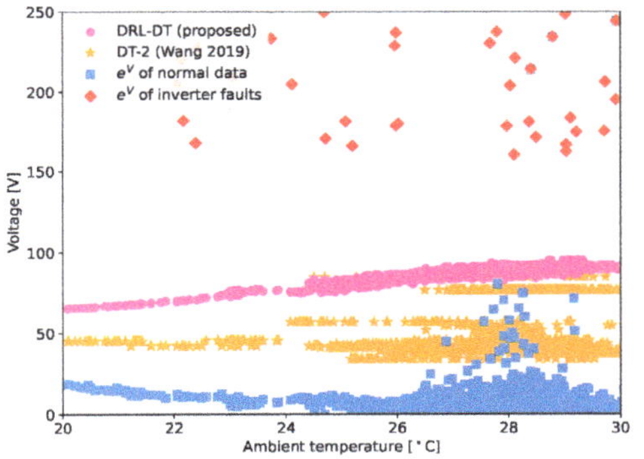

Figure 16. Scatter plots of voltage residuals and associated thresholds by the proposed and conventional schemes at high ambient temperatures [49].

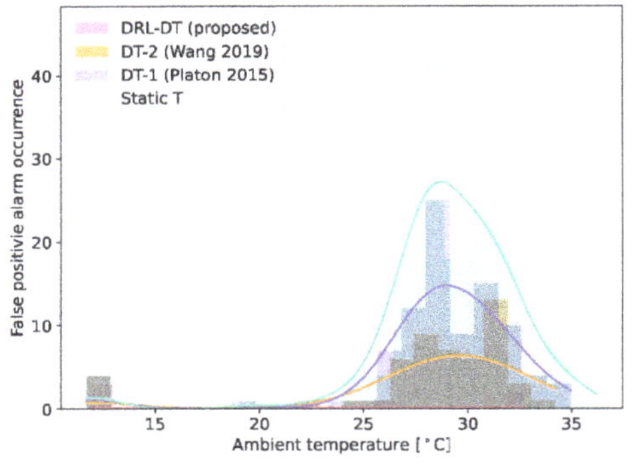

Figure 17. False positive alarm occurrences depending on ambient temperature [15,49].

As a result, various environmental variables, such as temperature, should also be considered to set more accurate threshold values. Both DT-1 and DT-2 used the threshold values of look-up tables, based on irradiance interval, which faces difficulties in considering various environmental variables. our proposed scheme could mitigate such difficulties and improve the accuracy of threshold values by using the DRL agent.

5. Conclusions

We presented a deep reinforcement learning (DRL)-based dynamic threshold scheme that operates in an unsupervised manner (without fault labels), and a reward function was designed to set appropriate threshold values under various weather conditions. The experimental results showed that the proposed scheme outperformed the existing unsupervised threshold schemes, based on statistical approaches, under varying weather conditions. Our scheme showed the closest performance to the performance of the supervised approach (with fault labels). In addition to PV plants, we expect that our scheme could be utilized in various fields, including monitoring systems and anomaly and fault detection systems, in which fault labels are sparse.

We applied the proposed scheme to a specific PV inverter fault. Our scheme should be considered for different PV faults or fault detection tasks in various fields. A potential limitation of the proposed method was that normal data was used in the training of the DRL agent. If the initial label did not exist and a large number of abnormal data was included for training, performance of the DRL-based approach might be degraded. As a future work, we are considering extending the proposed scheme for fully unlabeled datasets.

Author Contributions: G.S. overall devised the idea and led the research as the first author; S.Y. implemented benchmark models for performance comparison; J.S. conducted data analysis; E.S. reviewed previous studies; E.H. supervised the research as the corresponding author. All authors have read and agreed to the published version of the manuscript.

Funding: This research was funded by the Korea Government through the Ministry of Science and Information and Communication Technology (MSIT) (2021R1A2C1009803) and GIST Research Institute (GRI) grant funded by the GIST in 2023.

Institutional Review Board Statement: Not applicable.

Informed Consent Statement: Not applicable.

Data Availability Statement: Not applicable.

Acknowledgments: This research was supported, in part, by GIST Research Institute (GRI) grant funded by the GIST in 2023 and, in part, by the National Research Foundation of Korea (NRF) grant funded by the Korea government. (MSIT) (2021R1A2C1009803).

Conflicts of Interest: The authors declare no conflict of interest.

Abbreviations

PV	Photovoltaic
DNN	Deep neural network
DRL	Deep reinforcement learning
PPO	Proximal policy optimization
CNN	Convolution neural network
PCA	Principal component analysis
NKDE	Non-parametric kernel density estimation
MDP	Markov decision process
DQN	Deep Q-function network
A3C	Advantage actor-critic
DT	Dynamic threshold

References

1. Razmjoo, A.; Kaigutha, L.G.; Rad, M.V.; Marzband, M.; Davarpanah, A.; Denai, M. A Technical analysis investigating energy sustainability utilizing reliable renewable energy sources to reduce CO_2 emissions in a high potential area. *Renew. Energy* **2021**, *164*, 46–57. [CrossRef]
2. Algarín, C.R. An analytic hierarchy process based approach for evaluating renewable energy sources. *Int. J. Energy Econ. Policy* **2017**, *7*, 38–47.
3. Park, K.; Yoon, S.; Hwang, E. Hybrid load forecasting for mixed-use complex based on the characteristic load decomposition by pilot signals. *IEEE Access* **2019**, *7*, 12297–12306. [CrossRef]
4. Park, S.; Yoon, S.; Lee, B.; Ko, S.; Hwang, E. Probabilistic forecasting based joint detection and imputation of clustered bad data in residential electricity loads. *Energies* **2020**, *14*, 165. [CrossRef]
5. Yoon, S.; Hwang, E. Load guided signal-based two-stage charging coordination of plug-in electric vehicles for smart buildings. *IEEE Access* **2019**, *7*, 144548–144560. [CrossRef]
6. Seo, G.; Yoon, S.; Kim, M.; Mun, C.; Hwang, E. Deep Reinforcement Learning-Based Smart Joint Control Scheme for On/Off Pumping Systems in Wastewater Treatment Plants. *IEEE Access* **2021**, *9*, 95360–95371. [CrossRef]
7. Song, J.; Lee, Y.; Hwang, E. Time–frequency mask estimation based on deep neural network for flexible load disaggregation in buildings. *IEEE Trans. Smart Grid* **2021**, *12*, 3242–3251. [CrossRef]
8. Firth, S.K.; Lomas, K.J.; Rees, S.J. A simple model of PV system performance and its use in fault detection. *Sol. Energy* **2010**, *84*, 624–635. [CrossRef]
9. Hernández-Callejo, L.; Gallardo-Saavedra, S.; Alonso-Gómez, V. A review of photovoltaic systems: Design, operation and maintenance. *Sol. Energy* **2019**, *188*, 426–440. [CrossRef]
10. Correa-Betanzo, C.; Calleja, H.; Aguilar, C.; Lopez-Nunez, A.R.; Rodriguez, E. Photovoltaic-based DC microgrid with partial shading and fault tolerance. *J. Mod. Power Syst. Clean Energy* **2019**, *7*, 340–349. [CrossRef]
11. Spataru, S.; Sera, D.; Kerekes, T.; Teodorescu, R. Photovoltaic array condition monitoring based on online regression of performance model. In Proceedings of the 2013 IEEE 39th Photovoltaic Specialists Conference (PVSC), Tampa, FL, USA, 16–21 June 2013; IEEE: Piscataway, NJ, USA, 2013; pp. 0815–0820.
12. Muñoz, M.; Correcher, A.; Ariza, E.; García, E.; Ibañez, F. Fault detection and isolation in a photovoltaic system. In Proceedings of the International Conference on Renewable Energies and Power Quality, La Coruña, Spain, 25–27 March 2015; Volume 1, pp. 202–207.
13. Stauffer, Y.; Ferrario, D.; Onillon, E.; Hutter, A. Power monitoring based photovoltaic installation fault detection. In Proceedings of the 2015 International Conference on Renewable Energy Research and Applications (ICRERA), Palermo, Italy, 22–25 November 2015; IEEE: Piscataway, NJ, USA, 2015; pp. 199–202.
14. Andò, B.; Baglio, S.; Pistorio, A.; Tina, G.M.; Ventura, C. Sentinella: Smart monitoring of photovoltaic systems at panel level. *IEEE Trans. Instrum. Meas.* **2015**, *64*, 2188–2199. [CrossRef]
15. Platon, R.; Martel, J.; Woodruff, N.; Chau, T.Y. Online fault detection in PV systems. *IEEE Trans. Sustain. Energy* **2015**, *6*, 1200–1207. [CrossRef]
16. Chine, W.; Mellit, A.; Pavan, A.M.; Kalogirou, S.A. Fault detection method for grid-connected photovoltaic plants. *Renew. Energy* **2014**, *66*, 99–110. [CrossRef]
17. Kim, I.S. On-line fault detection algorithm of a photovoltaic system using wavelet transform. *Sol. Energy* **2016**, *126*, 137–145. [CrossRef]
18. Ali, M.H.; Rabhi, A.; El Hajjaji, A.; Tina, G.M. Real time fault detection in photovoltaic systems. *Energy Procedia* **2017**, *111*, 914–923. [CrossRef]
19. Spataru, S.; Sera, D.; Kerekes, T.; Teodorescu, R. Monitoring and fault detection in photovoltaic systems based on inverter measured string IV curves. In Proceedings of the 31st European Photovoltaic Solar Energy Conference and Exhibition, Hamburg, German, 14–18 September 2015; WIP Wirtschaft und Infrastruktur GmbH & Co Planungs KG: Munich, Germany, 2015; pp. 1667–1674.
20. Guerriero, P.; Piegari, L.; Rizzo, R.; Daliento, S. Mismatch based diagnosis of PV fields relying on monitored string currents. *Int. J. Photoenergy* **2017**, *2017*, 2834685. [CrossRef]
21. Stellbogen, D. Use of PV circuit simulation for fault detection in PV array fields. In Proceedings of the Conference Record of the Twenty Third IEEE Photovoltaic Specialists Conference-1993 (Cat. No. 93CH3283-9), Louisville, KY, USA, 10–14 May 1993; IEEE: Piscataway, NJ, USA, 1993; pp. 1302–1307.
22. Guasch, D.; Silvestre, S.; Calatayud, R. Automatic failure detection in photovoltaic systems. In Proceedings of the Proceedings of the 3rd World Conference on Photovoltaic Energy Conversion, Osaka, Japan, 11–18 May 2003; pp. 2269–2271.
23. Chao, K.H.; Ho, S.H.; Wang, M.H. Modeling and fault diagnosis of a photovoltaic system. *Electr. Power Syst. Res.* **2008**, *78*, 97–105. [CrossRef]
24. Hamdaoui, M.; Rabhi, A.; El Hajjaji, A.; Rahmoun, M.; Azizi, M. Monitoring and control of the performances for photovoltaic systems. In Proceedings of the International Renewable Energy Congress, Warsaw, Poland, 19–20 May 2009.
25. Jaskie, K.; Martin, J.; Spanias, A. PV fault detection using positive unlabeled learning. *Appl. Sci.* **2021**, *11*, 5599. [CrossRef]
26. Lu, F.; Niu, R.; Zhang, Z.; Guo, L.; Chen, J. A generative adversarial network-based fault detection approach for photovoltaic panel. *Appl. Sci.* **2022**, *12*, 1789. [CrossRef]

27. Nordmann, T.; Jahn, U.; Nasse, W. Performance of PV systems under real conditions. In Proceedings of the European Workshop on Life Cycle Analysis and Recycling of Solar Modules, The "Waste" Challenge, Brussels, Belgium, 18–19 March 2004; p. 1.
28. Takashima, T.; Yamaguchi, J.; Otani, K.; Oozeki, T.; Kato, K.; Ishida, M. Experimental studies of fault location in PV module strings. *Sol. Energy Mater. Sol. Cells* **2009**, *93*, 1079–1082. [CrossRef]
29. Lei, P.; Li, Y.; Chen, Q.; Seem, J.E. Extremum seeking control based integration of MPPT and degradation detection for photovoltaic arrays. In Proceedings of the Proceedings of the 2010 American Control Conference, Baltimore, Maryland, USA, 30 June–2 July 2010; IEEE: Piscataway, NJ, USA, 2010; pp. 3536–3541.
30. Roumpakias, E.; Stamatelos, T. Health Monitoring and Fault Detection in Photovoltaic Systems in Central Greece Using Artificial Neural Networks. *Appl. Sci.* **2022**, *12*, 12016. [CrossRef]
31. Wang, Y.; Li, X.; Ban, Y.; Ma, X.; Hao, C.; Zhou, J.; Cai, H. A DC Arc Fault Detection Method Based on AR Model for Photovoltaic Systems. *Appl. Sci.* **2022**, *12*, 10379. [CrossRef]
32. Boyd, M. Performance data from the nist photovoltaic arrays and weather station. *J. Res. Natl. Inst. Stand. Technol.* **2017**, *122*, 1. [CrossRef] [PubMed]
33. Park, S.; Park, S.; Kim, M.; Hwang, E. Clustering-based self-imputation of unlabeled fault data in a fleet of photovoltaic generation systems. *Energies* **2020**, *13*, 737. [CrossRef]
34. Qian, X.; Li, J.; Cao, J.; Wu, Y.; Wang, W. Micro-cracks detection of solar cells surface via combining short-term and long-term deep features. *Neural Netw.* **2020**, *127*, 132–140. [CrossRef]
35. Li, X.; Yang, Q.; Lou, Z.; Yan, W. Deep learning based module defect analysis for large-scale photovoltaic farms. *IEEE Trans. Energy Convers.* **2018**, *34*, 520–529. [CrossRef]
36. Deitsch, S.; Christlein, V.; Berger, S.; Buerhop-Lutz, C.; Maier, A.; Gallwitz, F.; Riess, C. Automatic classification of defective photovoltaic module cells in electroluminescence images. *Sol. Energy* **2019**, *185*, 455–468. [CrossRef]
37. Karimi, A.M.; Fada, J.S.; Hossain, M.A.; Yang, S.; Peshek, T.J.; Braid, J.L.; French, R.H. Automated pipeline for photovoltaic module electroluminescence image processing and degradation feature classification. *IEEE J. Photovoltaics* **2019**, *9*, 1324–1335. [CrossRef]
38. Du, B.; He, Y.; He, Y.; Duan, J.; Zhang, Y. Intelligent classification of silicon photovoltaic cell defects based on eddy current thermography and convolution neural network. *IEEE Trans. Ind. Inform.* **2019**, *16*, 6242–6251. [CrossRef]
39. Akram, M.W.; Li, G.; Jin, Y.; Chen, X.; Zhu, C.; Ahmad, A. Automatic detection of photovoltaic module defects in infrared images with isolated and develop-model transfer deep learning. *Sol. Energy* **2020**, *198*, 175–186. [CrossRef]
40. Herraiz, Á.H.; Marugán, A.P.; Márquez, F.P.G. Photovoltaic plant condition monitoring using thermal images analysis by convolutional neural network-based structure. *Renew. Energy* **2020**, *153*, 334–348. [CrossRef]
41. AbdulMawjood, K.; Refaat, S.S.; Morsi, W.G. Detection and prediction of faults in photovoltaic arrays: A review. In Proceedings of the 2018 IEEE 12th International Conference on Compatibility, Power Electronics and Power Engineering (CPE-POWERENG 2018), Doha, Qatar, 10–12 April 2018; IEEE: Piscataway, NJ, USA, 2018; pp. 1–8.
42. Pan, J.; He, W.; Shi, Y.; Hou, R.; Zhu, H. Uncertainty analysis based on non-parametric statistical modelling method for photovoltaic array output and its application in fault diagnosis. *Sol. Energy* **2021**, *225*, 831–841. [CrossRef]
43. Pillai, D.S.; Rajasekar, N. Metaheuristic algorithms for PV parameter identification: A comprehensive review with an application to threshold setting for fault detection in PV systems. *Renew. Sustain. Energy Rev.* **2018**, *82*, 3503–3525. [CrossRef]
44. Shimakage, T.; Nishioka, K.; Yamane, H.; Nagura, M.; Kudo, M. Development of fault detection system in PV system. In Proceedings of the 2011 IEEE 33rd International Telecommunications Energy Conference (INTELEC), Amsterdam, The Netherlands, 9–13 October 2011; IEEE: Piscataway, NJ, USA, 2011; pp. 1–5.
45. Bressan, M.; El-Basri, Y.; Alonso, C. A new method for fault detection and identification of shadows based on electrical signature of defects. In Proceedings of the 2015 17th European Conference on Power Electronics and Applications (EPE'15 ECCE-Europe), Geneva, Switzerland, 8–10 September 2015; IEEE: Piscataway, NJ, USA, 2015; pp. 1–8.
46. Dhimish, M.; Holmes, V. Fault detection algorithm for grid-connected photovoltaic plants. *Sol. Energy* **2016**, *137*, 236–245. [CrossRef]
47. Garoudja, E.; Harrou, F.; Sun, Y.; Kara, K.; Chouder, A.; Silvestre, S. Statistical fault detection in photovoltaic systems. *Sol. Energy* **2017**, *150*, 485–499. [CrossRef]
48. Rouani, L.; Harkat, M.F.; Kouadri, A.; Mekhilef, S. Shading fault detection in a grid-connected PV system using vertices principal component analysis. *Renew. Energy* **2021**, *164*, 1527–1539. [CrossRef]
49. Wang, H.; Zhao, J.; Sun, Q.; Zhu, H. Probability modeling for PV array output interval and its application in fault diagnosis. *Energy* **2019**, *189*, 116248. [CrossRef]
50. Sutton, R.S.; Barto, A.G. *Reinforcement Learning: An Introduction*; MIT Press: Cambridge, MA, USA, 2018.
51. Huang, C.; Wu, Y.; Zuo, Y.; Pei, K.; Min, G. Towards experienced anomaly detector through reinforcement learning. In Proceedings of the Thirty-Second AAAI Conference on Artificial Intelligence, New Orleans, LA, USA, 2–7 February 2018.
52. Yu, M.; Sun, S. Policy-based reinforcement learning for time series anomaly detection. *Eng. Appl. Artif. Intell.* **2020**, *95*, 103919. [CrossRef]
53. Khazaeli, S.; Nguyen, L.H.; Goulet, J.A. Anomaly detection using state-space models and reinforcement learning. *Struct. Control Health Monit.* **2021**, *28*, e2720. [CrossRef]

54. Malik, A.; Haque, A.; Kurukuru, V.B.; Khan, M.A.; Blaabjerg, F. Overview of Fault Detection Approaches for Grid Connected Photovoltaic Inverters. *e-Prime-Adv. Electr. Eng. Electron. Energy* **2022**, *2*, 100035.
55. Zaini, N.; Ab Kadir, M.; Izadi, M.; Ahmad, N.; Radzi, M.; Azis, N. The effect of temperature on a mono-crystalline solar PV panel. In Proceedings of the 2015 IEEE Conference on Energy Conversion (CENCON), Johor Bahru, Malaysia, 19–20 October 2015; IEEE: Piscataway, NJ, USA, 2015; pp. 249–253.
56. Dubey, S.; Sarvaiya, J.N.; Seshadri, B. Temperature dependent photovoltaic (PV) efficiency and its effect on PV production in the world–a review. *Energy Procedia* **2013**, *33*, 311–321. [CrossRef]
57. Koehl, M.; Heck, M.; Wiesmeier, S.; Wirth, J. Modeling of the nominal operating cell temperature based on outdoor weathering. *Sol. Energy Mater. Sol.* **2011**, *95*, 1638–1646. [CrossRef]
58. Schulman, J.; Wolski, F.; Dhariwal, P.; Radford, A.; Klimov, O. Proximal policy optimization algorithms. *arXiv* **2017**, arXiv:1707.06347.

Disclaimer/Publisher's Note: The statements, opinions and data contained in all publications are solely those of the individual author(s) and contributor(s) and not of MDPI and/or the editor(s). MDPI and/or the editor(s) disclaim responsibility for any injury to people or property resulting from any ideas, methods, instructions or products referred to in the content.

applied sciences

MDPI

Article

Pruning Quantized Unsupervised Meta-Learning DegradingNet Solution for Industrial Equipment and Semiconductor Process Anomaly Detection and Prediction

Yi-Cheng Yu [1,*], Shiau-Ru Yang [1], Shang-Wen Chuang [2], Jen-Tzung Chien [1] and Chen-Yi Lee [3,*]

[1] College of Electrical and Computer Engineering, National Yang Ming Chiao Tung University, Hsinchu City 300, Taiwan; zuzuyang0912.ee08@nycu.edu.tw (S.-R.Y.); jtchien@nycu.edu.tw (J.-T.C.)

[2] Advance Tech Co., Taipei City 114, Taiwan; swchuang@gmail.com

[3] Department of Electronics Engineering, National Yang Ming Chiao Tung University, Hsinchu City 300, Taiwan

* Correspondence: yichengyu.ee06@nycu.edu.tw (Y.-C.Y.); cylee@nycu.edu.tw (C.-Y.L.)

Abstract: Machine- and deep-learning methods are used for industrial applications in prognostics and health management (PHM) for semiconductor processing and equipment anomaly detection to achieve proactive equipment maintenance and prevent process interruptions or equipment downtime. This study proposes a Pruning Quantized Unsupervised Meta-learning DegradingNet Solution (PQUM-DNS) for the fast training and retraining of new equipment or processes with limited data for anomaly detection and the prediction of various equipment and process conditions. This study utilizes real data from a factory chiller host motor, the Paderborn current and vibration open dataset, and the SECOM semiconductor open dataset to conduct experimental simulations, calculate the average value, and obtain the results. Compared to conventional deep autoencoders, PQUM-DNS reduces the average data volume required for rapid training and retraining by about 75% with similar AUC. The average RMSE of the predictive degradation degree is 0.037 for Holt–Winters, and the model size is reduced by about 60% through pruning and quantization which can be realized by edge devices, such as Raspberry Pi. This makes the proposed PQUM-DNS very suitable for intelligent equipment management and maintenance in industrial applications.

Keywords: anomaly detection; deep learning; prediction; pruning; edge computing; intelligent equipment management; meta-learning; retraining; semiconductor; unsupervised learning; vibration; quantization

1. Introduction

Research related to prognostics and health management (PHM) indicates that anomaly detection and prediction are important approaches to monitoring equipment faults and semiconductor process abnormalities. However, this type of detection often relies on subjective assessments performed by operators with prior experience. Automating the detection of equipment faults or semiconductor process anomalies is essential for reliable predictive maintenance and can potentially eliminate the need for manual monitoring. Moreover, interconnected intelligent monitoring systems play a crucial role in Industry 4.0, which focuses on artificial intelligence (AI)-driven factory automation.

In recent years, various deep-learning techniques have been introduced for anomaly detection and prediction in PHM, focusing on equipment faults (vibration and current anomalies) and semiconductor processes [1–10].

The related techniques include dense autoencoders (AEs) [11], convolutional AEs [12], and pretrained convolutional neural networks [13]. Although these deep-learning approaches exhibit excellent performance in anomaly detection, their widespread adoption in real factory settings remains limited. One major reason for the slow adoption is the high computational resource requirements of many deep-learning-driven anomaly detection methods. These methods often lack practical considerations and system integration, ignoring factors such as multi-machine or multi-production line deployment, model retraining, optimization, and systematic integration. Consequently, their applicability in real factory environments is hindered.

Many studies have been conducted in the field of PHM. Pradeep et al. proposed that machine-learning techniques could be used to predict wafer defects with a random forest classifier, achieving an accuracy of over 93.62% [14]. This predictive maintenance approach enhanced the semiconductor manufacturing productivity. Nuhu et al. introduced synthetic data generation techniques that combined two missing value imputation methods and feature selection techniques [15]. This approach achieved an accuracy ranging from 99.5% to 100% when paired with the proposed machine-learning (ML) methods. Mao et al. introduced a novel deep AE (DAE) method that fused discriminative information with a gradient descent optimization approach [16]. This technique enhanced the numerical stability of the model in cases with limited training data. Abbasi et al. presented a series of highly compact deep convolutional AE network architectures that reduced the model size while maintaining a detection accuracy comparable to that of structures with over four million parameters [17]. Givnan et al. proposed an ML method for modeling and detecting anomalies during the operation of rotating machinery. This ML approach learned and generalized based on the fault severity to generate threshold values for anomaly detection [18].

A DAE model specifically designed for factory scenarios involving chillers was introduced that effectively distinguished between normal and abnormal vibration signals based on reconstruction differences [19]. Additionally, meta-learning was employed to improve the accuracy of the new sensor models with limited vibration data. For the new sensor model with fewer vibration data, the accuracy increased by about 33.50%. However, this method is mainly oriented to anomaly detection; it has not yet considered model retraining, anomaly prediction, lightweight models, edge computing, and integration with a complete intelligent management system.

Considering the aforementioned issues, this study proposes a Pruning Quantized Unsupervised Meta-learning DegradingNet System (PQUM-DNS) to address the real-world conditions of practical factories. This approach integrates five key features based on the actual needs of factories as illustrated in Figure 1.

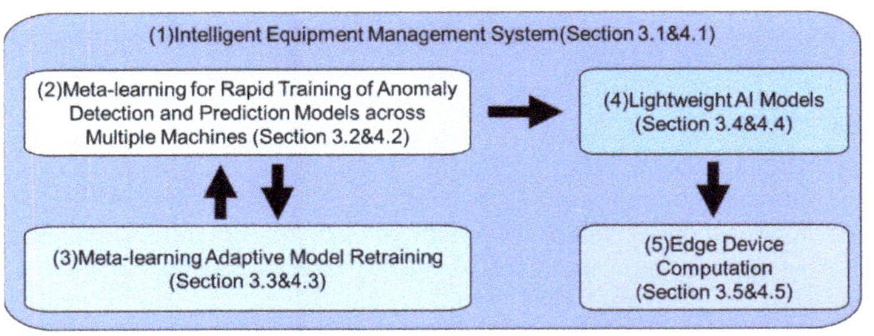

Figure 1. Overview of the proposed Intelligent Equipment Management System with related techniques highlighted in this paper.

(1) Intelligent Equipment Management System (Sections 3.1 and 4.1)

This system includes automated methods for vibration signal sensing, data transmission, data preprocessing, model training and retraining, anomaly detection, and prediction. Visual results are presented through dashboards and alert notifications are sent to onsite personnel and managers to facilitate timely problem solutions.

(2) Meta-learning for Rapid Training of Anomaly Detection and Prediction Models across Multiple Machines (Sections 3.2 and 4.2)

This approach rapidly establishes models for new machines or production lines with limited data by leveraging meta-learning and unsupervised learning through AEs, thereby achieving anomaly detection and prediction objectives.

(3) Meta-learning Adaptive Model Retraining (Sections 3.3 and 4.3)

Machine-specific models are adaptively retrained by employing meta-learning, quickly adjusting to the slow-changing characteristics of the machine over time and enabling long-term anomaly detection and prediction.

(4) Lightweight AI Models (Sections 3.4 and 4.4)

The proposed pruning and quantization compression model significantly reduces model size and conserves computational resources.

(5) Edge Device Computation (Sections 3.5 and 4.5)

Substituting traditional AI inference engines (IPC) with embedded Raspberry Pi systems enhances lightweight deployment, resource savings, cost reduction, and large-scale deployment feasibility.

2. Related Study

The following fundamental hardware, software techniques, and specifications are commonly used in semiconductor process and industrial scenarios for intelligent equipment management.

2.1. Vibration Signal Data Acquisition

(1) Device Sensors

One study collected real-time vibration data from equipment in an actual factory using an Advantech WISE-2410 LoRaWAN wireless sensor that integrated an ARM™ Cortex-M4 processor, LoRa transceiver, three-axis accelerometer, and temperature sensor [20]. It operated within a temperature range of −20 °C to 85 °C and was powered via USB.

(2) Vibration Data Feature Transformation

The raw vibration signals received from the sensors undergo feature transformation. The transformed features are critical for determining the vibration state and mainly include values such as velocity root mean square (RMS), acceleration RMS, acceleration peak, displacement kurtosis, displacement crest factor, displacement skewness, displacement peak to peak, and displacement deviation.

2.2. Vibration Signal Data of Chiller

Chillers are primarily utilized to build air conditioning systems. Chiller motors drive compressors and facilitate the exchange of heat and cold. In this study, vibration sensors were installed on a chiller motor to detect vibration values, as shown in Figure 2. Key vibration characteristics representing actual vibration measurements in a factory field were obtained through feature transformation [19].

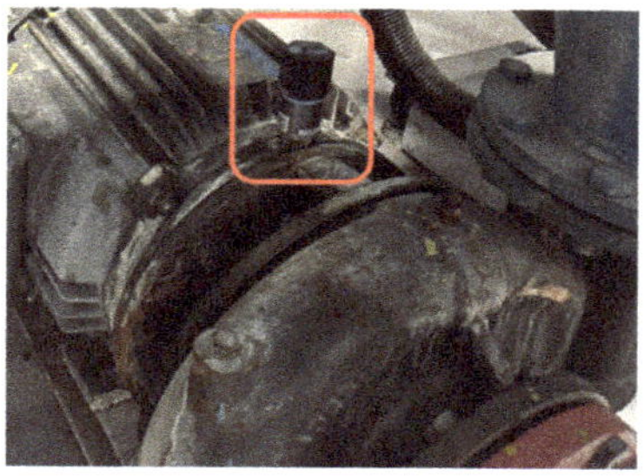

Figure 2. Vibration sensor (in the red square) installed on the motor of a chiller machine.

Machine condition monitoring (MCM) was used to convert vibration signals into 44 key time- and frequency-domain physical and statistical feature values, as listed in Table 1 [21].

Table 1. Key time- and frequency-domain physical and statistical feature values of vibration signals transformed by MCM.

Feature Types	Feature
Time Domain (22 Features)	Clearance factor, max, mean, median, min, coefficient of variation, crest factor, local max, local min, max in range, frequency, impulse factor, kurtosis, percentile, peak to peak, RMS, shape factor, skewness, standard deviation, variance, X of max, X of min
Frequency Domain (22 Features)	Clearance factor, coefficient of variation, crest factor, frequency, impulse factor, kurtosis, local max, local min, max in range, max, mean, median, min, percentile, peak to peak, RMS, shape factor, skewness, standard deviation, variance, X of max, X of min

This study refers to ISO10816 [22] as shown in Figure 3 to determine the normal and abnormal vibration data of the chiller host motor. Results may vary depending on the size of the equipment where the sensor is installed.

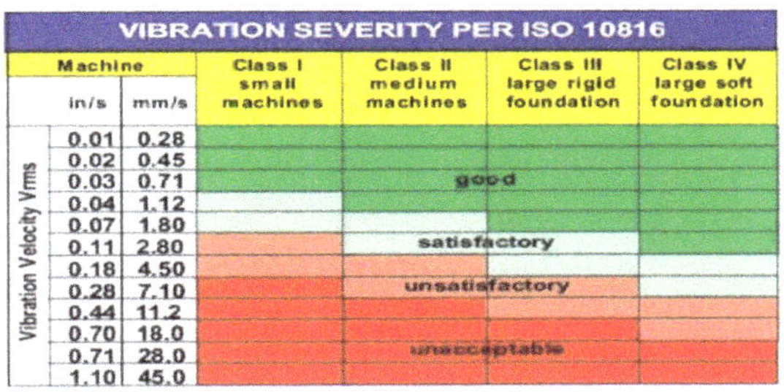

Figure 3. Vibration Severity defined in ISO10816 [22].

2.3. *Paderborn University Bearing Dataset*

A condition-monitoring experimental bearing dataset based on vibration and motor current signals was used from Germany's Paderborn University, as shown in Figure 4 [23]. Experimental datasets were generated by installing different types of damaged ball bearings in the bearing test module. The setup consisted of healthy bearings with 1920 and 1600 entries for outer and inner ring damage, respectively. The most common bearing damage analysis [24] involves using motor current signals (MCSs) to convert time-domain signals into frequency-domain signals to observe the spectrum differences between normal and abnormal bearings as shown in Figure 5.

Feature engineering was applied to extract features from raw data, including common statistical, signal factor-related, fast Fourier transform (FFT), power spectral density (PSD)-related, and wavelet packet decomposition (WPD)-related features, as listed in Table 2.

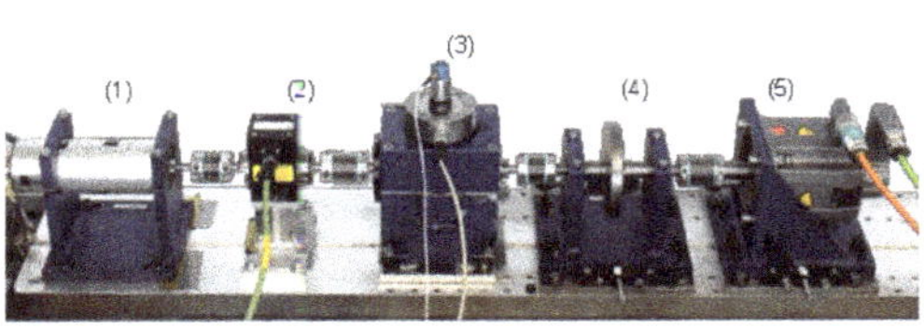

Figure 4. Germany's Paderborn University Bearing Dataset, consisting of: (1) motor, (2) torque measurement shaft, (3) rolling bearing test module, (4) flywheel, and (5) load motor.

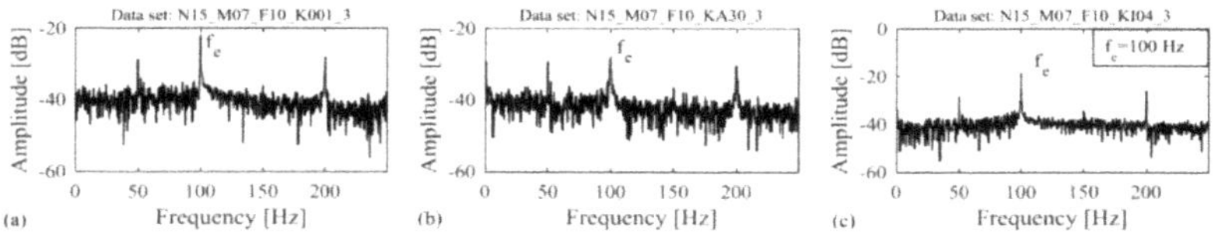

Figure 5. Frequency spectrum from MCS for (**a**) healthy bearing, (**b**) outer ring damage, and (**c**) inner ring damage [23].

Table 2. Feature values of vibration and current features after feature extraction.

Feature Types	Feature
Statistical feature	Mean, median, min, max, peak to peak, Std, variance, RMS, Absmean, Abslogmean, Meanabsdev, Medianabsdev, coefficient of variance, midrange
Signal factor-related feature	Shape factor, impulse factor, crest factor, clearance factor, skewness, kurtosis, entropy
FFT, PSD-related feature	1. Highest 3 FFT frequencies and values (6) 2. Highest 5 PSD frequencies (5) 3. Energy sum of FFT and PSD (2)
Wavelet packet decomposition (WPD)-related feature	Variance of 3 levels of WPD coefficient: Level 1 (2) Level 2 (4) Level 3 (8)

2.4. SECOM Semiconductor Analysis Dataset

The semiconductor manufacturing process utilizes semiconductor wafers as substrates and processes them through a series of steps. The main steps include: cleaning the wafer, depositing the film, cleaning after film formation, exposure, development, etching, inserting impurities, generating semiconductor properties, activation, assembly, and packaging. At each stage, sensors measure relevant parameters, including film thickness, size, resistance, temperature, etc. Through data analysis and machine learning, predictive maintenance, fault detection, process monitoring, and yield improvement are performed. These methods can also extract useful information from limited event records. Large amounts of data help solve predictive maintenance issues and build fault detection and diagnosis models. Through timely fault detection and diagnosis, downtime is reduced, costs are lowered, and product quality is enhanced [25].

Data were collected from a complex modern SECOM semiconductor manufacturing process that was under consistent surveillance through the monitoring of signals/variables collected from sensors or process measurement points. The dataset contained 1567 examples, 591 features, and 104 failures. Random forests were used to extract the 16 most important features for subsequent analysis to filter the key features because the SECOM dataset contained 591 features, as shown in Figure 6 [26–28]. For the SECOM dataset, the target column "−1" corresponds to a pass and "1" corresponds to a fail.

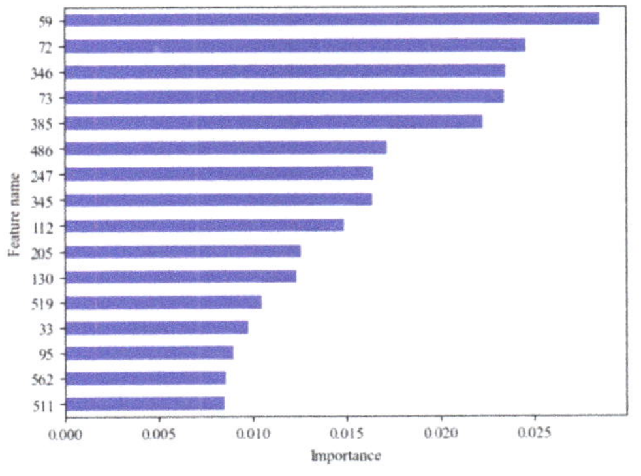

Figure 6. Extracting 16 key features from the SECOM dataset's 591 features using random forests.

2.5. Data Preprocessing

Data often encounter issues in practical scenarios, such as incomplete or missing data, noise, and outliers. These issues can disrupt the proper functioning of models. Hence, data preprocessing is essential for adjusting and manipulating data before applying analytical algorithms and preventing inaccurate judgments owing to flaws in the data. In this context, missing value deletion and statistical analysis were utilized to identify and eliminate values exceeding three standard deviations. This resulted in cleaner data, which facilitated subsequent analyses.

2.6. Equipment Process Degradation Level

The algorithm employed was an AE, which is a technique based on unsupervised learning. This approach involves computing the root mean square error (RMSE) by comparing the reconstructed output values of the model with the numerical input values. The calculated RMSE serves as an indicator of the equipment process degradation level, where smaller and larger values indicate healthier and poorer states, respectively [7,8].

2.7. Equipment Process Data Storage

A real-time PostgreSQL database was used to store the data collected from the sensors, including sensor names, equipment names, sensor registration times, operational data, and access to AI model training and inference. The inference results were stored and the database offered interfaces for other platforms to access the required data.

2.8. Area under Curve

Area under curve (AUC) is the area under the receiver operating characteristics (ROC) curve. AUC is a widely used evaluation metric in ML that assesses the performance of binary or multi-class classifiers [6,29]. Values range from 0 to 1, with a higher AUC value indicating better classifier performance. A notable advantage of AUC is its immunity to threshold variations, providing a comprehensive evaluation across different thresholds. This attribute makes AUC particularly robust in scenarios with imbalanced datasets. Figure 7 is an introduction to statistical principles related to AUC [30].

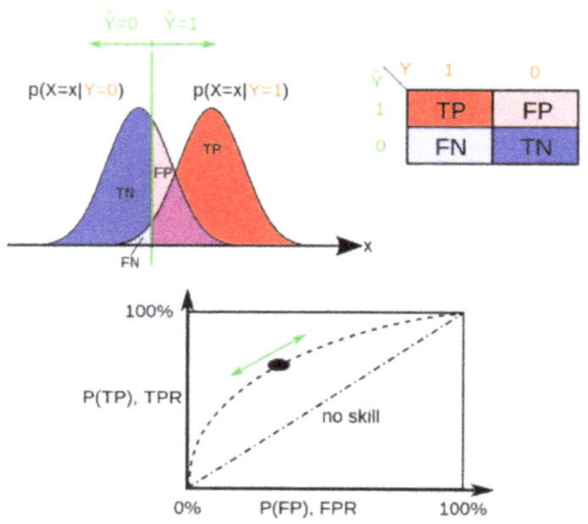

Figure 7. The results obtained from negative sample (left curve) overlap with the results obtained from positive samples (right curve). By moving (green arrow) the result cutoff value (vertical bar), the rate of false positives (FP) can be decreased, at the cost of raising the number of false negatives (FN), or vice versa (TP = True Positives, TPR = True Positive Rate, FPR = False Positive Rate, TN = True Negatives).

3. Proposed Method

3.1. Intelligent Equipment Management System

The proposed PQUM-DNS was integrated into an intelligent equipment management system for practical field applications. Signals were initially collected from the device sensors within the system before feature extraction transformation and data preprocessing, as shown in Figure 8. The feature extraction transformation involved converting raw vibration feature values into multiple key parameters related to machine health. Data preprocessing involved eliminating unnecessary empty and abnormal values to retain only the normal values required for unsupervised learning. The processed data were stored in a database (PostgreSQL) until there were sufficient accumulated data (e.g., 3000 records, adjustable). Subsequently, both AE and Holt–Winters algorithms were used to train the anomaly detection and prediction models, respectively. An inference was performed using anomaly detection and prediction models to generate results that indicated the degree of equipment or process degradation. The inference results were stored in a database and abnormal detection and prediction outcomes were sent to the visual dashboard of the intelligent management system, providing users with insights into the machine conditions. The system issued alert notifications to relevant personnel if the results of the abnormal detection and prediction exceeded the threshold of the AI equipment or process degradation. When retraining the model with a small amount of data using PQUM-DNS, the anomaly detection model can adapt after the first training of the Pretrain and Metatrain models, and then use a small amount of data to fine-tune retraining to obtain the new model.

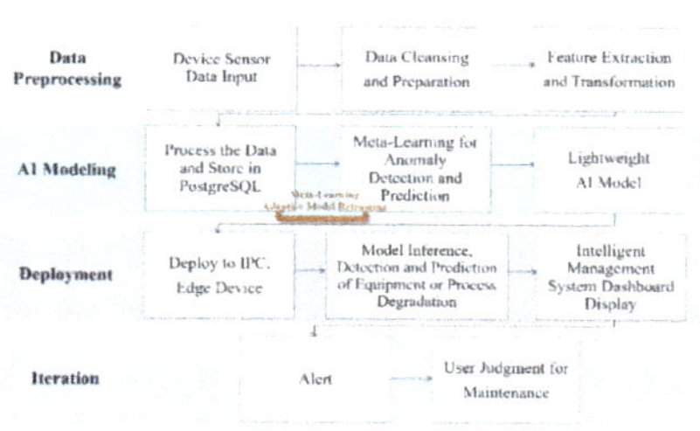

Figure 8. Flowchart of the proposed intelligent equipment management system.

3.2. Meta-Learning for Rapid Training of Multi-Machine Models for Anomaly Detection and Prediction

The rapid training of meta-learning multi-machine models for anomaly detection and prediction used in the PQUM-DNS was introduced in a previous study [19]. This method utilizes abundant data from numerous machines, trains the Metatrain model using the AE + meta-learning approach, and fine-tunes the Metatrain model with a small amount of data from new machines. This process yields a model adapted to a new machine, facilitating the inference for anomaly detection. In this study, the anomaly detection results were combined with prediction models to forecast future anomalies based on past anomaly detection outcomes. The following provides a brief introduction to the techniques employed.

(1) Meta-learning

Meta-learning is a technique aimed at enabling machine-learning systems to swiftly adapt to new tasks or environments [4]. Traditional machine-learning algorithms often require large amounts of labeled data to train models. In addition, it is necessary to collect and label substantial data for retraining when faced with new tasks. In contrast, the goal of meta-learning is to train a "learner" that is capable of rapidly learning new tasks from a small amount of labeled data. This approach typically relies on prior experience with numerous similar tasks and applies this experience to new tasks. These tasks can be expressed as:

$$\text{Learn } \theta \text{ such that } \Phi_i = f_\theta(D_i^{tr}) \text{ is good for } D_i^{ts} \quad (1)$$

$$\theta^* = \underset{\theta}{argmax} \sum_{i=1}^{n} logP(\phi_i | D_i^{ts}) \quad (2)$$

where

$$\Phi_i = f_\theta(D_i^{tr}) \quad (3)$$

$$D_{meta-train} = \{(D_1^{tr}, D_1^{ts}), (D_n^{tr}, D_n^{ts})\} \quad (4)$$

$$Ti \begin{cases} D_i^{tr} = \{(x_1^i, y_1^i), \cdots, (x_k^i, y_k^i)\} \\ D_i^{ts} = \{(x_1^i, y_1^i), \ldots, (x_l^i, y_l^i)\} \end{cases} \quad (5)$$

where Ti is a (meta-learning) task.

An illustration of the meta-learning method is shown in Figure 9 [31].

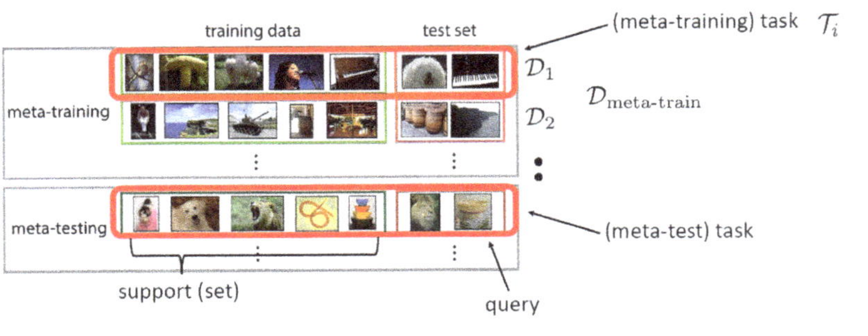

Figure 9. Illustration of the meta-learning method.

Various meta-learning models have been proposed for deep learning that are generally categorized as follows: learning good weight initializations, metamodels that generate the parameters of other models, and learning transferable optimizers. Model-agnostic meta-learning belongs in the first category and learns a good initial weight initialization to achieve fast adaptation to new tasks, enabling rapid convergence and fine-tuning of small-scale training samples [32].

(2) Anomaly Detection

The AI degradation level index was used to detect anomalies and the model was established using an unsupervised AE algorithm, as shown in Figure 10. This method involves constructing a model with normal data and applying an AE to compute the root mean square error (RMSE) between the input and output data, referred to as the reconstruction error. Here, the reconstruction error was defined as the AI degradation level index. A smaller value indicates a closer alignment between the model input and output values, leading to a better data reconstruction capability and a higher likelihood of normal equipment or processes.

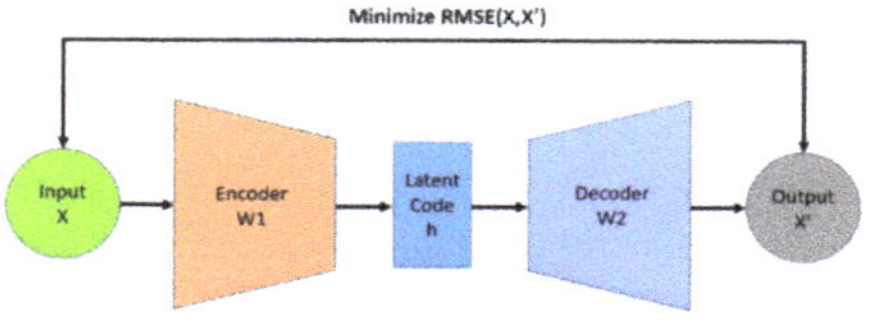

Figure 10. Structure of the autoencoder.

In practical applications, suitable threshold values are defined based on the conditions of the equipment or processes used for anomaly determination. These are expressed as:

$$h = g(W_1X + b_1) \tag{6}$$

$$X' = f(W_2h + b_2) \tag{7}$$

$$Minimize\ RMSE\left(X, X'\right) \tag{8}$$

(3) Anomaly Prediction

Time-series algorithms can use historical data from the past to predict future trends. In this study, historical records of the health status of semiconductor manufacturing processes and machine equipment in the past can be used to predict future health status using time-series algorithms. The proposed anomaly detection models trained on the anomaly detection dataset could only detect current and past data anomalies. Therefore, an additional anomaly prediction model was required for future anomaly states. The historical data of the AI degradation level index obtained from anomaly detection were compared with various commonly used prediction models. Finally, the Holt–Winters algorithm was applied to the anomaly prediction model [7].

Various anomaly prediction algorithms are introduced below and the overview is in Table 3.

Table 3. Overview of time-series algorithms.

Algorithm	Overview (or Main Features)
Simple exponential smoothing (SES)	By placing a large weight on the most recent data, only two values and constants are needed to predict the next period of value
Holt (Holt's linear trend)	Holt's linear trend, which predicts trends in data, consists of a prediction equation and two smoothing equations for sequences that have a linear trend and no seasonality
Holt–Winters (Holt–Winters seasonal method)	The Holt–Winters method forecasts time series with both trends and seasonality
Autoregressive (AR)	The historical data of the variable itself are used to predict its own data, and the autoregression must meet the requirements of stationarity
Moving Average (MA)	A simple smoothing forecasting technique that calculates the sequence average of a certain number of items in turn according to the time-series data and the passage of time items to reflect the long-term trend
Autoregressive integrated moving average (ARIMA)	In the case analysis of non-stationary time series, the originally non-stationary time series becomes a stationary time series after many differences
Seasonal ARIMA (SARIMA)	ARIMA (differentially integrated moving average autoregressive) time-series-forecasting method with seasonal periodicity

(a) Simple Exponential Smoothing (SES)

This algorithm is used when there is no clear trend or seasonal pattern in the predictive data [33,34]. The prediction is calculated using weighted averages, meaning that the largest and smallest weights are associated with the most recently and least recently observed values, respectively. This is expressed as:

$$\hat{y}_{T+1|T} = \alpha y_T + \alpha(1-\alpha)y_{T-1} + \alpha(1-\alpha)^2 y_{T-2} + \ldots \tag{9}$$

where $\hat{y}_{T+1|T}$ denotes the one-step-ahead forecast for time $T + 1$, y_T denotes the most recent observation, and $0 \leq \alpha \leq 1$ denotes the smoothing parameter.

(b) Holt (Double Exponential Smoothing Method)

The Holt double exponential smoothing method is an extension of the simple exponential smoothing method that predicts trends in data [9]. This method is suitable for linear trending sequences without seasonal patterns and consists of one prediction and two smoothing equations that represent the level and trend components (l_t, b_t), which are respectively expressed as:

$$\hat{y}_{t+h|t} = l_t + hb_t \tag{10}$$

$$l_t = \alpha y_t + (1-\alpha)(l_{t-1} + b_{t-1}) \tag{11}$$

$$b_t = \beta^*(l_t - l_{t-1}) + (1-\beta^*)b_{t-1} \tag{12}$$

where y_t and l_t denote the observed value and level at time t, respectively. b_t, h, α, and β^* denote the trend at time t, weight for the level ($0 \leq \alpha \leq 1$), and weight for the trend ($0 \leq \beta^* \leq 1$), respectively.

(c) Holt–Winters Forecasting (Triple Exponential Smoothing)

Holt–Winters forecasting, also known as triple exponential smoothing, is a method used to predict the behavior of time-series data that includes trends and seasonality. This algorithm considers three factors: the level l_t, trend b_t, and seasonal component s_t. It is effective for forecasting time-series data with seasonality patterns. There are two variations of this method: the additive and multiplicative models.

In the additive model, the components are expressed as:

$$\hat{y}_{t+h|t} = l_t + hb_t + s_{t+h-m(k+1)} \tag{13}$$

$$l_t = \alpha(y_t - s_{t-m}) + (1-\alpha)(l_{t-1} + b_{t-1}) \tag{14}$$

$$b_t = \beta^*(l_t - l_{t-1}) + (1-\beta^*)b_{t-1} \tag{15}$$

$$s_t = \gamma(y_t - l_{t-1} - b_{t-1}) + (1-\gamma)s_{t-m} \tag{16}$$

In the multiplicative model, the components are expressed as:

$$\hat{y}_{t+h|t} = (l_t + hb_t)s_{t+h-m(k+1)} \tag{17}$$

$$l_t = \alpha\frac{y_t}{s_{t-m}} + (1-\alpha)(l_{t-1} + b_{t-1}) \tag{18}$$

$$b_t = \beta^*(l_t - l_t - 1) + (1-\beta^*)b_{t-1} \tag{19}$$

$$s_t = \gamma\frac{y_t}{(l_{t-1} + b_{t+1})} + (1-\gamma)s_{t-m} \tag{20}$$

where s_t, k, m, α, β^*, and y denote the season at time t, integer part of $(h-1)/m$, number of cycles/frequency of seasonality (e.g., four for quarterly data), level smoothing parameter, trend smoothing parameter, and seasonal smoothing parameters/weight for the season ($0 \leq \gamma \leq 1$), respectively.

In the additive model, the forecast value for each data element is the sum of the baseline, trend, and seasonality components. However, a multiplicative model is preferred when seasonal variations change proportionally to the level of the series.

(d) Autoregressive Model

The autoregressive (AR) model is a statistical method used to analyze time-series data that predict the future value of a variable using its own historical data [35]. AR is an evolution of linear regression analysis, where it analyzes the relationship between parameter x and its own past value instead of analyzing the relationship between the parameter x and dependent variable y. This is expressed as:

$$y_t = C + \phi_1 y_{t-1} + \phi_2 y_{t-2} + \cdots + \phi_P y_{t-p} + \varepsilon_t \tag{21}$$

where y_t, C, p, ϕ_i, and ε_t denote the stationary time series, constant term, autoregressive order, non-zero autocorrelation coefficients, and independent error term, respectively.

(e) Moving Average

Moving average (MA) is a simple smoothing prediction technique used for time-series data that calculates a moving average over a certain number of terms to reflect long-term trends [36]. However, it is difficult to discern the development trend when time-series data are influenced by periodic and random variations causing large fluctuations. Using MAs can eliminate these influences and reveal the direction and trend of the events, which is expressed as:

$$y_t = \mu + \varepsilon_t + \theta_1 \varepsilon_{t-1} + \theta_2 \varepsilon_{t-2} + \ldots + \theta_q \varepsilon_{t-q} \tag{22}$$

where y_t, μ, q, and θ_i denote the stationary time series, mean of the sequence, moving average order, and non-zero autocorrelation coefficients, respectively.

(f) Autoregressive Integrated Moving Average

Autoregressive integrated moving average (ARIMA) is an evolution of the AR, MA, and autoregressive moving average models. This approach is used to analyze non-stationary time-series data by transforming them into stationary data through differencing [37]. This method is employed when dealing with non-stationary time-series data that exhibit a changing mean and variance over time. A new stationary time series can be obtained by using the differences in the data, and a suitable probabilistic model can be derived from historical data to represent the dependence between time and data. ARIMA can be expressed as ARIMA (p, d, q), where p, d, and q denote the autoregressive, differencing, and moving average orders, respectively. Furthermore:

$$y'_t = c + \phi_1 y'_{t-1} + \ldots + \phi_P y'_{t-P} + \theta_1 \varepsilon_{t-1} + \ldots + \theta_q \varepsilon_{t-q} + \varepsilon_t \tag{23}$$

where ϕ_i and θ_i denote non-zero autocorrelation coefficients.

(g) Seasonal Autoregressive Integrated Moving Average

The seasonal autoregressive integrated moving average (SARIMA) model incorporates seasonal factors into the ARIMA model [38]. Generally, the SARIMA model is denoted as SARIMA (p, q, d)(P, Q, D)s, where s, P, Q, D denote the seasonal period and seasonal autoregressive, seasonal moving average, and seasonal differencing orders, respectively. This is expressed as:

$$\begin{aligned} &\left(1 - \sum_{i=1}^{p} \varphi_i B^i\right)\left(1 - \sum_{i=1}^{p} \Phi_i B^{im}\right)(1-B)^d \left(1 - B^{Dm}\right) y_t = \\ &\left(1 + \textstyle\sum_{i=1}^{q} \theta_i B^i\right)\left(1 + \textstyle\sum_{i=1}^{q} \Theta_i B^{im}\right)\epsilon_t \end{aligned} \tag{24}$$

where B denotes the lag operator, and Φ_i and Θ_i denote non-zero constants.

3.3. Meta-Learning Adaptive Model Retraining

Meta-learning was employed to develop an adaptive method for retraining machine-specific models [4]. This approach enables models to quickly adjust to the slow changes observed in each machine over time, thereby achieving prolonged anomaly detection. The concept of rapid training models for different machine devices was extended to different time segments by combining the background technique of AEs with meta-learning.

The data were segmented into three intervals based on chronological order: Pretrain (older and long-running machine data), Metatrain (newer and long-running machine data), and Fine-tune (latest operational data). Leveraging the principles of meta-learning, Pretrain and Metatrain data were used to train a generalized anomaly detection model, whereas Fine-tune adapted the model to the most recent machine conditions. This approach automatically trains models that adapt to data changes over time.

Model retraining was based on the operational conditions of the factory to maintain the effectiveness of the anomaly detection model. The historical data were also divided into Pretrain, Metatrain, and Fine-tune segments by utilizing seven days of equipment operation data, as shown in Figure 11. This process led to the training of an anomaly detection model that could assess the degree of equipment degradation, as shown in Figure 12. Additionally, predictive algorithms were applied to forecast equipment degradation over the next seven days. The ultimate goal was to achieve continuous automatic updates for anomaly detection and prediction.

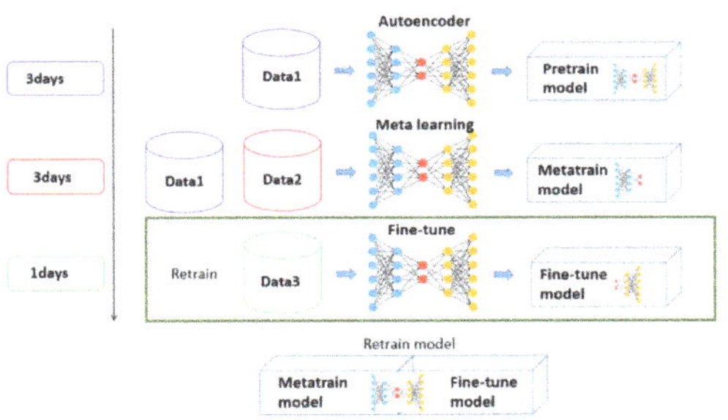

Figure 11. Historical data of seven days segmented into Pretrain, Metatrain, and Fine-tune for training an anomaly detection model to detect equipment degradation.

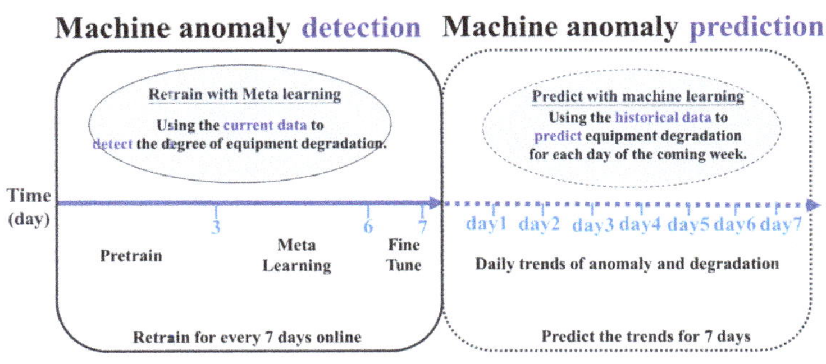

Figure 12. Automatic model retraining based on time intervals, with the retrained model used for predicting anomalies in the next seven days.

3.4. Lightweight AI Model

A pruning and quantization-based meta-learning anomaly detection model was introduced owing to the possibility of certain neural neuron weights being small or negligible during the retraining process of the neural network of the proposed model. This approach significantly reduced the model size and enhanced the computational speed.

(1) Model Pruning

Deep-learning neural network models often contain redundant parameters, with many neuron weights approaching zero. Model pruning involves removing these neurons while preserving the same model expressive capability. Model pruning retains the essential weights and parameters, reducing the number of connections between the neural network layers, as shown in Figure 13 [39]. This reduction helps to decrease the number of parameters involved in the calculations, thereby lowering the computation requirements. By maintaining the performance of the model, this approach reduces the storage space, lowers computational costs, and accelerates the training process.

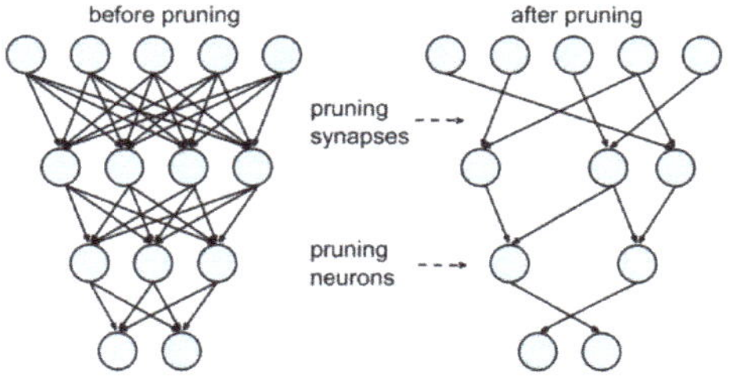

Figure 13. Model pruning retains important weights and parameters while reducing the number of connections between neural network layers.

Pruning algorithms typically employ a three-stage pipeline: training, pruning, and fine-tuning. The weight adjustment process in the three-step training pipeline for pruning is shown in Figure 14. The model's weights are trained, pruning techniques are applied to remove neurons with weights approaching zero, and the model is fine-tuned to adjust the remaining weights to approximate the performance of the original model. This iterative process helps align the performance of the model with that of the original model.

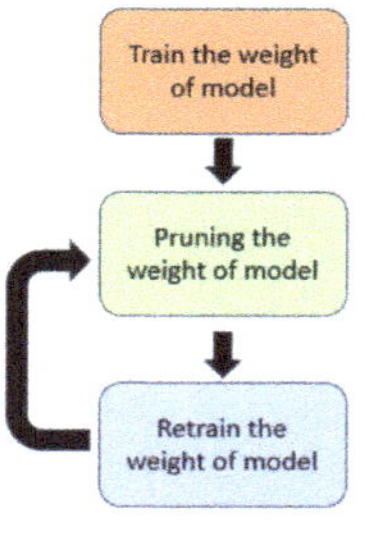

Figure 14. Weight adjustment process in the three-step training pipeline for pruning.

(2) Model Quantization

The principle of quantization involves reducing the precision of the bits used to represent model parameters (typically 32-bit floating-point (float32) numbers) [40]. This approach results in smaller model sizes and faster computations. Model quantization involves approximating the continuous values (or a large number of possible discrete values) of the floating-point model weights with a limited set of discrete values (usually 8-bit integer (int8) numbers) at a lower inference accuracy loss, as shown in Figure 15 [40]. A lower-bit data type is used to approximate the finite-range floating-point data, which leads to a reduced model size, decreased memory consumption, and faster inference speed. The calculations are expressed as:

$$Q = \frac{R}{s} + z \quad (25)$$

$$s = (R_{max} - R_{min})/(S_{max} - S_{min}) \quad (26)$$

where R, Q, z, and s denote the real floating-point value, fixed-point value after quantization, fixed-point value after quantization of the 0 floating-point value, and minimum scale that can be represented after fixed-point quantization, respectively.

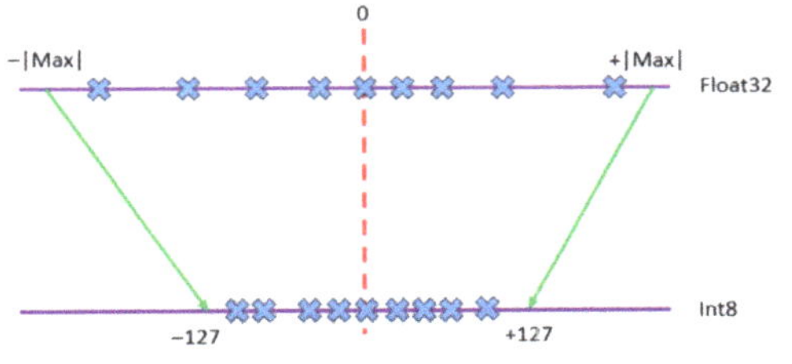

Figure 15. Model quantization: conversion from float32 to int8.

The model-pruning approach is applied using weight sparsity, where weights close to zero are removed from the original model. Subsequently, the model is retrained to adjust its performance. Additionally, quantization techniques are combined to convert the weights from float32 to int8, thereby significantly reducing the model size and enhancing the computational speed.

3.5. Edge Device Computing

The pruned and quantized models are deployed on embedded systems, such as a Raspberry Pi, replacing traditional IPCs. This lightweight approach conserves resources, reduces costs, and facilitates a large-scale deployment. Edge devices can analyze data and promptly alert onsite maintenance personnel in industrial scenarios where certain equipment is located in hard-to-reach locations [10,18]. This enables real-time responsiveness to the equipment conditions, allowing immediate intervention and preventing downtime.

Benefits of edge computing are as follows:

- Provides rapid real-time reflection of situations, enabling onsite personnel to detect anomalies promptly and take immediate action.
- Solves bandwidth issues in cloud and edge transmissions because edge devices only need to send inference results back to the control centers.
- Addresses cybersecurity concerns, protecting against network attacks that could lead to factory shutdowns.
- Reduces energy consumption because lightweight edge-computing models conserve power.

4. Results and Discussion

4.1. Intelligent Equipment Management System

Automated methods are used for vibration signal sensing, data transmission, data preprocessing, model training, and retraining. The results are visually presented through dashboards. The system utilizes AI degradation level values to detect anomalies and sends warning notifications for timely handling by managers and onsite personnel. A comparison between this system and traditional methods is presented in Table 4.

Table 4. Comparison of intelligent equipment management system and traditional approaches.

Issue	Intelligent Management System	Traditional Management System
Intelligent management	Integrate and automate equipment alarms, display visualization results in a Kanban style, and send warning notices to facilitate timely handling of problems by managers on site.	A manager will only be notified of a situation by onsite personnel when an abnormality occurs in the equipment or the production line stops, thus not dealing with the situation in a more timely manner.
Meta-learning anomaly detection and prediction	Apply meta-learning to quickly train AI models for automatic detection and prediction of new equipment anomalies for preventive equipment maintenance.	New machine models require a great deal of data to train, personnel need to confirm the condition of the equipment from time to time, and preventative maintenance cannot be performed in advance.
Meta-learning adaptive modeling with retraining	Meta-learning can be used to quickly adapt to the characteristics of machines that change slowly over time, thus realizing the purpose of model updating over a long period of time.	The model is retrained by AI analysts when an abnormality occurs in the model, which is labor-intensive and increases risk to the equipment.
Lightweight quantitative AI models	Dramatically reduces the size of the model and increases the speed of the operation.	Larger models consume more hardware space for storage and run more slowly.
Edge computing	It can be lightweight, save resources, reduce cost, and achieve the purpose of large-scale parts.	Larger PC computing devices are bulky, heavy, costly, energy-intensive, lack mobility, and are difficult to deploy on a large scale.

4.2. Meta-Learning for Rapid Training of Multi-Machine Models for Anomaly Detection and Prediction

For the sake of versatility, meta-learning is applied to multi-machine anomaly detection and prediction models with different datasets, such as factory chiller vibration data, publicly available datasets for analyzing the vibration and current of Paderborn bearings, and SECOM semiconductor analysis.

(1) Rapid Training of Multi-Machine Models for Anomaly Detection

The proposed PQUM-DNS method is compared with the general DAE method. PQUM-DNS achieves similar AUC values with minimal data compared to those of the DAE method when used to train models for new machines. This is because PQUM-DNS used meta-learning to train a versatile model applicable to various conditions (Metatrain model). Therefore, training a new machine model only requires a small amount of data for fine-tuning, resulting in a rapidly adaptable anomaly detection model. Test results for different data types, such as chiller vibration, SECOM, and Paderborn current and vibration datasets are presented in Table 5 and Figure 16. Compared with the DAE, PQUM-DNS reduces the required data for training new machine models on average by approximately 75%, with a decrease in AUC performance of only 0.35%, which is a very close AUC value.

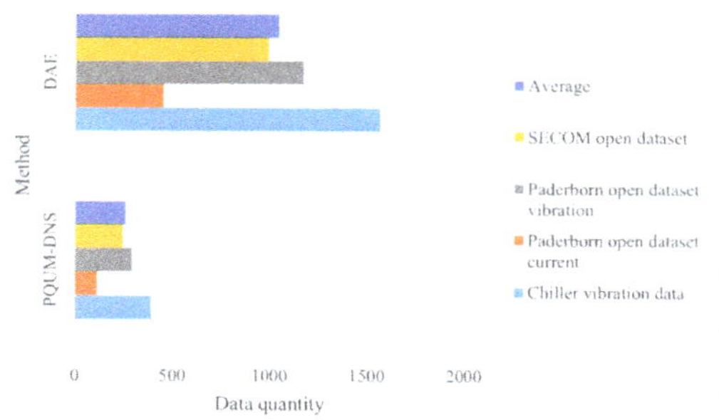

Figure 16. Comparison of PQUM-DNS and DAE training data.

Table 5. Comparison of PQUM-DNS and DAE training performance for different data types.

	Training				
	Chiller Vibration Data	**Paderborn Open Dataset Current**	**Paderborn Open Dataset Vibration**	**SECOM Open Dataset**	**Average**
	PQUM-DNS				
AUC (%) Data quantity	99.99 392	92.73 113	99.81 293	68.97 248	90.38 262
	DAE				
AUC (%) Data quantity	99.99 1569	97.03 452	99.73 1172	66.14 993	90.73 1047
	Improvement				
AUC Increase (%)	0.00	−4.3	0.07	2.83	−0.35
Data quantity Decrease (%)	75.02	75	75	75.03	75.01

(2) Anomaly Prediction:

PQUM-DNS detects machine degradation levels through anomaly detection and evaluated these levels using various prediction algorithms. Seven prediction algorithms are compared and one based on RMSE calculations is selected based on the predicted and actual values. The performances of these algorithms on different datasets are presented in Table 6 and Figure 17. The Holt–Winters algorithm demonstrates the best performance, with the lowest RMSE value of approximately 0.037, making it the chosen algorithm for anomaly prediction in the PQUM-DNS.

Table 6. Comparison of PQUM-DNS prediction algorithm performance on different datasets.

	RMSE				
Algorithm	**Chiller Vibration Data**	**Paderborn Open Dataset Current**	**Paderborn Open Dataset Vibration**	**SECOM Open Dataset**	**Average**
SES	0.03350	0.05153	0.02934	0.03547	0.03746
Holt	0.03330	0.05824	0.03348	0.07614	0.05029
Holt–Winters	0.03315	0.05161	0.02928	0.03395	0.03700
AR	0.03855	0.06250	0.03052	0.03500	0.04164
MA	0.03853	0.06289	0.03057	0.03505	0.04176
ARIMA	0.03349	0.05062	0.02933	0.03606	0.03737
SARIMA	0.03347	0.05088	0.02932	0.03626	0.03748

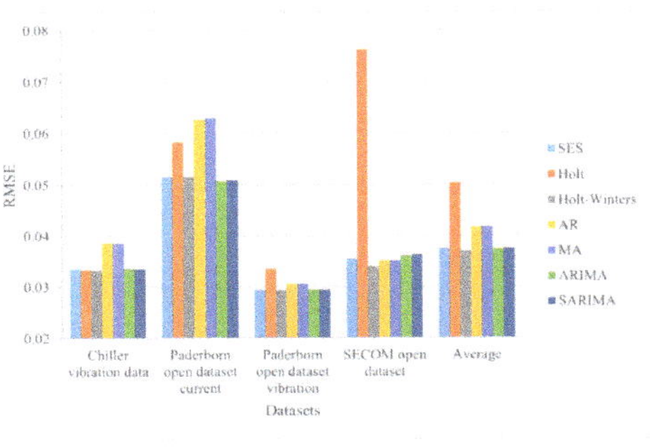

Figure 17. Comparison of PQUM-DNS prediction algorithm performance on different datasets.

4.3. Meta-Learning Adaptive Model Retraining

An adaptive method is employed for retraining the machine model using meta-learning. This enabled the model to adapt quickly to gradual changes over time, thereby facilitating long-term anomaly detection. PQUM-DNS chronologically segments data from the same machine and fine-tunes the model using the latest data, thereby achieving a model suited to the machine's latest condition. Unlike the general DAE method that requires retraining with all data, PQUM-DNS significantly reduces the amount of data needed for retraining, as shown in Table 7. This is because PQUM-DNS already trains a versatile meta-learning model using past data (a meta-trained model), enabling efficient fine-tuning with a small amount of new machine data.

Table 7. Comparison of PQUM-DNS and DAE retraining performance for different data types.

	Retraining				
	Chiller Vibration Data	**Paderborn Open Dataset Current**	**Paderborn Open Dataset Vibration**	**SECOM Open Dataset**	**Average**
	PQUM-DNS				
Retraining data quantity (fine-tune)	586	54	50	248	235
AUC	99.99	94.68	99.99	62.6	89.32
	DAE				
Retraining data quantity	2346	219	204	993	941
AUC	99.99	96.17	99.99	66	90.54
	Improvement				
Retraining data quantity decrease	75.02	75.34	75.49	75.03	75.07
AUC increase	0.00	−1.49	0.00	5.15	1.35

4.4. Lightweight AI Model

PQUM-DNS drastically reduces the model size while maintaining a similar AUC performance to that of the non-lightweight DAE model. This is achieved by removing the near-zero weights from the original DAE-trained model and compressing the model data format from float32 to int8. Consequently, the model size is significantly reduced. The application of PQUM-DNS to various datasets demonstrates that the lightweight model size is reduced by approximately 60% with AUC performance maintained at similar levels, as shown in Table 8 and Figure 18.

Table 8. Comparison of PQUM-DNS and DAE model sizes for different data types.

	Model Size				
	Chiller Vibration Data	**Paderborn Open Dataset Current**	**Paderborn Open Dataset Vibration**	**SECOM Open Dataset**	**Average**
	PQUM-DNS				
Model size (bytes)	5776	6969	5973	5866	6146
AUC	99.31	93.16	99.34	69.69	90.38
	DAE				
Model size (bytes)	18767	14765	19466	11866	16216
AUC	99.31	93.16	99.34	69.69	90.38
	Improvement				
Model size decrease	69	53	69	51	60
AUC increase	0.00	0.00	0.00	0.00	0.00

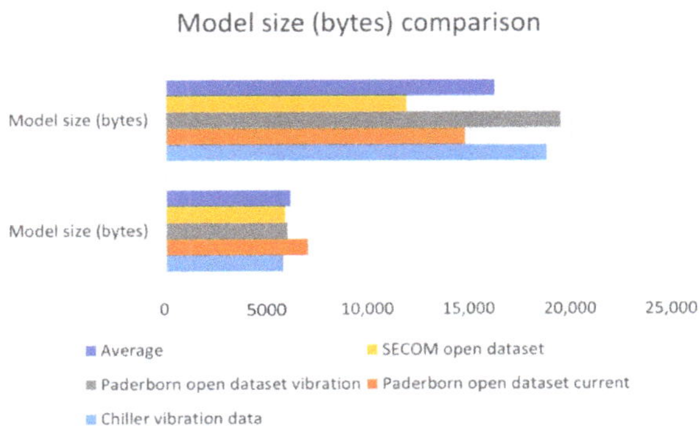

Figure 18. Comparison of PQUM-DNS and DAE model sizes for different data types.

4.5. Edge Device Computing

The PQUM-DNS, with its reduced and compressed model size, is well suited for lightweight edge-computing devices. Therefore, it is applied to replace traditional IPCs with embedded systems, such as a Raspberry Pi. This substitution reduces the size and weight, conserves resources, lowers costs, and supports large-scale deployments.

5. Conclusions

This study proposes a new PQUM-DNS model, which is an intelligent device management system that combines pruning, quantization, meta-learning, anomaly detection, prediction using AEs, adaptive model retraining, and edge inference. This system effectively reduces the manual labor, provides fault notifications, prevents downtime, decreases model computational resources, accelerates the model inference speed, and enables edge inference.

The system is suitable for various factory scenarios and types of machine equipment and process states. Compared with general DAEs, the system achieves a similar AUC while reducing the training data by approximately 75%. The average RMSE of the predictive degradation degree is 0.037 for Holt–Winters, retraining is conducted using 75% fewer data with similar AUC performance, and the model size is reduced by approximately 60% through pruning and quantization. The proposed system can be deployed on lightweight edge devices, such as a Raspberry Pi, enabling real-time anomaly detection and prediction. The system demonstrates superior performance, thereby realizing intelligent equipment management and maintenance.

Author Contributions: Conceptualization, Y.-C.Y.; methodology, Y.-C.Y. and S.-W.C.; software, Y.-C.Y.; validation, Y.-C.Y.; formal analysis, Y.-C.Y.; investigation, Y.-C.Y.; resources, C.-Y.L. and S.-W.C.; data curation, Y.-C.Y.; writing—original draft preparation, Y.-C.Y.; writing—review and editing, C.-Y.L. and S.-R.Y.; visualization, Y.-C.Y.; supervision, C.-Y.L., J.-T.C. and S.-W.C.; project administration, C.-Y.L., Y.-C.Y. and S.-R.Y.; funding acquisition, Y.-C.Y. All authors have read and agreed to the published version of the manuscript.

Funding: This research received no external funding.

Data Availability Statement: Publicly available datasets were analyzed in this study. This data can be found here: https://mb.uni-paderborn.de/kat/forschung/kat-datacenter/bearing-datacenter/data-sets-and-download, accessed on 1 June 2023 (Paderborn University Bearing Dataset) and https://archive.ics.uci.edu/dataset/179/secom, accessed on 1 June 2023 (SECOM Dataset).

Acknowledgments: Thanks for the data support from Advantech Taipei headquarters and the technical support from SI2 Laboratory at National Yang Ming Chiao Tung University.

Conflicts of Interest: The authors declare no conflict of interest.

Nomenclature

AE	Autoencoder
AI	Artificial Intelligence
AR	Autoregressive
ARIMA	Autoregressive Integrated Moving Average
AUC	Area under Curve
DAE	Deep AE
FFT	Fast Fourier Transform
IPC	Industrial Personal Computer
LoRaWAN	Long-Range Wide-Area Network
MA	Moving Average
MCM	Machine Condition Monitoring
MCS	Motor Current Signal
ML	Machine Learning
PHM	Prognostics and Health Management
PQUM-DNS	Pruning Quantized Unsupervised Meta-learning DegradingNet Solution
PSD	Power Spectral Density
RMS	Root Mean Square
RMSE	Root Mean Square Error
ROC	Receiver Operating Characteristics
SARIMA	Seasonal Autoregressive Integrated Moving Average
USB	Universal Serial Bus
WPD	Wavelet Packet Decomposition

References

1. Tung, F.; Mori, G. Clip-q: Deep network compression learning by in-parallel pruning-quantization. In Proceedings of the IEEE Conference on Computer Vision and Pattern Recognition, Salt Lake City, UT, USA, 18–22 June 2018; pp. 7873–7882.
2. Frantar, E.; Alistarh, D. Optimal brain compression: A framework for accurate post-training quantization and pruning. *Adv. Neural Inf. Process. Syst.* **2022**, *35*, 4475–4488.
3. Hu, P.; Peng, X.; Zhu, H.; Aly, M.M.S.; Lin, J. Opq: Compressing deep neural networks with one-shot pruning-quantization. *Proc. AAAI Conf. Artif. Intell.* **2021**, *35*, 7780–7788. [CrossRef]
4. Hospedales, T.; Antoniou, A.; Micaelli, P.; Storkey, A. Meta-learning in neural networks: A survey. *IEEE Trans. Pattern Anal. Mach. Intell.* **2021**, *44*, 5149–5169. [CrossRef] [PubMed]
5. Givnan, S.; Chalmers, C.; Fergus, P.; Ortega-Martorell, S.; Whalley, T. Anomaly detection using autoencoder reconstruction upon industrial motors. *Sensors* **2022**, *22*, 3166. [CrossRef]
6. Bradley, A.P. The use of the area under the ROC curve in the evaluation of machine learning algorithms. *Pattern Recognit.* **1997**, *30*, 1145–1159. [CrossRef]
7. Holt, C.C. Forecasting seasonals and trends by exponentially weighted moving averages. *Int. J. Forecast.* **2004**, *20*, 5–10. [CrossRef]
8. Jiang, W.; Xu, Y.; Chen, Z.; Zhang, N.; Xue, X.; Liu, J.; Zhou, J. A feature-level degradation measurement method for composite health index construction and trend prediction modeling. *Measurement* **2023**, *206*, 112324. [CrossRef]
9. Lehmann, A. Joint modeling of degradation and failure time data. *J. Stat. Plan. Inference* **2009**, *139*, 1693–1706. [CrossRef]
10. Bellavista, P.; Della Penna, R.; Foschini, L.; Scotece, D. Machine learning for predictive diagnostics at the edge: An IIoT practical example. In Proceedings of the ICC 2020–2020 IEEE International Conference on Communications (ICC), Dublin, Ireland, 7–11 June 2020; pp. 1–7.
11. Li, P.; Pei, Y.; Li, J. A comprehensive survey on design and application of autoencoder in deep learning. *Appl. Soft Comput.* **2023**, *138*, 110176. [CrossRef]
12. Singh, S.A.; Desai, K. Automated surface defect detection framework using machine vision and convolutional neural networks. *J. Intell. Manuf.* **2023**, *34*, 1995–2011. [CrossRef]
13. Tang, T.W.; Hsu, H.; Li, K.M. Industrial anomaly detection with multiscale autoencoder and deep feature extractor-based neural network. *IET Image Process.* **2023**, *17*, 1752–1761. [CrossRef]
14. Pradeep, D.; Vardhan, B.V.; Raiak, S.; Muniraj, I.; Elumalai, K.; Chinnadurai, S. Optimal Predictive Maintenance Technique for Manufacturing Semiconductors using Machine Learning. In Proceedings of the 2023 3rd International Conference on Intelligent Communication and Computational Techniques (ICCT), Jaipur, India, 19–20 January 2023; pp. 1–5.
15. Nuhu, A.A.; Zeeshan, Q.; Safaei, B.; Shahzad, M.A. Machine learning-based techniques for fault diagnosis in the semiconductor manufacturing process: A comparative study. *J. Supercomput.* **2023**, *79*, 2031–2081. [CrossRef]
16. Mao, W.; Feng, W.; Liu, Y.; Zhang, D.; Liang, X. A new deep auto-encoder method with fusing discriminant information for bearing fault diagnosis. *Mech. Syst. Signal Process.* **2021**, *150*, 107233. [CrossRef]

17. Abbasi, S.; Famouri, M.; Shafiee, M.J.; Wong, A. OutlierNets: Highly compact deep autoencoder network architectures for on-device acoustic anomaly detection. *Sensors* **2021**, *21*, 4805. [CrossRef]
18. Yazici, M.T.; Basurra, S.; Gaber, M.M. Edge machine learning: Enabling smart internet of things applications. *Big Data Cogn. Comput.* **2018**, *2*, 26. [CrossRef]
19. Yu, Y.-C.; Chuang, S.-W.; Shuai, H.-H.; Lee, C.-Y. Fast Adaption for Multi Motor Anomaly Detection via Meta Learning and deep unsupervised learning. In Proceedings of the 2022 IEEE 31st International Symposium on Industrial Electronics (ISIE), Anchorage, AK, USA, 1–3 June 2022; pp. 1186–1189.
20. Advantech. WISE-2410—LoRaWAN Wireless Vibration Sensor—Advantech. 2023. Available online: https://www.advantech.com/en/products/b7e2306f-d561-4ca9-b0e3-33f7057e185f/wise-2410/mod_25018dc7-355c-40b4-bf9b-c93f6c73f1a0 (accessed on 1 June 2023.).
21. Advantech. WebAccess_MCM_DS(07.18.17)—Advantech Support—Advantech. 2023. Available online: https://www.advantech.com/en/support/details/datasheet?id=b5660e1c-d223-40ed-86bd-bdab7be541d7 (accessed on 1 June 2023.).
22. Artono, B.; Susanto, A.; Hidayatullah, N.A. Design of Smart Device for Induction Motor Condition Monitoring. *J. Phys. Conf. Ser.* **2021**, *1845*, 012035. [CrossRef]
23. Lessmeier, C.; Kimotho, J.K.; Zimmer, D.; Sextro, W. Condition monitoring of bearing damage in electromechanical drive systems by using motor current signals of electric motors: A benchmark data set for data-driven classification. In Proceedings of the European Conference of the PHM Society, Bilbao, Spain, 5–8 July 2016; Volume 3.
24. Huang, S.R.; Huang, K.H.; Chao, K.H.; Chiang, W.T. Fault analysis and diagnosis system for induction motors. *Comput. Electr. Eng.* **2016**, *54*, 195–209. [CrossRef]
25. Salem, M.; Taheri, S.; Yuan, J.S. An experimental evaluation of fault diagnosis from imbalanced and incomplete data for smart semiconductor manufacturing. *Big Data Cogn. Comput.* **2018**, *2*, 30. [CrossRef]
26. Ho, T.K. The random subspace method for constructing decision forests. *IEEE Trans. Pattern Anal. Mach. Intell.* **1998**, *20*, 832–844.
27. Ho, T.K. Random decision forests. In Proceedings of the 3rd International Conference on Document Analysis and Recognition, Montreal, QC, Canada, 14–16 August 1995; Volume 1, pp. 278–282.
28. Dua, D.; Graff, C. *UCI Machine Learning Repository*; University of California, School of Information and Computer Science: Irvine, CA, USA, 2019; Available online: https://archive.ics.uci.edu/ml/datasets.php (accessed on 1 June 2023.).
29. Huang, J.; Ling, C.X. Using AUC and accuracy in evaluating learning algorithms. *IEEE Trans. Knowl. Data Eng.* **2005**, *17*, 299–310. [CrossRef]
30. Type I Error and Type II Error. Available online: https://explorable.com/type-i-error (accessed on 12 February 2024).
31. ICML. ICML 2019 Meta-Learning Tutorial. 2019. Available online: https://sites.google.com/view/icml19metalearning (accessed on 1 June 2023.).
32. Finn, C.; Abbeel, P.; Levine, S. Model-agnostic meta-learning for fast adaptation of deep networks. In Proceedings of the International Conference on Machine Learning, Sydney, Australia, 6–11 August 2017; pp. 1126–1135.
33. Gardner, E.S., Jr. Exponential smoothing: The state of the art. *J. Forecast.* **1985**, *4*, 1–28. [CrossRef]
34. Ostertagova, E.; Ostertag, O. Forecasting using simple exponential smoothing method. *Acta Electrotech. Et Inform.* **2012**, *12*, 62. [CrossRef]
35. Amo-Salas, M.; López-Fidalgo, J.; Pedregal, D.J. Experimental designs for autoregressive models applied to industrial maintenance. *Reliab. Eng. Syst. Saf.* **2015**, *133*, 87–94. [CrossRef]
36. Zhao, Z.; Liu, F. Industrial monitoring based on moving average PCA and neural network. In Proceedings of the 30th Annual Conference of IEEE Industrial Electronics Society, IECON 2004, Busan, Republic of Korea, 2–6 November 2004; Volume 3, pp. 2168–2171.
37. Box, G.E.; Jenkins, G.M.; Reinsel, G.C.; Ljung, G.M. *Time Sries Analysis: Forecasting and Control*; John Wiley & Sons: Hoboken, NJ, USA, 2015.
38. Liang, Y.-H. Combining seasonal time series ARIMA method and neural networks with genetic algorithms for predicting the production value of the mechanical industry in Taiwan. *Neural Comput. Appl.* **2009**, *18*, 833–841. [CrossRef]
39. Han, S.; Pool, J.; Tran, J.; Dally, W. Learning both weights and connections for efficient neural network. *Adv. Neural Inf. Process. Syst.* **2015**, *28*.
40. Jacob, B.; Kligys, S.; Chen, B.; Zhu, M.; Tang, M.; Howard, A.; Adam, H.; Kalenichenko, D. Quantization and training of neural networks for efficient integer-arithmetic-only inference. In Proceedings of the IEEE Conference on Computer Cision and Pattern Recognition, Salt Lake City, UT, USA, 18–22 June 2018; pp. 2704–2713.

Disclaimer/Publisher's Note: The statements, opinions and data contained in all publications are solely those of the individual author(s) and contributor(s) and not of MDPI and/or the editor(s). MDPI and/or the editor(s) disclaim responsibility for any injury to people or property resulting from any ideas, methods, instructions or products referred to in the content.

applied sciences

MDPI

Article

Enhancing Reliability in Wind Turbine Power Curve Estimation

Pere Marti-Puig [1,*], Jose Ángel Hernández [2], Jordi Solé-Casals [1] and Moises Serra-Serra [1]

[1] Data and Signal Processing Group, University of Vic-Central University of Catalonia, 08500 Vic, Spain; jordi.sole@uvic.cat (J.S.-C.); moises.serra@uvic.cat (M.S.-S.)

[2] Advanced Signal Processing, Open University of Catalonia, 08018 Barcelona, Spain

* Correspondence: pere.marti@uvic.cat

Abstract: Accurate power curve modeling is essential to continuously evaluate the performance of a wind turbine (WT). In this work, we characterize the wind power curves using SCADA data acquired at a frequency of 5 min in a wind farm (WF) consisting of five WTs. Regarding the non-parametric methods, we select artificial neural networks (ANNs) to make curve estimations. Given that, we have the curves provided by the manufacturer of the WTs given by some very precisely measured pair of wind speed and power points. We can evaluate the difference between the manufacturer characterization and the ones estimated with the data provided by the SCADA system. Before the estimation, we propose a method of filtering the anomalies based on the characteristics provided by the manufacturer. We use three-quarters of the available data for curve estimation and one-quarter for the test. One WT suffered a break in the test part, so we can check how the test estimates reflect this problem in its wind-power curve compared to the estimations obtained in the WTs that worked adequately.

Keywords: wind power curve modeling; artificial neural networks (ANN); wind turbine (WT); SCADA data; industrial AI

1. Introduction

The world's needs for cleaner and renewable energy show no signs of slowing down, leaving markets seeking immediate solutions. Researchers face a daunting task: discovering ways to meet this ever-growing demand. Wind energy offers a promising possibility, particularly with the proliferation of wind farms.

While advances in wind technology are encouraging, many existing farms were built hastily, neglecting crucial long-term maintenance considerations.

Given that an essential part of the costs of generating wind energy originates in wind farms' operation and maintenance stage, the most relevant line of action is making wind energy more competitive by reducing these costs. Once the WFs are built, O&M costs are the few parts of the process involved in generating wind power that can be addressed. To do so, condition-based maintenance (CBM) is considered one of the best solutions [1]. Supervisory control and data acquisition (SCADA) systems are used for data collection in wind farms, providing operational and condition information about wind turbines. These systems collect information from sensors distributed over different parts of the WT and provide the recorded captures' maximum, minimum, average, and standard deviation values every 5/10 min. Although the SCADA systems were not originally designed to perform WTs prognosis, due to their wide availability and low cost (since we found them installed), much research has been generated to take advantage of their data.

As is well known, WTs are highly non-linear systems. One of the simplest and most efficient ways used by the manufacturers of WTs to characterize their performance is through the curve that relates the power generated as a function of the wind speed. These curves have four zones. The wind turbines do not generate power in zone 1, corresponding to low speeds. In zone 2, above a wind speed value, the generated power increases as the wind increases. In zone 3, the wind turbine generates its nominal power, and even if the wind speed continues to increase, the generated power remains constant. In zone 4, from a critical wind speed, the wind turbine stops for safety.

Given a wind speed profile, the power curve is used for the planning and development of wind farms and, as it is known that the malfunction of a wind turbine can drastically reduce its power generation capacity, and that the fact is reflected in this curve, it is also used to diagnose its health [2,3]. WT manufacturers certify these curves for specific wind conditions and turbine types, including air density and temperature at tower height. The International Electrotechnical Commission (IEC) standard 61400-12-1 [4] describes the certification process for measuring and recording a single turbine's power performance characteristics. In practice, the power curve is constructed using historical data from the supervisory control and data acquisition (SCADA) system, with several SCADA-based models proposed for this purpose [5–7]. It is not surprising, therefore, that among the various magnitudes captured by SCADA, wind speed and active power output have attracted the attention of researchers [7]. When the pair of wind speed and power points are plotted directly from the SCADA data, the points are distributed around the theoretical curve, exhibiting some dispersion due to measurement errors and showing variations from the curve of the manufacturer, among other causes, in the 5 min averages made by the SCADA system. In addition, from the representation of the power curves from the raw data, a set of points appears outside the expected area, far from the manufacturer-given points. In this context, we call such points anomalies. To correctly estimate the power curves, it is necessary to clean the data anomalies [8–10]. These anomalies remain after typical outlier detection based primarily on statistical techniques. Most of the applied filtering and anomaly detection techniques are still based on statistical criteria and apply rules based on interquartile ranges (IQRs), or other statistical measures such as skewness or kurtosis, or more sophistical methods based on ML techniques such as self-organized maps (SOMs) and linear mixtures (LSOMs), which are unsupervised methods.

According to the literature [2], power curve modeling methods can be classified into discrete, deterministic/probabilistic, parametric/non-parametric, and based on the data used for modeling methods [2,3,11]. This work has focused on modeling power curves with non-parametric methods, as they are suitable for extracting models from large amounts of data. So, concerning modeling wind turbines' power curves, the main non-parametric methods used explore artificial neural networks (ANN) [12], adaptive neuro-fuzzy inference system (ANFIS) [13,14], clustering [15], regression models, as well as different Machine Learning (ML) methods, such as support vector machine (SVM) and Gaussian Process (GP), due to their capabilities to modeling nonlinearities [16]. In addition, and in a very particular way to the estimation of power curves, copula models focus on learning the joint probability of wind speed and wind power [17].

As the estimation of power curves in wind turbines using SCADA data remains an active field, contributions are constantly made. Recently, improved estimations have been achieved mainly by the incremental sophistication of ML techniques. For example, in [18], super vector regressors are developed using a set of multivariate input features in which, in addition to mean variables, maximums, minimums and standard deviations provided by the SCADA are also considered. SVMs continue to be applied for a long time, as recently as in [19]. In [20], a deep learning technique based on convolutional neural networks (CNN) is introduced. To be able to train big networks, in this case, the authors first combine the action of multiple extreme learning machines (ELM) to increase the data dimensionality, followed by techniques to reduce redundant mapping before the application of CNNs. In general, data-driven ML-based methods exhibit a superior performance; however, they lack explainability, so a relevant line of research seeks to interpret the results in explainability, as in [21], or physical uncertainty quantification, as in [22]. Although most of the proposed estimators are data-driven, there are also contributions concerning parametric estimators, as in [23]. Due to the multiple contributions addressing the problem from different angles, it is always interesting to be aware of the latest state-of-the-art compilations, e.g., [24].

This work characterizes wind power curves using SCADA data, and is based on a new anomalies filtering technique approach and small artificial neural networks (ANNs) to estimate the curves. The filtering method introduced directly applies graphical information expressed as a condition to each pair of wind speed and power points. Removing such points using this method is fast. So, before making the estimation, a continuous power curve is obtained from the information provided by the WT manufacturer. The method we present is simple but effective, and especially practical compared to proposals appearing in the recent state of the art, where the dominant trend seems to add complexity to the methods. Although more complex methods can perform optimally in optimum conditions, they can face problems in real wind farm (WF) applications. Our method directly relates the two variables involved, the power and the wind speed, in the form of SCADA averages. A small-sized ANN performs well in the test without overfitting and requires only a few parameters to be tuned. That also requires less data to obtain a power curve estimation from new data, which helps to visualize possible changes faster. So, from our point of view, that is a valuable fact.

As for this work, we also have the curves of the particular WT model used in the experiments provided directly by the manufacturer; we can compare the characterization of the manufacturer concerning the estimated curves obtained from their estimation using the SCADA.

This work will be organized into Materials and Methods (Section 2), where we provide detailed descriptions of the following components: data used, Fuhrländer FL 2500/100 power curve provided by the manufacturer, anomaly filtering, and Wind Power curves estimation. The results are presented in Section 3, where we provide detailed descriptions of the following components: filtering and isolation of anomalies, using the filtered SCADA signals to estimate power curves, the training data, using the filtered SCADA signals to estimate power curves, and the test. Section 4 has a Discussion, and the main Conclusions are summarized in Section 5.

2. Materials and Methods

2.1. Data Used

This study utilizes a comprehensive three-year SCADA log from five 2.5 MW Fuhrläender FL2500 wind turbines. The SCADA system employs the IEC 61400-25 standard [25] communication protocol for transferring the wind turbine's data and storing it in a MySQL database (DB). The database encompasses 312 analogue variables sourced from 78 different sensors. The initial DB entry commences on 1 January 2012 at 01:00, and concludes on 8 December 2014 at 06:00. The Fuhrläender FL2500 comprises various components, including the grid, transmission, generator, converter, nacelle, hydraulic system, rotor, meteorological unit, turbine, and tower. Signal categorization is organized into nine groups based on the specific system to which the signal pertains. Additionally, events and statistical parameters such as mean, standard deviation, maximum, and minimum values are documented at five-minute intervals. The used data are extracted from the database freely available at https://github.com/alecuba16/fuhrlander (accessed on 15 February 2024).

Of all the measurements provided by the SCADA system, we mainly use two signals that collect the wind and the power measured for each WT, with an additional signal to confirm whether the WT is active or not. We use the wgdc_avg_TriGri_PwrAt signal provided by the generator subsystem (wgdc) of each WT for active power measurements. For the wind speed, the signals wnac_avg_WSpd1 are obtained from each WT's sensor in their Nacele system (wnac). Additionally, we use the rotation speed of the central axis of the turbine wgen_avg_RtrSpd_IGR in those cases where we need to verify that a WT is stopped despite high wind registers being measured. In all three signals, we use 5 min average (avg) values. We identify the five WTs with the following names: WT80, WT81, WT82, WT83 and WT84. Figure 1 shows the training portion of the wgdc_avg_TriGri_PwrAt signals, corresponding to the active power of each WT in the same temporal reference. The representation collects two-thirds of the total available signals. The third part is reserved for the test. Notice the strong similarity between signals, although it is occasionally observed that some WTs are not producing while the rest are.

Figure 1. Active power of each WT measured from the SCADA wgdc_avg_TriGri_PwrAt variable under the same temporal reference.

Figure 2 shows the training portion of the wnac_avg_WSpd1 signals, corresponding to the wind speed measured in the nacelle of each WT in the same temporal reference. As in Figure 1, the representation collects two-thirds of the available signals. We note the close similarities of the envelopes of the wind signals, much greater than the powers generated for each WT.

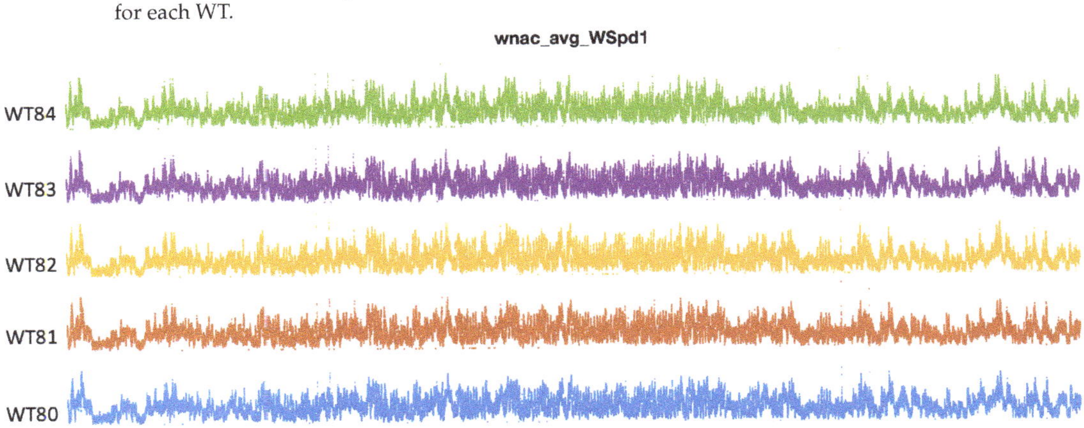

Figure 2. Wind speed measured in the nacelle of each WT provided by the SCADA signals wnac_avg_WSpd1.

In Figure 3, for each WT, we superimpose over the signals wgdc_avg_TriGri_PwrAt the rotation speed of the rotors provided by wgen_avg_RtrSpd_IGR, which are represented scaled and in black color, to confirm that in the intervals where power is not produced, the WT is indeed also stopped. So, the data are consistent and reliable.

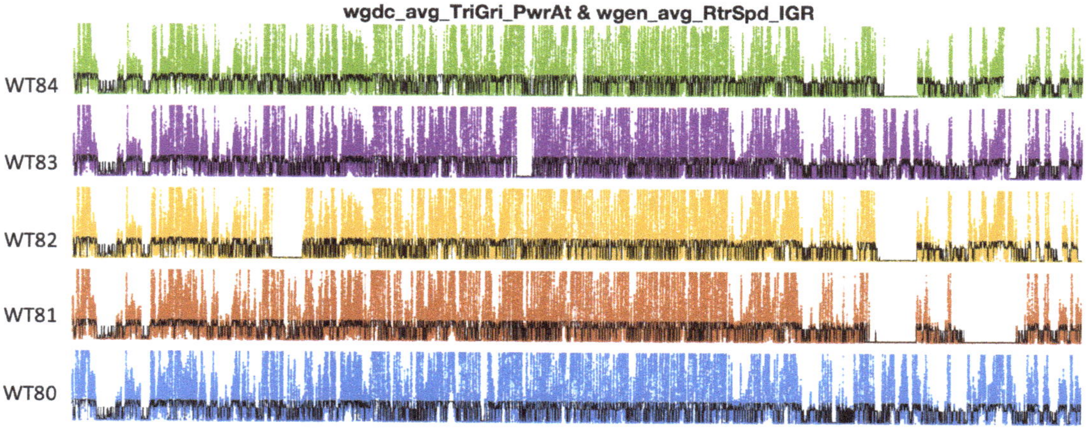

Figure 3. Active power of each WT (wgdc_avg_TriGri_PwrAt) in different colors and a scaled version of their main axes rotation speed (wgen_avg_RtrSpd_IGR) in black, to verify the consistency o the SCADA data.

2.2. Fuhrländer FL 2500/100 Power Curve Provided by the Manufacturer

The Fuhrländer FL 2500/100 manufacturer characterizes the power generated with the wind speed of its WT model at increments of 0.5 m/s. One can find such data at https://www.thewindpower.net/turbine_en_154_fuhrlander_fl-2500-100.php (accessed on 15 February 2024). Notice that the manufacturer provides a set of 50 pairs of wind-power points measured precisely following the recommendations of the measurement protocol, but does not provide the intermediate values. Indeed, carrying out these measurements is expensive. In the left graph of Figure 4, one can see those points represented by small black squares. So, the function on the left of Figure 4, shown in green, is made by connecting the points provided by the manufacturer with straight lines. The green function is, therefore, a function defined by parts. To carry out the filtering we propose, we need a continuous function defined for any wind value, not only for the values provided by the manufacturer. If the interest was a continuous function defined by parts, a better alternative solution could be to use the third-order splines which, in addition to providing continuity at the points, would give continuity for the derivatives in such points. In this case, to manipulate functions defined by such a large number of parts makes them few operational. In the representation on the right of Figure 4, in blue, we show the continuous function used to have a power value for the wind speeds between the points provided by the manufacturer from the sum of only seven sinusoidal functions, so it is very compact. Previously, to decide the used curve, we compared techniques such as polynomial approximations of different orders, developments in the Fourier series, and even third-order splines to make this approximation. We discarded the third-order splines because we did not want to handle a piecewise function with such many parts. Of the contrasted approximations, the decomposition into sinusoidal functions is the one that provides better results in terms of RMSE, better even than the Fourier series for the same number of terms. Using sinusoidal functions, we have verified that seven sinusoidal terms obtain a better approximation than nine. For the approximation of the curve, we use the following expression:

$$P_N(w) = \sum_{i=1}^{N} A_i sin(a_i w + \phi_i). \tag{1}$$

where $P_N(w)$ provides the power at the wind speed w. So, we particularize $P_N(w)$ for $N = 7$. We represent the values of amplitudes A_i, frequencies a_i, and phases ϕ_i in Table 1.

Table 1. Coefficients of $P_7(w)$.

i	A_i	a_i	ϕ_i
1	4131.2000	0.1487	−0.2597
2	2199.6000	0.1909	2.8832
3	385.6073	0.5622	1.5075
4	86.5623	1.1268	1.9258
5	19.0806	2.1933	−3.3415
6	30.4685	1.6828	−4.1747
7	12.4781	2.7778	−3.2777

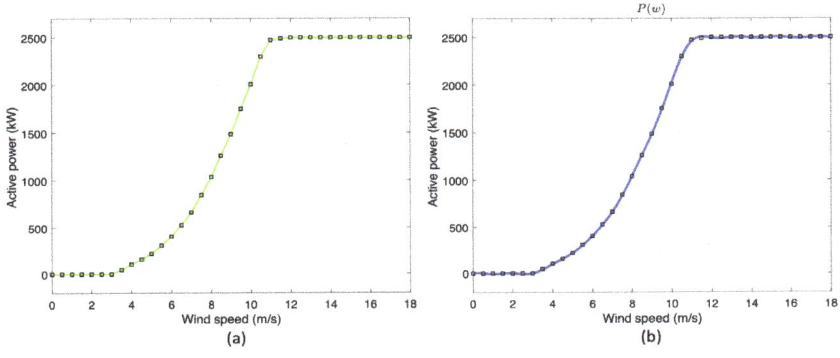

Figure 4. (**a**) Points in (black squares) of the power curve provided by the manufacturer. (**b**) Continuous function $P_7(w)$ in blue with the square points given by the manufacturer.

2.3. Anomaly Filtering

According to the distribution presented by these points (the anomalies), there are authors such as those of [9] who classify them according to three typologies: (1) power sensor failures, (2) wind speed sensors failure, or (3) due to the starting and stopping effect of the WT.

In Figure 5, we see represented the power curves of each WT resulting from directly crossing the wnac_avg_WSpd1 and wgdc_avg_TriGri_PwrAt signals provided by the SCADA. A total of 3/4 of the available signals are used. In some of these curves, anomalies appear, forming horizontal lines, which are attributed to failures of type (1), which are related to the power sensors. The anomalies distributed randomly inside the curves are of type (3), mainly attributed to the phenomenon of starting and stopping the WTs.

In any case, the anomalies introduce errors in estimating the wind power curves. Before the estimation of the power curves of each WT, it is necessary to filter them.

When estimating the power curve from the data, what is evident from the related literature is the need to filter out anomalies to obtain decent estimates. This filtering determines the rest of the processing chain. The work in [24], which compiles recent state-of-the-art literature on estimating wind-power curves, devotes a section to anomaly classification and determines up to seven different types of anomalies, five of which are systematically treated in the texts. The other two appear less referenced. When discussing anomalies in the context of power curves, the texts systematically explain them graphically, and do so based on the graphical representation of the curve. However, the filtering process is usually based on more or less complicated statistical criteria although, we repeat, they are visually identified clearly in a graph.

Our approach may be surprisingly simple, but is very effective. It uses such graphical information. Since we only have the manufacturer's information, we have an idea of the region of the graph where the points must appear. For the first filtering, we use such information. Since we know that the noise in the sensors, the different environmental conditions around the turbine (temperature, turbulence, humidity, etc.) and the averaging performed by the SCADA system generate some dispersion in the points around the curve, we select the candidate points in a margin around the nominal curve provided by the manufacturer. Since, as far as we know, the power performances, in practice, are usually slightly lower than those provided by the curve, we establish the criterion of selecting all the points that enter the region in the upper part represented in red in Figure 6. This red curve is made from the points provided by the manufacturer by shifting to the right and down a given distance.

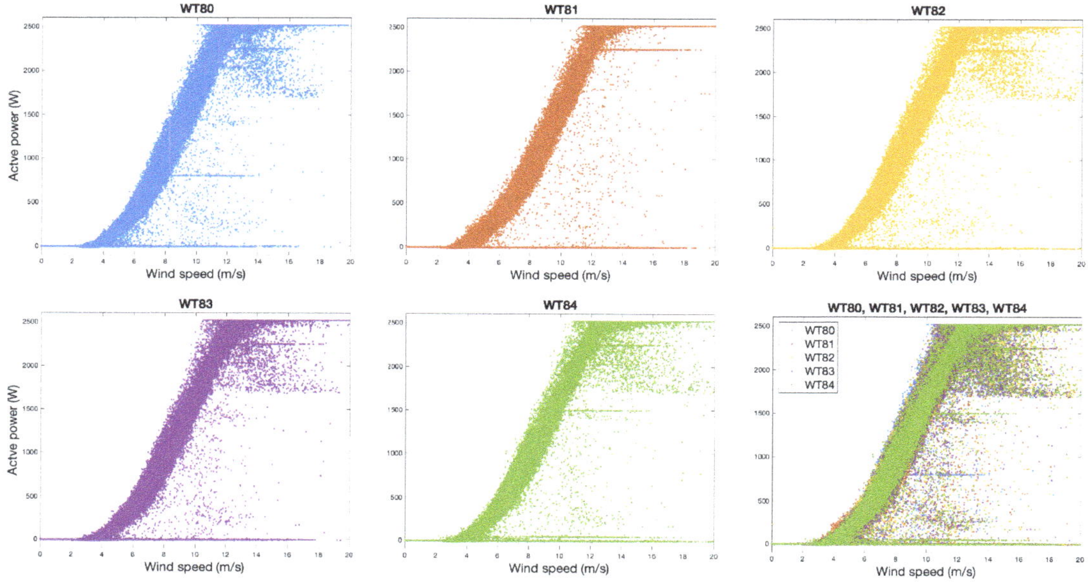

Figure 5. Power curves of each WT resulting from directly crossing the wnac_avg_WSpd1 and wgdc_avg_TriGri_PwrAt signals. The last subplot is the superimposition of all the data of the 5 WTs.

So, to filter the data of each WT, we use the continuous manufacturer's curve approximated by $P_7(w)$ given in (1) that we move horizontally to the right w_{off} and vertically down a distance p_{off}. Considering the signals wnac_avg_WSpd1 and wgdc_avg_TriGri_PwrAt at the point k to be w_k, p_k, respectively, then if the following inequality is fulfilled, we establish that p_k is an anomaly:

$$p_k < P_7(w_k - w_{off}) - p_{off} \tag{2}$$

We initially determine w_{off} and p_{off} empirically to 1m/s and 100W, respectively. Later on, we realize that small variations in such values do not affect the estimations.

Note that if w_{off} increases, we collect more points around the curve, but we can also enter some anomalies, mainly of type 3. Otherwise, maintaining a value of p_{off} of 100 allows us to collect the points of the curve that present slight measurement errors when the WTs are stopped or working at total capacity.

In Figure 6, we can see the graphic interpretation of the filtering rule where superimposed on the WT82 points, in red, is represented the function $P_7(w_k - 1.3) - 120$.

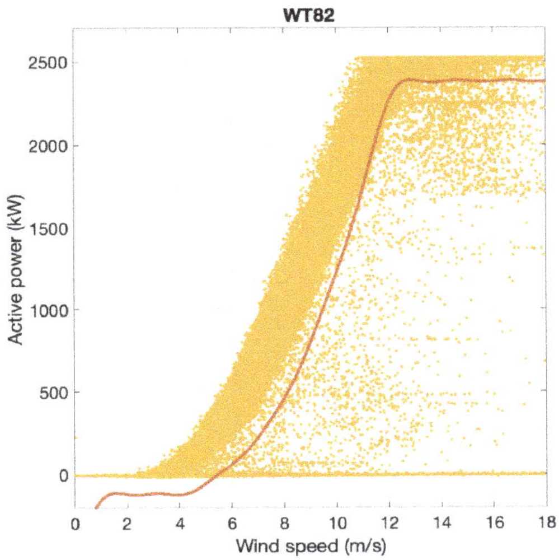

Figure 6. Graphical interpretation of the anomaly filtering rule. The wind speed-active power pairs measured by the SCADA system are represented in yellow dots, and the line marking the filtering decision is in red.

2.4. Wind Power Curves Estimation

Once the SCADA data anomalies have been filtered according to the rule described in the previous section, we estimate the wind power curve. Of all the existing non-parametric methods, we choose to use a classical neural network of relatively small dimensions, since we understand that the filtering quality used to eliminate anomalies and clean the data used for the training has more influence on the results than the chosen method or increase the dimensions of the network.

Then, considering s the wind speed and $p(s)$ the power for such s, the ANN model N hidden neurons takes the form:

$$p(s) = \mathbf{w}_o^T \sigma(\mathbf{w}_i s + \mathbf{b}_i) + b_o \quad (3)$$

where the column vectors $\mathbf{w}_i$ and $\mathbf{w}_o$ of size N contain the input and output weights, respectively. The same size vector $\mathbf{b}_i$ contains the input biases while the scalar b_o contains the output bias. Function $\sigma()$ stands for the activation function, which, in this case, applies the hyperbolic tangent sigmoid to all the elements of its input vector.

We use the Levenberg–Marquardt algorithm for training with 2/3 of the total available data. We chose N = 10. The test will use the last third of the available data.

Once the network weights are calculated, p(s) provides the curve directly.

3. Results

3.1. Filtering and Isolation of Anomalies

This section presents the result of filtering anomalies according to the expression in (2). The same filtering values of $w_{off} = 1.3$ m/s and $p_{off} = 120$ W have been used for all WTs. In Figures 7–11, we show the same results WT to WT.

So, each figure shows the results for each of the five WTs of the WF individually. Notice that different colors identify the data of each WT. The points detected as anomalies in the left subplot are marked in black. On the color points, we present the curve, also in black, provided by the manufacturer. In the subplot on the right, we represent the points captured as anomalies in isolation.

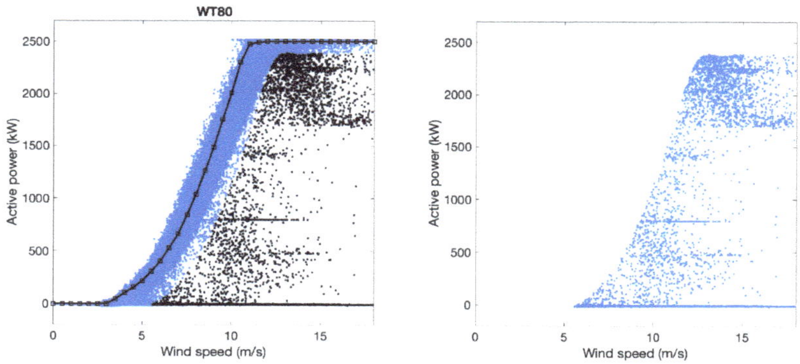

Figure 7. On the left, in black, the anomaly points of WT80 and the power curve provided by the manufacturer. On the right, the anomalies of WT80 are isolated.

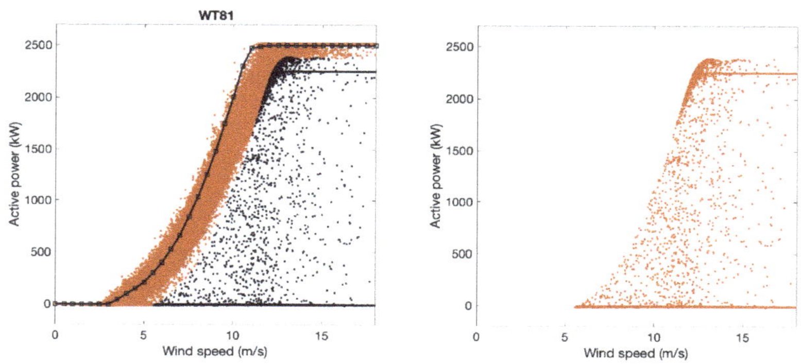

Figure 8. On the left, in black, the anomaly points of WT81 and the power curve provided by the manufacturer. On the right, the anomalies of WT81 are isolated.

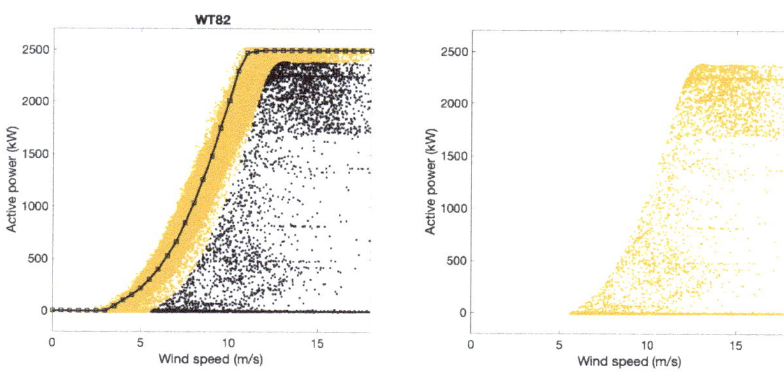

Figure 9. On the left, in black, the anomaly points of WT82 and the power curve provided by the manufacturer. On the right, the anomalies of WT82 are isolated.

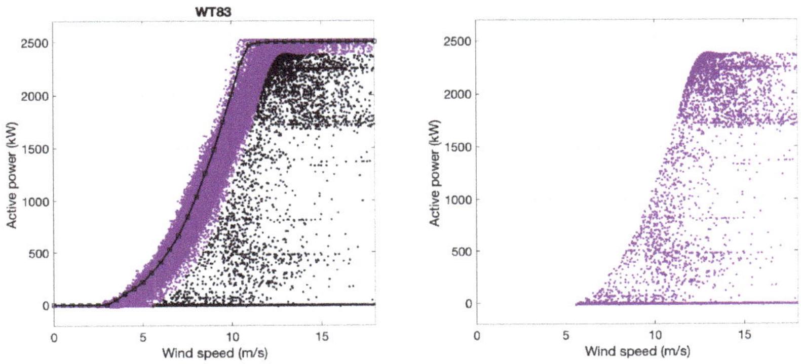

Figure 10. On the left, in black, the anomaly points of WT83 and the power curve provided by the manufacturer. On the right, the anomalies of WT83 are isolated.

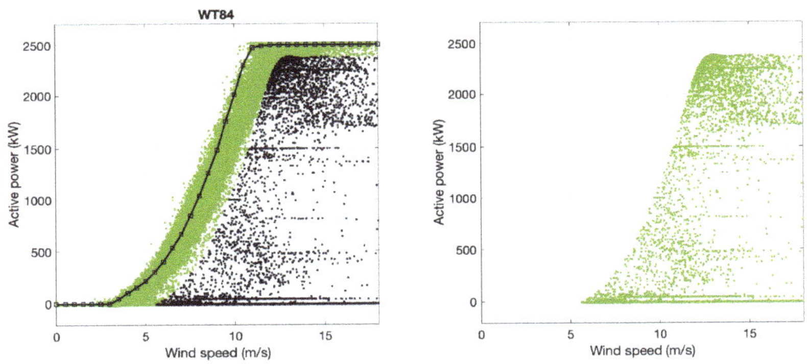

Figure 11. On the left, in black, the anomaly points of WT84 and the power curve provided by the manufacturer. On the right, the anomalies of WT84 are isolated.

3.2. Using the Filtered SCADA Signals to Estimate Power Curves: Training Data

For each WT, we estimate the wind-power curve using the ANN architecture described in (3) by using 3/4 of the total available data for the training by selecting the filtered wind speeds (wnac_avg_WSpd1) as the predictor and the filtered active power (wgdc_avg_TriGri_PwrAt) as the target. For all WTs, we have used the same ANN size consisting of $N = 10$ hidden nodes (neurons). Figure 12 shows all the estimated curves of each WT together with the power curve calculated from the data provided by the manufacturer in black indicating, in small squares, the specific points provided.

At this point, we notice how the estimated power-wind curves obtained from SCADA data differ from the one provided by the manufacturer, especially in the rang that goes from 8 to 12 m/s.

Comparing the estimated curves of each WT, we also notice that they are slightly different.

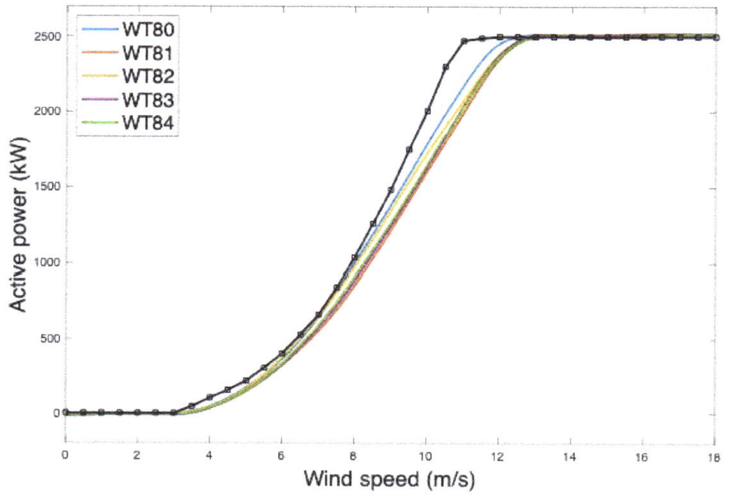

Figure 12. Power curves of each WT estimated from the wind and power data reported by the SCADA system after removing the anomalies. The black line is the power curve provided by the manufacturer.

3.3. Using the Filtered SCADA Signals to Estimate Power Curves: Test Data

Using the same anomaly filtering and estimation method, in this section, we estimate the same curves from the last quarter of available data that we reserved for the test. Once estimated, we compare them with the curves obtained in the training part. We provide the results in Figure 13.

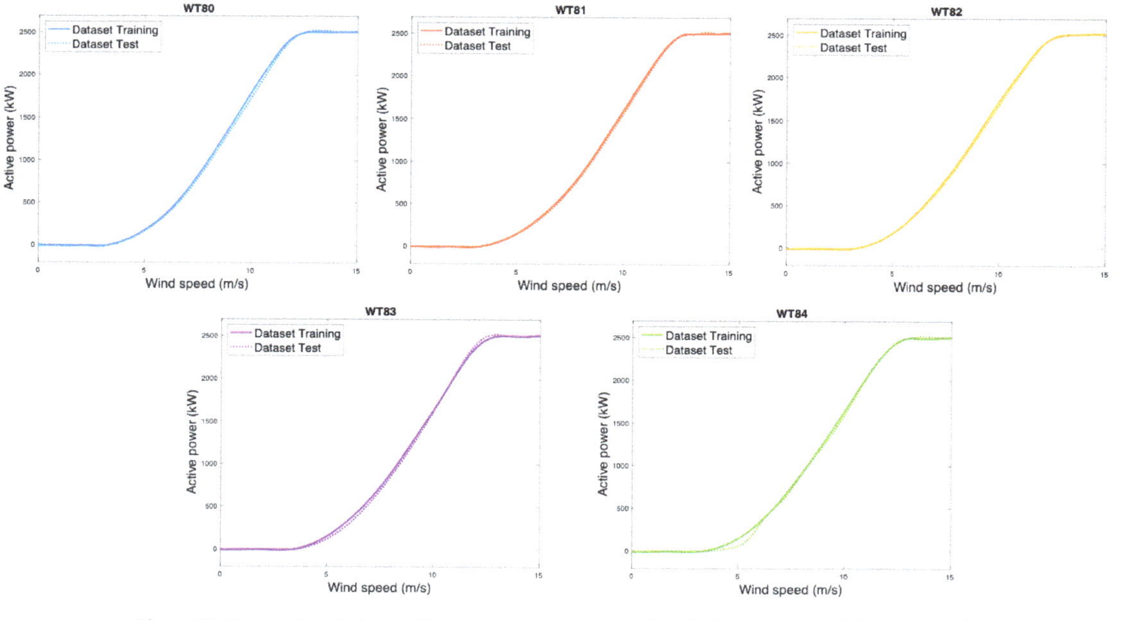

Figure 13. Comparison between the power curves estimated with the training and the test data for each of the WTs of the WF.

It is interesting to check that the estimated power curves of each machine in the test part fit very well with those obtained using the training data. The most notable difference is the WT84 machine, which suffered a major breakdown, causing it to be stopped for a long time. In this case, we see that the differences between the curves are observed in the part of the initial start-up of the WT when little power is still generated.

3.4. Experiment 1

Next, we experiment to check whether the size of the ANNs works correctly and whether the results depend on the network size.

In the following experiment, for each WT, we use the filtered data according to the proposed filter to train a series of models of different sizes and evaluate them with the test data. We evaluate the models by calculating the RMSE of each model's estimation error concerning the measured power. In the experiment, we evaluate ANNs with only five hidden nodes and progressively increase their hidden layer. In Table 2, we present the RMSEs obtained in the training phase, and in Table 3, the RMSEs obtained with the test data.

Table 2. RMSE in the training (experiment 1).

Hidden Layer Size:	5	7	10	15	20	25
WT80	86.07	86.03	86.02	85.98	85.98	85.96
WT81	78.22	77.80	77.94	77.78	77.77	77.74
WT82	82.10	82.15	81.89	82.01	81.94	81.76
WT83	74.87	75.05	75.57	74.82	74.84	75.13
WT84	75.63	74.62	74.56	74.77	74.50	74.60
mean:	79.38	79.13	79.19	79.07	79.00	79.04

Table 3. RMSE in the test (experiment 1).

Hidden Layer Size:	5	7	10	15	20	25
WT80	248.51	248.57	248.60	248.61	248.54	248.61
WT81	240.44	240.80	240.74	240.77	240.78	240.81
WT82	324.74	324.97	324.70	324.29	324.32	324.81
WT83	308.16	308.09	308.69	308.10	308.16	306.95
WT80	386.87	388.01	387.84	388.31	387.77	388.41
mean:	301.75	302.09	302.11	302.02	301.92	301.92

Notice that, in the test, which is the relevant part of the experiment and serves to size the network, all the ANNs evaluated work practically similarly, with the smallest configuration achieving the best result. Notice also that the magnitudes of RMSE are high, but it must be considered that the WTs are 2500 kW, and therefore, high values are handled.

3.5. Experiment 2

Once we have estimated the power curve of each WT, we have an estimate of the power curve of each WT that fits the data better than the curve determined from the points provided by the manufacturer, which has been used in the first filter. The question at this point is whether the filtering rule can be improved with the knowledge of this new curve. That is why we conduct this second experiment. We use the network of five hidden nodes to filter the row data and eliminate some points that in the improved rule may be considered anomalies, but that with the original rule may not.

In Figure 14, we show graphically, for the WT80, the power curve obtained from the ANN on the filtered data. In this experiment, for all WTs, values of $w_{off} = 1.1$ and $P_{off} = 55$ had been used, which are smaller than those used in the original filtering rule and therefore are more selective. In this figure, the points considered are shown in lighter blue, and those discarded as anomalies are shown in darker blue. With this new filtering, the previous experiment was repeated by training models of different sizes and evaluating them in the test. The results are again given in terms of the RMSE. In Table 4 for the training and in Table 5 for the test.

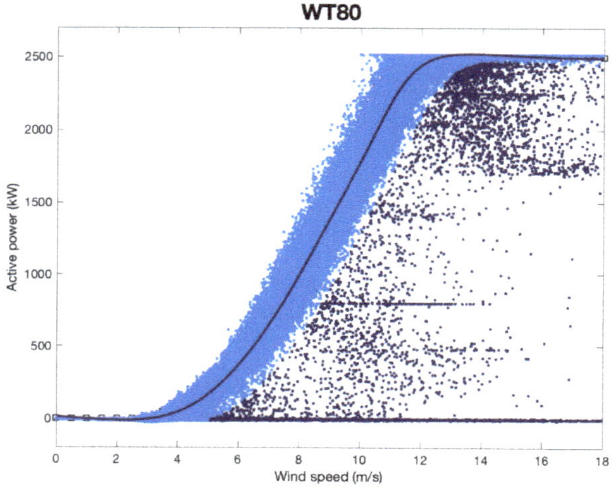

Figure 14. Representation on the SCADA data (points in lighter blue) of the WT80 of the modification of the filtering rule by changing the manufacturer's curve to the power curve (darker blue) estimated with an ANN of 5 hidden nodes.

Table 4. RMSE in the training (experiment 2).

Hidden Layer Size:	5	7	10	15	20	25
WT80	82.57	82.59	82.44	82.45	82.40	82.40
WT81	75.36	74.90	74.84	74.87	74.84	74.85
WT82	77.73	77.35	77.37	77.32	77.28	77.42
WT83	72.63	72.59	72.58	72.53	72.56	72.82
WT84	70.66	70.69	71.18	70.61	70.58	70.54
mean:	75.79	75.63	75.68	75.56	75.53	75.61

Table 5. RMSE in the test (experiment 2).

Hidden Layer Size:	5	7	10	15	20	25
WT80	247.96	247.94	247.91	247.88	247.88	247.88
WT80	235.62	235.68	235.71	235.71	235.68	235.45
WT80	323.82	323.83	323.80	323.99	323.90	324.01
WT80	309.51	309.35	309.59	309.64	309.37	309.97
WT80	392.62	392.39	392.95	393.58	393.03	392.53
mean: 301.91	301.84	301.99	302.16	301.97	301.97	301.96

Based on the results of this experiment, improving the filtering rule improves the results obtained in the training, but leaves the test results practically unchanged.

4. Discussion

We want to point out different aspects. In the context of this work, we understand by anomalies all those points that, due to failures in the sensors that measure the power or the wind speed or due to the starting and stopping processes of the WTs, fall in places far from the theoretical curve. Therefore, to obtain a good curve estimation, we note the need to filter anomaly data before the estimation to remove anomalies, as otherwise supervised ML-based methods also attempt to capture them into the models. So, when working with

SCADA data, estimating the power curves of each WT from the data (cleaned of anomalies) is necessary, since there are significant differences from the power curve provided by the turbine manufacturer. The cause of such significant differences seems to be the nature of the SCADA data, which consists of temporal averages, in our case, of 5 minutes. Instead, the manufacturer's values model instantaneous relations of wind and power.

In this work, we use a filtering rule that depends on the choice of w_{off} and p_{off} values, which can slightly determine the estimates of the curves in the regions where the curve meets the flat areas. In any case, what is vital is to consider the same rule in both the training and test phases to clean anomalies in the same way. This rule can be refined by changing the estimate of the curve provided by the manufacturer, $P_7(w)$, to the estimate of the curve measured in each WT. As noted in our work, we have kept a general rule for all WTs.

Once the estimates are made, we see that, despite presenting significant similarities, each WT has its particular curve. In this sense, it is also interesting to see that, for each WT, the estimates made with the training and test data are highly similar for all the WTs that work correctly and for the one we know has a failure in the test part, the WT84, the malfunction is captured in its curve.

The curves (to be precise, the points provided by the manufacturer) have been obtained according to a measurement protocol that is very clearly defined to be comparable between WTs from different manufacturers. Taking measurements under these conditions is expensive, so the manufacturer only provides points every 0.5 m/s. We understand, therefore, that the measurements supplied by the manufacturer are exact. In working conditions, things change slightly. First, of all, environmental conditions are continuously evolving. For example, it is known that air turbulence affects the generation conditions or that sometimes the orientation of the blades is not optimal either. These phenomena, in itself, cause observable differences between WTs of the same WF. However, one of the most significant things in this context is that the estimates are made from data provided by the SCADA system. Apart from the fact that the sensors are subject to noise, what is also very relevant is that these data are obtained from temporal averages of usually 10 min (although, in our case, they are 5 min). Temporal averages combined with wind variations in this time interval explain these differences between the estimated and manufacturer curves.

5. Conclusions

Since the malfunction of WTs directly impacts their power production capacity, characterizing the relationship between wind and power has attracted much interest. This dependence, however, is complex and non-linear since factors of a very varied nature (mechanical, electromagnetic, aerodynamic, control systems, etc.) are involved. One powerful way to express this relationship is through the wind-power curve. From the result of this work, after analyzing five WTs of a small WF, we note, after the estimation of their power curves from SCADA data, that:

1. Estimated curves substantially differ from the curve that one can deduce from the data provided by the manufacturer.
2. Given that the locations of the WTs are not identical, the environmental conditions of each turbine are also slightly different. Consequently, the estimated curves for each WT reflect such differences and have slight differences.
3. Once the curve of one WT is determined from the training data, it is reproduced very accurately when re-estimated from the test data.
4. The anomaly filtering methodology introduced permits consistently estimating the power curves using simple ANNs.
5. When there is a problem in one of the WTs, such is the case of the WT84, where a fault is documented at one point in the test phase, the problem becomes visible in its power curve when estimated by test data.

Author Contributions: Conceptualization, P.M.-P. and M.S.-S.; Methodology, P.M.-P.; Software, P.M.-P., J.Á.H. and M.S.-S.; Validation, P.M.-P.; Investigation, P.M.-P., J.Á.H. and M.S.-S.; Resources, J.S.-C.; Writing—original draft, P.M.-P. and M.S.-S.; Visualization, J.Á.H.; Supervision, P.M.-P. and J.S.-C.; Project administration, J.S.-C. All authors have read and agreed to the published version of the manuscript.

Funding: This research was supported by the Ministerio de Ciencia e Innovación of the Spanish Government (ref: PID2020-120314RB-I00/AEI/10.13039/501100011033).

Institutional Review Board Statement: Not applicable.

Informed Consent Statement: Not applicable.

Data Availability Statement: The data can be extracted from the freely available database at: https://github.com/alecuba16/fuhrlander (accessed on 15 February 2024).

Acknowledgments: The author would like to thank the Smartive company (https://gust.com/companies/smartive (accessed 13 March 2024)) for providing the data used in the experimental part.

Conflicts of Interest: The authors declare no conflicts of interest.

Abbreviations

The following abbreviations are used in this manuscript:

ANFIS	Adaptive network fuzzy inference system
ANN	Artificial neural network
GP	Gaussian Process
IQR	Interquartile range
LSE	Least-square estimate
LSOM	Linear Self-organized map
RMSE	Root mean squared error
SCADA	Supervisory control and data acquisition
SOM	Self organized map
WF	Wind farm
WT	Wind turbine
WT80	Wind turbine 80
WT81	Wind turbine 81
WT82	Wind turbine 82
WT83	Wind turbine 83
WT84	Wind turbine 84

References

1. Yang, W.; Court, R.; Jiang, J. Wind turbine condition monitoring by the approach of SCADA data analysis. *Renew. Energy* **2013**, *53*, 365–376. [CrossRef]
2. Sohoni, V.; Gupta, S.; Nema, R. A critical review on wind turbine power curve modelling techniques and their applications in wind based energy systems. *J. Energy* **2016**, *2016*, 8519785. [CrossRef]
3. Wang, Y.; Hu, Q.; Li, L.; Foley, A.M.; Srinivasan, D. Approaches to wind power curve modeling: A review and discussion. *Renew. Sustain. Energy Rev.* **2019**, *116*, 109422. [CrossRef]
4. Woebbeking, M. IEC TS 61400-22 (First Revision of IEC WT 01). *Citeseer* **2008**, *22*, 1–11.
5. Al-Quraan, A.; Al-Masri, H.; Al-Mahmodi, M.; Radaideh, A. Power curve modelling of wind turbines-A comparison study. *IET Renew. Power Gener.* **2022**, *16*, 362–374. [CrossRef]
6. Bandi, M.M.; Apt, J. Variability of the wind turbine power curve. *Appl. Sci.* **2016**, *6*, 262. [CrossRef]
7. Lydia, M.; Selvakumar, A.I.; Kumar, S.S.; Kumar, G.E.P. Advanced algorithms for wind turbine power curve modeling. *IEEE Trans. Sustain. Energy* **2013**, *4*, 827–835. [CrossRef]
8. Wang, S.; Huang, Y.; Li, L.; Liu, C. Wind turbines abnormality detection through analysis of wind farm power curves. *Measurement* **2016**, *93*, 178–188. [CrossRef]
9. Moreno, S.R.; Coelho, L.d.S.; Ayala, H.V.; Mariani, V.C. Wind turbines anomaly detection based on power curves and ensemble learning. *IET Renew. Power Gener.* **2020**, *14*, 4086–4093. [CrossRef]
10. Morrison, R.; Liu, X.; Lin, Z. Anomaly detection in wind turbine SCADA data for power curve cleaning. *Renew. Energy* **2022**, *184*, 473–486. [CrossRef]

11. Pandit, R.K.; Infield, D.; Kolios, A. Comparison of advanced non-parametric models for wind turbine power curves. *IET Renew. Power Gener.* **2019**, *13*, 1503–1510. [CrossRef]
12. Li, S.; Wunsch, D.C.; O'Hair, E.A.; Giesselmann, M.G. Using neural networks to estimate wind turbine power generation. *IEEE Trans. Energy Convers.* **2001**, *16*, 276–282.
13. Jang, J.S. ANFIS: Adaptive-network-based fuzzy inference system. *IEEE Trans. Syst. Man Cybern.* **1993**, *23*, 665–685. [CrossRef]
14. Karaboga, D.; Kaya, E. Adaptive network based fuzzy inference system (ANFIS) training approaches: A comprehensive survey. *Artif. Intell. Rev.* **2019**, *52*, 2263–2293. [CrossRef]
15. Raj, M.M.; Alexander, M.; Lydia, M. Modeling of wind turbine power curve. In Proceedings of the ISGT2011-India, Kollam, India, 1–3 December 2011; IEEE: Piscataway, NJ, USA, 2011; pp. 144–148.
16. Ouyang, T.; Kusiak, A.; He, Y. Modeling wind-turbine power curve: A data partitioning and mining approach. *Renew. Energy* **2017**, *102*, 1–8. [CrossRef]
17. Stephen, B.; Galloway, S.J.; McMillan, D.; Hill, D.C.; Infield, D.G. A copula model of wind turbine performance. *IEEE Trans. Power Syst.* **2010**, *26*, 965–966. [CrossRef]
18. Astolfi, D.; Castellani, F.; Lombardi, A.; Terzi, L. Multivariate SCADA data analysis methods for real-world wind turbine power curve monitoring. *Energies* **2021**, *14*, 1105. [CrossRef]
19. Pandit, R.; Kolios, A. SCADA data-based support vector machine wind turbine power curve uncertainty estimation and its comparative studies. *Appl. Sci.* **2020**, *10*, 8685. [CrossRef]
20. Wang, Y.; Duan, X.; Zou, R.; Zhang, F.; Li, Y.; Hu, Q. A novel data-driven deep learning approach for wind turbine power curve modeling. *Energy* **2023**, *270*, 126908. [CrossRef]
21. Letzgus, S.; Müller, K.R. An explainable AI framework for robust and transparent data-driven wind turbine power curve models. *Energy AI* **2024**, *15*, 100328. [CrossRef]
22. Mclean, J.H.; Jones, M.R.; O'Connell, B.J.; Maguire, E.; Rogers, T.J. Physically meaningful uncertainty quantification in probabilistic wind turbine power curve models as a damage-sensitive feature. *Struct. Health Monit.* **2023**, *22*, 3623–3636. [CrossRef]
23. Saint-Drenan, Y.M.; Besseau, R.; Jansen, M.; Staffell, I.; Troccoli, A.; Dubus, L.; Schmidt, J.; Gruber, K.; Simões, S.G.; Heier, S. A parametric model for wind turbine power curves incorporating environmental conditions. *Renew. Energy* **2020**, *157*, 754–768. [CrossRef]
24. Bilendo, F.; Meyer, A.; Badihi, H.; Lu, N.; Cambron, P.; Jiang, B. Applications and modeling techniques of wind turbine power curve for wind farms—A review. *Energies* **2022**, *16*, 180. [CrossRef]
25. Giebel, G.; Gehrke, O.; McGugan, M.; Borum, K. Common access to wind turbine data for condition monitoring the IEC 61400-25 family of standards. In Proceedings of the 27th Risø International Symposium on Materials Science: Polymer Composite Materials for Wind Power Turbines, Roskilde, Denmark, 4–7 September 2006.

Disclaimer/Publisher's Note: The statements, opinions and data contained in all publications are solely those of the individual author(s) and contributor(s) and not of MDPI and/or the editor(s). MDPI and/or the editor(s) disclaim responsibility for any injury to people or property resulting from any ideas, methods, instructions or products referred to in the content.

applied sciences

MDPI

Article

Main Factors on Effect of Precise Measurement and Precise Tamping Based on BP Neural Network

Jianjun Qu [1,*], Pan Liu [1,2], Yiyu Long [1] and Fei Xu [1]

[1] Infrastructure Inspection Research Institute, China Academy of Railway Sciences, Beijing 100081, China
[2] Graduate Department, China Academy of Railway Sciences, Beijing 100081, China
* Correspondence: qujianjun@rails.cn

Abstract: With the continuous development of precise measurement and precise tamping (PMPT) technology on Chinese railway conventional speed lines, the efficiency of machinery tamping operation and the quality of the track have been effectively improved. A variety of PMPT modes have been tried in the field operation, however there are some differences in the operation effect. The quality of the tamping operation is affected by multiple factors. In order to identify the key factors affecting the operation quality and to further improve the tamping operation effect, this paper establishes both the database of PMPT operation modes and the selection index system for evaluating the operation effect. Based on mega multi-source heterogeneous data and track geometry inspection data, this paper adopts the Back Propagation Neural Network (BPNN) prognosis model to quantify and sort the main factors affecting the effect of PMPT. The research results show that the initial quality of the track before tamping, whether the stabilizing operation or the tamping modes have great influence weights. It can scientifically guide the field operation to control the key factors and put forward some practical suggestions for promoting the field application of PMPT and the optimization of operation modes on the conventional speed lines.

Keywords: ballast track; BP neural network; precise measurement and precise tamping; tamping effect; influencing factors; track geometry inspection data

1. Introduction

The application of heavy-duty track maintenance machinery to realize digitally-controlled tamping operation and repairing the tracks by sections is an effective means to eliminate the ballasted track irregularities at the longitudinal level and the alignment [1–3]. Owing to the high mechanization, excellent operation quality, intelligent control and other advantages, it has become a popular maintenance practice accepted by railways all over the world. In 2021, in order to effectively improve the operation effect of heavy-duty tamping machines on ballasted tracks to remove track geometry irregularity, the technology of track PMPT was promoted and rolled out in China. The PMPT technology uses measuring equipment to accurately measure the track profile, optimizes to eliminate the track irregularity and formulate the tamping operation scheme based on the measurement results, inputs the tamping scheme to the heavy-duty machine before maintenance, and carries out precise digital tamping operations.

In the PMPT of a ballasted track, precise track measurement and precise tamping with the heavy-duty machine are two important links. Precise track measurement can quickly and accurately obtain the static track geometry state and control position information with the track geometry measurement trolley (TGMT), such as the three-dimensional coordinates of the track center line and the left and right rails [4]. It guides the track geometry adjustment performance, which is the premise to ensure the high smoothness and stability of the track. In line with the different measurement principles and methods, it can be classified into three modes of measurement: absolute measurement, relative measurement [5], and "relative + absolute" combined measurement. There are certain differences in measurement accuracy. Among them, the representative measuring equipment are the inertial navigation system (INS), the trolley and total station (TS) [6], the INS trolley, and the INS trolley and GNSS [7]. The INS trolley with TS configuration can provide 3D tamping surveying solutions for ballasted track works, as is shown in Figure 1a. Its 3D positioning data acquisition adopts the IMS measuring module and TS. The TS is used for leveling and free station setting on the track, thereby observing the surrounding CPIII control points (generally 4~8) and obtaining the 3D coordinates of the station setting position. The INS trolley with GNSS configuration adopts position and attitude measurements with a multi-antenna combination and INS real-time data fusion technology, thereby possessing a non-stop continuous measurement function, as is shown in Figure 1b.

(a) (b)

Figure 1. (**a**) INS trolley with TS configuration; (**b**) INS trolley with GNSS configuration.

Precise tamping with a heavy-duty machine is to guide heavy-duty track maintenance machinery to carry out digitally controlled tamping operations on ballasted tracks by using the measurement results. The tamping machine conducts track lifting, leveling, and alignment lining so as to improve the compactness of the track bed [8–10], and make the track geometry parameters meet the requirements of track design standards or maintenance codes [11], such as longitudinal level, cross level, and alignment of the track. At present, the main models of tamping machines used in China are the DCL-32 and DWL-48, which have some differences in the operating system, operation accuracy, and heavy-duty tamping machine parameter settings. DCL-32 [12] is a double-sleeper continuous tamping machine, which needs to be equipped with a dynamic track stabilizer, as is shown in Figure 2a. DWL-48 is a three-sleeper continuous tamping machine that integrates tamping and stabilization [13] and features high efficiency, high precision, and low energy consumption [14], as is shown in Figure 2b. DCL-32 and DWL-48 are produced using two tamping models produced by Plasser & Theurer as prototypes, which respectively include the 09-32 and 09-3X. Both the DCL-32 and DWL-48 models adopt Automatischer Leit Computer (ALC) control system and can achieve precise tamping. The stabilizing operation is used to simulate the stress state of the track under the action of the train, so as to achieve the purpose of re-compacting the ballast and avoiding uneven settlement of the track [15]. On-site maintenance personnel arrange the tamping machines and dynamic stabilizers appropriately to form different tamping modes for track maintenance operations.

(a) (b)

Figure 2. The main models of tamping machines: (**a**) DCL-32; (**b**) DWL-48.

The operation effect of different PMPT modes is affected by multiple factors. The degree of track geometry improvement depends on the track state before maintenance [16,17]. Different tamping operation modes have different effects on track geometry and track bed mechanical properties. Stabilizing operation will improve the track stability and extend the track quality warranty period [15]. Qu J. summed up the main factors affecting the tamping effect, which include the initial state of the tamping track, the measurement method, the tamping mode, the stabilizing operation, the models of the tamping machine, and the amount of lifting [1]. The condition of the track before the tamping operation affects the outcome of the tamping operation [18,19]. At present, there is no quantitative analysis method to determine the weight of the factors affecting the effect on the PMPT. The influence of different factors can only be qualitatively evaluated through field operation experience. We must use an optimal calculation method to quantify the influence weights of different factors and diagnosis the main factors affecting the effect on the PMPT. Scientifically guiding on-site construction operations and effectively improving the quality of operations are the pressing issues to be solved in the PMPT of ballast railways in China.

With the popularization and application of track inspection vehicles in China, a large amount of multi-source track inspection data has been accumulated, providing an important basis for track status evaluation. Neural networks can handle and analyze a huge amount of data, and even extract more data information from nonlinear data [20]. Machine learning technology has been widely applied in different fields, such as track status assessment [21], track geometry quality prediction [22,23], and railway fault diagnosis. In this study, the BPNN is selected because it is able to easily explore the importance of input features on output parameter. Therefore, having collected and sorted out the national-wide ballasted track PMPT maintenance statistics, the track inspection vehicle provides mega track geometry inspection data for the operation section. Based on the mega multi-source heterogeneous data and track geometry detection data, the BPNN is used to establish the main factor model that affects the PMPT operation effect. The influence weight of different factors on the PMPT effect is quantized for the first time, which is used to scientifically guide the control of key factors in the field operations. Some practical suggestions for the on-site application and operation mode optimization of the PMPT of the normal speed railway tracks are proposed, which provide a basis for reducing costs and increasing efficiency for heavy-duty track maintenance machinery operations, which has important practical significance and great economic benefits.

2. Related Work

2.1. The PMPT Operation Mode Database

From 2021 to 2022, a survey was carried out on the PMPT operation of national normal speed railway lines, and a database of the PMPT operation mode was established, which contains many actual on-site operations and track geometry inspection data, totaling about 27,000 km. The text data of the PMPT operation mode database includes the measurement method used before tamping, the model of the tamping machine, the tamping mode, the operation frequency of the stabilizer, etc. The conditional data includes whether it is with stabilizing operation, etc., and the digital data includes track geometry inspection data before and after tamping operation, improvement rate (IR), etc., among which various types of data contain data of different levels. The measurement methods mainly include INS trolley, INS trolley and CPIII, and INS trolley and GNSS. The tamping modes mainly include single tamping, single tamping and stabilizing, single pass double tamping and stabilizing, and single tamping and single tamping and stabilizing. The tamping machine models mainly include the DCL-32 and DWL-48. The operation frequency of the stabilizer mainly includes 0 Hz, 0–25 Hz and 25–40 Hz. On this basis, a scientific indicator system representing the main factors of PMPT operation effect was constructed.

2.2. Operation Effect Evaluation Index

The precise measuring and precise tamping operation are mainly used to improve the longitudinal level and the alignment irregularity of the track section by section so that the track profile of the line is constantly improved closer to the designed track profile. The study found that there was almost no change in track gauge before and after tamping operation, as is shown in Figure 3. In order to further improve the effectiveness of the operation effect evaluation index on the quality effect evaluation, the machine track quality index (*MTQI*) given in Equation (1) was used to the effect evaluation index. The cumulative distribution of the *MTQI* before and after tamping is shown as in Figure 4. It shows that there are obvious differences before and after the operation. The track geometry improvement before and after tamping is characterized by the track geometry improvement rate—which is expressed in Equation (2).

$$MTQI = \sum_{i=1}^{6} \sigma_i \tag{1}$$

$$IR = \frac{MTQI_{bf} - MTQI_{af}}{MTQI_{bf}} \tag{2}$$

where σ_i is the standard deviation of the 200 m inspection data of six track irregularities ($i = 1, 2, \ldots, 6$), including the left and right longitudinal level, the left and right rail alignment lining, the cross level, and the twist; $MTQI_{bf}$ is the *MTQI* before tamping, $MTQI_{af}$ is the *MTQI* after tamping, and *IR* is the improvement rate of the *MTQI*.

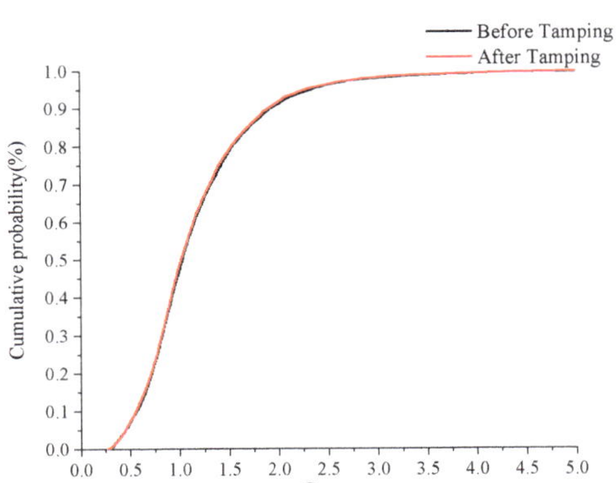

Figure 3. Cumulative distribution of track gauge before and after tamping.

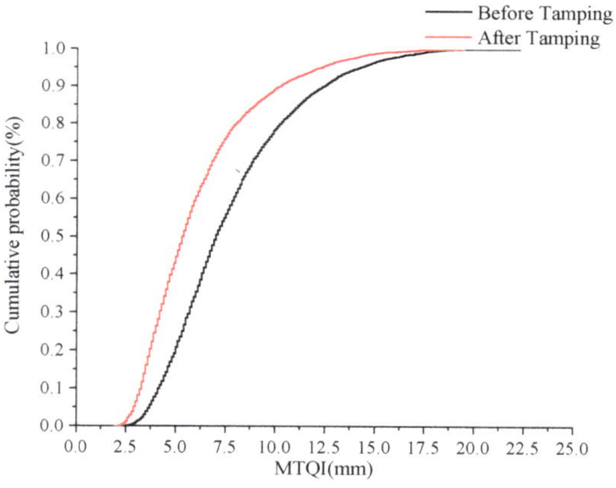

Figure 4. Cumulative distribution of *MTQI* before and after tamping.

2.3. Selection of Main Influencing Factors

Taking into consideration the actual operation and investigation on site, the main influencing factors that may exist in the whole process of PMPT operation were preliminarily determined. The quality of the track before tamping, $MTQI_{bf}$, the measurement method before tamping, the models of the tamping machine, the tamping mode, the frequency of stabilizing operation, and whether it had a stabilizer or not were input as the influencing factors. The input set X of the main factor model is defined as follows:

$$X = [X_1, X_2, X_3, X_4, X_5, X_6] \tag{3}$$

where X_1 is the measurement method before tamping. With the continuous progress of precise measurement technology, based on the different control network types, a variety of fast precise measurement methods were formed.

X_2 is the tamping mode. There are four main tamping modes used in field operation, and different tamping modes have different quality effects after operation.

X_3 is the model of the tamping machine. The main models of heavy-duty tamping machine used in China are the DWL-48 and DCL-32.

X_4 is whether it has a stabilizer or not, which refers to whether there is stabilizing operation during the tamping operation.

X_5 is the stabilizing operation frequency, which refers to the vibration frequency used in the stabilizing operation, which is mainly classified into 0 Hz (without stabilizing), 0~25 Hz, and 25~40 Hz. The compactness of the ballast bed after operation varies with different stabilizing operation frequencies [12].

X_6 is track quality before tamping, $MTQI_{bf}$.

3. Model Building

The BPNN [24] is a multi-layer feedforward network trained based on the error back propagation algorithm. It uses the steepest descent method to continuously adjust the weights and thresholds of the network through back propagation to minimize the sum of squared errors of the network. The neural network structure diagram is shown in Figure 5.

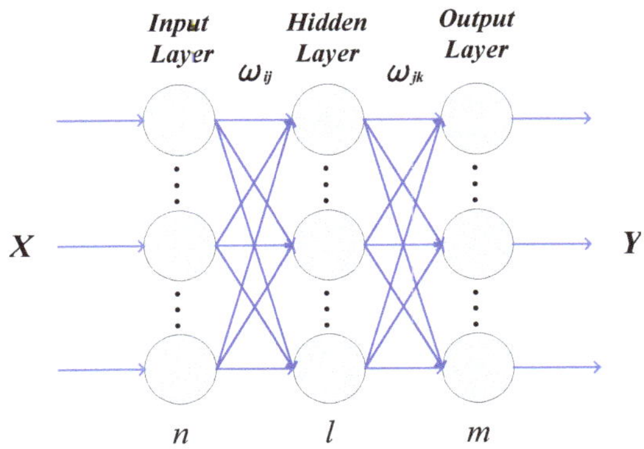

Figure 5. BPNN structure.

3.1. BPNN Training Model

Based on the data of different PMPT operation modes, a BPNN model was constructed by Matlab. By taking different influencing factors as inputs, the weights between each layer were calculated, and the weights of the influencing factors were obtained by statistics.

The main factor analysis model of the PMPT was characterized by the selection index system based on the influence of the PMPT effect. A huge amount of multi-source heterogeneous data was used as the input for the multi-layer feedforward neural network model. Through the transmission of signals between layers, the weighted connection of each neuron between the influencing factors of the input layer, the evaluation indicators of the hidden layer and the output layer were realized. The network weights and thresholds were adjusted based on the error of the predicted inspection data, and the cycle was repeated until a result meeting the error requirements was obtained. The network structure of the main factor analysis method that affected the PMPT operation effect is shown in Figure 6.

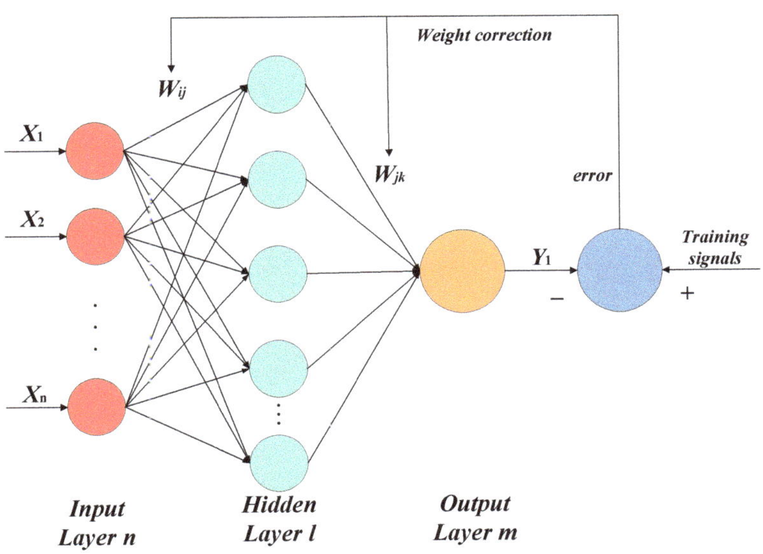

Figure 6. Neural network structure diagram by the main factor analysis method.

3.1.1. Statistical Index

MSE is the mean square error, as is shown in Equation (4). It calculates the expectation of the square of the difference between the true value y_i and the estimated value $\hat{y}_i$. The smaller the MSE is, the better the model is. R is the correlation coefficient between the expected output and the actual output, as is shown in Equation (5). The closer the fitting slope of the expected output and the actual output is to 1, the better. k is the expected output and actual output fit slope, as is shown in Equation (6).

$$\text{MSE} = \frac{1}{m}\sum_{i=1}^{m}(y_i - \hat{y}_i)^2 \tag{4}$$

$$R = \frac{\sum_{i=1}^{n}(y_i - \overline{y})(\hat{y}_i - \overline{\hat{y}})}{\sqrt{\sum_{i=1}^{n}(y_i - \overline{y})^2}\sqrt{\sum_{i=1}^{n}(\hat{y}_i - \overline{\hat{y}})^2}} \tag{5}$$

$$k = \frac{\sum_{i=1}^{n} y_i\hat{y}_i - n\overline{y}\overline{\hat{y}}}{\sum_{i=1}^{n} {y_i}^2 - n\overline{y}^2} \tag{6}$$

3.1.2. Selecting Model Parameters

In the original data, the initial quality of the track ($MTQI_{bf}$), the measurement method before tamping, the model of the tamping machine, the tamping mode, whether it had a stabilizer, and the frequency of stabilizing operation were processed as the input layer. The number of nodes in the input layer of the network was six (n = 6). The track quality after tamping ($MTQI_{af}$) or the IR was used as the output layer. The number of nodes in the output layer of the network was one (m = 1).

A network, with bias and at least one sigmoid hidden layer plus a linear output layer, is capable of approximating any rational function. Increasing the number of layers can further reduce the error and improve the accuracy; however, it also complicates the network. The parameter configuration of the neural network requires multiple attempts to adjust different parameters to determine the activation function of the hidden layer and the number of neurons in the hidden layer. By evaluating the network indicators with the correlation coefficient R, the fitting slope k and the value of MSE, a relatively good combination method can meet the principal factor analysis of the PMPT model was selected. The relatively optimal combination is shown as in Tables 1 and 2.

Table 1. Both input and output normalized fitting results.

Output Layer	Hidden Layer	The Number of Neurons in Hidden Layer	R	k	MSE
purelin	sigmoid	10	0.80841	0.655	1.38917
		20	0.81167	0.665	1.36820
		30	0.81265	0.667	1.36610
		50	0.81387	0.669	1.35414
	poslin	10	0.79767	0.636	1.45859
		20	0.80408	0.646	1.41714
		30	0.80706	0.649	1.40782
		50	0.81117	0.662	1.37169
	purelin	10	0.74900	0.561	1.76048
		20	0.74897	0.555	1.76073
		30	0.74902	0.557	1.76050
		50	0.74906	0.560	1.75990

Table 1. *Cont.*

Output Layer	Hidden Layer	The Number of Neurons in Hidden Layer	R	k	MSE
	tansig	10	0.80639	0.652	1.40226
		20	0.81199	0.665	1.36627
		30	0.81201	0.665	1.36635
		50	0.81211	0.666	1.36640

Table 2. Input normalized, output unnormalized fitting result.

Output Layer	Hidden Layer	The Number of Neurons in Hidden Layer	R	k	MSE
purelin	sigmoid	10	0.81044	0.659	1.37633
		20	0.81307	0.667	1.35926
		30	0.81328	0.667	1.35872
		50	0.81342	0.670	1.35720
	poslin	10	0.79514	0.635	1.47453
		20	0.80393	0.647	1.41840
		30	0.80452	0.650	1.40821
		50	0.80960	0.654	1.38164
	purelin	10	0.74893	0.561	1.76071
		20	0.74890	0.563	1.76096
		30	0.74893	0.561	1.76080
		50	0.74895	0.562	1.76067
	tansig	10	0.80783	0.657	1.39301
		20	0.81267	0.662	1.36202
		30	0.81398	0.665	1.35420
		50	0.81427	0.665	1.35219

Tables 1 and 2 show that, when the training results of the neural network are relatively good, the parameters of the neural network are set as follows (as is shown in Figure 7):

a. The number of neurons in the hidden layer: $l = 30$. Selecting 30 hidden layers not only makes the results more accurate, but also has high computational efficiency.
b. Input normalized; output not normalized;
c. Hidden layer activation function: tansig;
d. Output layer activation function: purelin;
e. The learning rate of the BPNN is set to 0.01;
f. The epoch of the BPNN is set to 100.

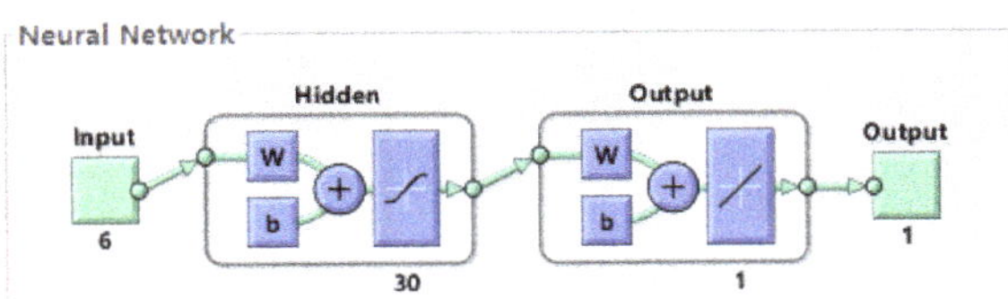

Figure 7. Neural network training structure.

3.1.3. Model Training Step

The training steps of the model are as follows:

Step 1: Calculate the output value H of the hidden layer according to the influencing factors of the input layer X, calculate the connection weight ω_{ij} between the input layer and the hidden layer, and calculate the threshold α of the hidden layer.

$$H_j = f\left(\sum_{i=1}^{n} \omega_{ij} x_i - \alpha_j\right) \quad j = 1, 2, \cdots, l \tag{7}$$

where l is the number of hidden layers at 30; and f is the activation function tansig of the hidden layer.

Step 2: According to the output value H of the hidden layer, the connection weight ω_{jk} and threshold b between the hidden layer and the output layer, calculate the O of the BPNN to predict the output of the $MTQI_{af}$ or IR.

$$O_k = \sum_{j=1}^{l} H_j \omega_{jk} - b_k \quad k = 1, 2, \cdots, m \tag{8}$$

Step 3: Calculate the network prediction error e according to the network prediction output O and the expected output value Y.

$$e_k = o_k - y_k \quad k = 1, 2, \cdots, m \tag{9}$$

Step 4: According to the network prediction error e, update the weights ω_{ij} and ω_{jk} between the layers of the network, as well as the new network node thresholds α and b.

$$\omega_{ij} = \omega_{ij} + \eta H_j (1 - H_j) x(i) \sum_{k=1}^{m} \omega_{jk} e_k \quad i = 1, 2, \cdots, n; j = 1, 2, \cdots, l \tag{10}$$

$$\omega_{jk} = \omega_{jk} + \eta H_j e_k \quad j = 1, 2, \cdots, l; k = 1, 2, \cdots, m \tag{11}$$

Step 5: Determine whether the iteration of the model algorithm has reached its end; if not, go back to Step 1.

3.2. Abnormal Data Cleaning

Due to the random error in the data set, in order to further improve the fitting effect of the model, on the basis of the established BPNN prognosis model, it was necessary to clean the data of the PMPT modes. We calculated the standard deviation of the actual error value of the predicted output value (estimated by the 1/2 power of the sample variance), removed the samples exceeding 3σ, and retrained until the data MSE stabilized.

$$|e_i - \mu| > 3\sigma \tag{12}$$

where μ is the mean error; and σ is the error standard deviation.

We iteratively cleaned n times until the number of cleanings had little effect on the R, MSE, and the amount of data that was cleaned.

3.3. Calculation Method of Influence Factor Weight

The weight associated with each feature conveys the importance of the feature in predicting the output value, so the weight is used to represent the influence of each factor on the quality of the track work after tamping.

Note that the weight w of the hidden layer is iiw in Equation (13), and the weight of the output layer is llw in Equation (13). The input data matrix is $n \times m$ dimensional: m is the amount of data. There are 30 hidden layer neurons, so the hidden layer weight iiw is a $n \times 30$ dimensional matrix, and the output layer weight llw is a 30×1 dimensional matrix. Use the product of iiw and llw (n $\times$ 1 dimensional matrix) as the weight for the n factors in the corresponding input. The weight formula of influencing factors is shown in Equation (14).

$$\begin{cases} iiw = net.IW\{1,1\} \\ llw = net.LW,\{2,1\} \end{cases} \quad (13)$$

$$W_1 = iiw \times llw \quad (14)$$

Due to the defects of the standard BP algorithm, it is easy to form a local minimum that does not obtain a global optimal solution. Therefore, the neural network model with the same structure was repeatedly trained 10,000 times on the cleaned training data, and the network weight in Equation (14) or its related form each time, as well as the ranking value of each factor weight each time, which was recorded for subsequent statistical processing.

4. Case Analysis

Taking the PMPT data of the national conventional speed lines as an example, the overall process of using the BPNN to study the effect of different influencing factors on the effect of the heavy-duty tamping machine is mainly classified into the following steps, as shown in Figure 8:

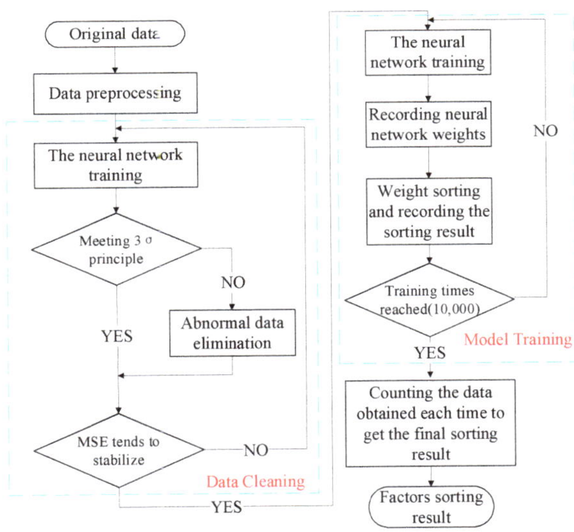

Figure 8. Overall flowchart.

4.1. Original Data Source

Local railway administrations in our country provide the operation information of the ballasted track PMPT maintenance section on the conventional speed railway every month, including the line, line type, operation start and end mileage, tamping date, measurement method, tamping mode, tamping machine model, stabilizer vibration frequency etc. Track inspection vehicles or comprehensive inspection trains are used to inspect the track geometry of the operating railway in China. Different speed grade lines have different inspection frequencies. The more important the line grade is, the higher the inspection frequency is. It is mainly divided into once a month, once every 20 days, or twice a month. Therefore, the track geometry inspection data of national railway lines were possessed. According to the maintenance information, the latest inspection data before and after tamping operation are matched from the inspection database as the $MTQI_{bf}$ and $MTQI_{af}$, respectively. Part of the original data is shown in Table 3.

Table 3. Part of original data.

Measurement Method	Tamping Machine Model	Tamping Mode	Stabilized Track	Stabilizer Vibration Frequency	$MTQI_{bf}$/mm	$MTQI_{af}$/mm	IR (%)
INS Trolley	DCL-32	single tamping and single tamping and stabilizing	Yes	0~25 Hz	4.9	4.7	4.1
INS Trolley	DWL-48	single pass double tamping and stabilizing	Yes	25~40 Hz	8.6	7.1	17.4
INS Trolley and GNSS	DCL-32	single tamping	No	0 Hz	3.5	2.5	28.6
INS Trolley and GNSS	DWL-48	single tamping and stabilizing	Yes	25~40 Hz	3.4	2.5	26.5
INS Trolley and CPIII	DCL-32	single pass double tamping and stabilizing	Yes	0~25 Hz	5.9	4.4	25.4
INS Trolley and CPIII	DWL-48	single tamping	No	0 Hz	6.3	5.3	15.9

4.2. Data Preprocessing

In the data preprocessing stage, each factor level number and combined data screening are performed on the original data (Table 3). The specific process is as follows:

Using the neural network for regression analysis, it identifies the variables with significant influence from multiple variables that affect a variable and uses the obtained relational expression to estimate or predict another specific variable based on the value of one variable or multiple variables. The influence of each factor on the quality of the track after tamping is studied, so the numbering method is used to number each factor and each level, and each factor is used as the input, as is shown in Table 4. The data form of *MTQI* or IR after tamping is the target value.

Table 4. Input factor corresponding coding example.

Code	1	2	3	4
Tamping mode	single tamping	single tamping + stabilizing	single pass double tamping and stabilizing	single tamping and single tamping and stabilizing
Measurement method	INS Trolley	INS Trolley and GNSS	INS Trolley and CPIII	-
Stabilizer vibration frequency	0 Hz	25~40 Hz	0~25 Hz	-
Tamping machine model	DCL-32	DWL-48	-	-
Stabilized track	No	Yes	-	-

4.3. Data Cleaning

Figure 9 shows the R, MSE, and the data that was cleaned away with the change in the number of cleanings. The total amount of known data was 30,022. A total of about 1800 pieces of data were cleaned. As the number of cleaning times increased, the correlation coefficient became bigger, and the MSE gradually became smaller and stable. The number of cleaning times was set 100. If the cleaning effect was not ideal, and, moreover, Figure 8 shows that increasing the cleaning number did not improve the cleaning effect, then the range of effective data could be accurately determined by reducing 3σ in Equation (12).

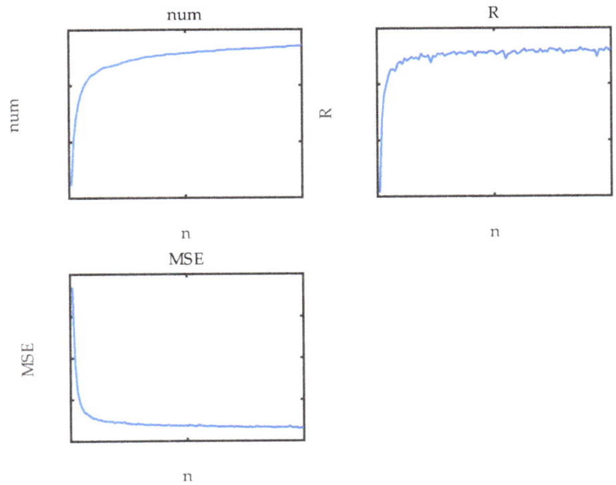

Figure 9. Data changes with the number of cleaning.

4.4. Neural Network Training

4.4.1. Splitting the Dataset

We split the dataset into a training set, validation set, and test set. Among them, the training set was used to train the neural network model; the validation set came from the subdivision of the training set, which was used for model selection and parameter adjustment; the test set was used to test the identification ability of the learner for new samples, and the test error was used as the generalization error. The approximate value of the data set was divided into 70% for the training set, 15% for the validation set, and 15% for the test set.

4.4.2. Model Training

The maximum number of iterations allowed in the model epoch was set to 100. The system stops when it reaches the training goal or reaches the maximum number of iterations. Figure 10 respectively shows the performance of the MSE metrics for the BP training, validation, and testing processes in each generation. The figure shows that the MSE decreased with the increase in the number of iteration rounds. The best verification performance was 0.44533 in the 7th round, and it tended to be stable. The training ended after 13 rounds.

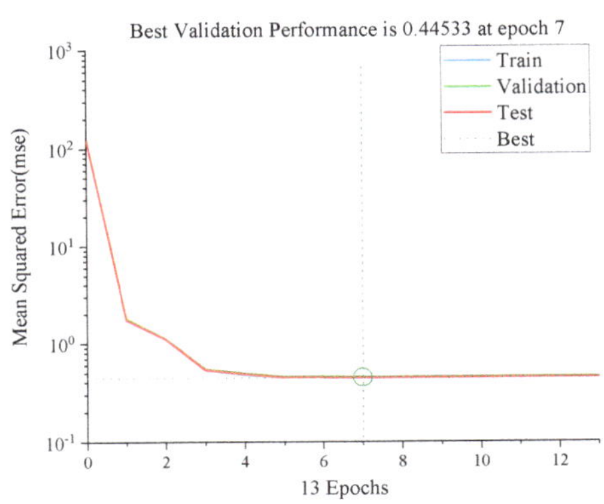

Figure 10. Performance of MSE indicators in each generation.

4.4.3. Evaluation Index

In this paper, the coefficient of determination R (Equation (5)) was adopted to evaluate the effectiveness of the model. The coefficient of determination R was close to 1, indicating that the trained network model was trained well. Meanwhile, the median absolute deviation (*MAD*) and mean absolute percentage error (*MAPE*) evaluation indicators were used to aid the evaluation of the prognosis effectiveness of the target value. The foregoing define the deviation between the predictive value and true value, thus reflecting the prognosis capacity. The lower the error value, the better the prognosis effect and the higher the accuracy of the target data. The calculation formula is:

$$\begin{cases} MAD = median(|v_i - v_m|) \\ MAPE = \frac{100\%}{n}\sum_{i=1}^{n}\left|\frac{\hat{y}_i - y_i}{y_i}\right| \end{cases} \quad (15)$$

where v_i is the i th value of $\hat{y} - y$, v_m is the median value of v, and n is the amount of data.

Figure 11 is the result of fitting the target value of each dataset with the estimated value of the neural network. After experimentation, the data results after cleaning were trained based on the above structural neural network model. The correlation coefficient R could reach about 0.85 when the target value was the $MTQI_{af}$. When the target value was the IR, R was only about 0.5. It shows that the neural network can fit $MTQI_{af}$ well, but the processing effect of IR is not satisfactory. Therefore, the $MTQI_{af}$ was selected as the only output of the neural network model. After 10,000 trainings, the mean of the MSE was 0.432, as is shown in Figure 12; the mean of the *MAD* was 0.64, as is shown in Figure 13; and the mean of the *MAPE* was 13.2%, as is shown in Figure 14. It can be seen that the deviation was small and the prediction effect was good. They met our precision needs for this study.

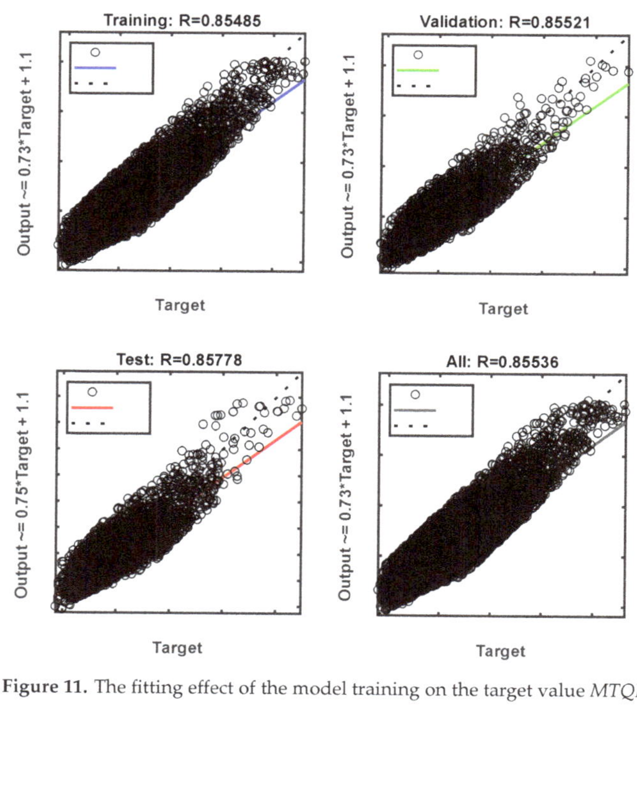

Figure 11. The fitting effect of the model training on the target value $MTQI_{af}$.

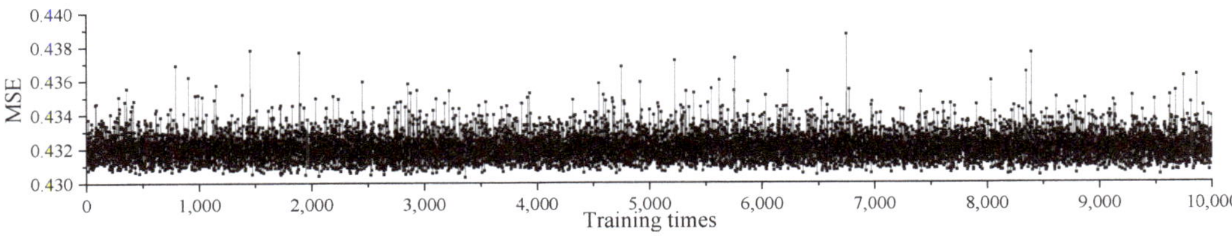

Figure 12. The MSE of $MTQI_{af}$.

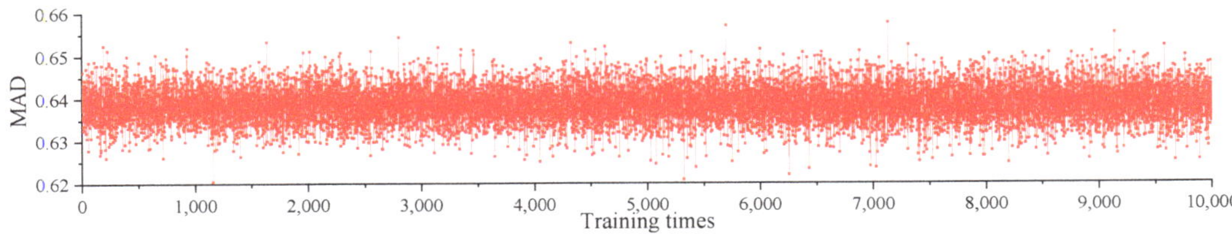

Figure 13. The *MAD* of $MTQI_{af}$.

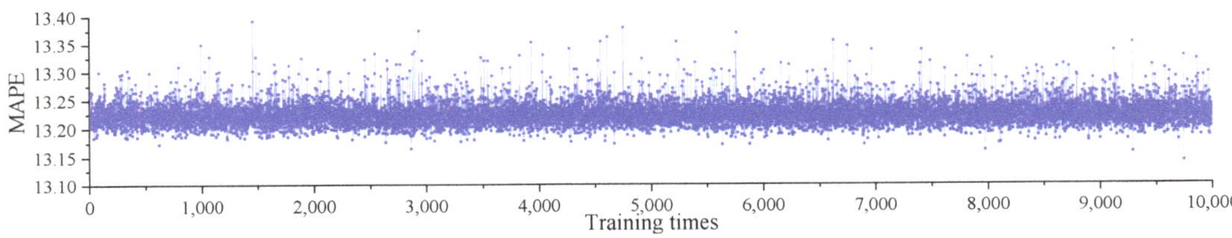

Figure 14. The *MAPE* of $MTQI_{af}$.

4.5. Result

Statistical methods for the research results were affected by various factors and include the follow steps:

Sort the weights of each factor obtained by each training model, record the sorting value of each training 10,000 times, find the mean of the sorting value of each factor, and finally sort the obtained sorting mean to get the final sorting result of each factor.

As is shown in Figure 15, the horizontal axis represents the ranking of weights (from small to big), and the vertical axis represents the probability of the corresponding weights of each influencing factor during 10,000 times of training. For example, the $MTQI_{bf}$ was 0.0341 at weight 1 and 0.3847 at weight 6. The weighted average value of the weighted sorting value (Equation (16)) was taken as the weight of each influencing factor, which was sorted from the biggest to the smallest, as is shown in Table 5. The statistical results show that the order of factors influencing the effect of tamping was the $MTQI_{bf}$, stabilized track or not, tamping mode, stabilizer vibration frequency, measurement method, and tamping machine model.

$$AS = \sum_{i=1}^{6} \omega_i S_i \tag{16}$$

where AS is the average sort of influencing factors, ω_i is the statistical probability, and S_i is the weight sorting (S_i = 1, 2, ... , 6).

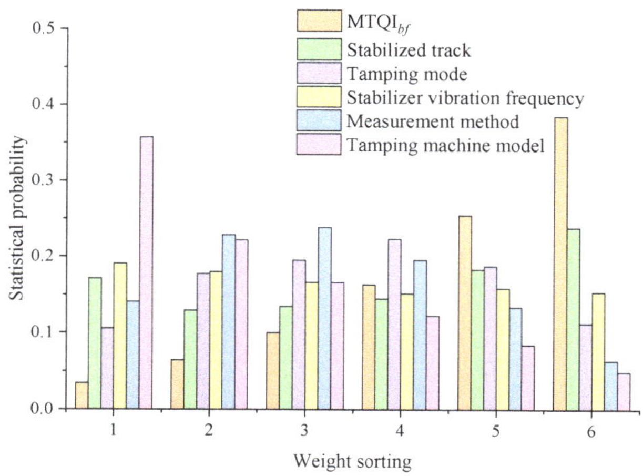

Figure 15. Distribution of weight ranking of each influencing factor.

Table 5. Influencing factor weighted statistics results.

Weight Sorting \ Factors	$MTQI_{bf}$	Stabilized Track	Tamping Mode	Stabilizer Vibration Frequency	Measurement Method	Tamping Machine Model
1	0.0341	0.1713	0.1055	0.1909	0.1410	0.3572
2	0.0641	0.1291	0.1772	0.1798	0.2282	0.2216
3	0.1001	0.1345	0.1950	0.1661	0.2382	0.1661
4	0.1631	0.1446	0.2230	0.1515	0.1955	0.1223
5	0.2539	0.1829	0.1872	0.1585	0.1336	0.0839
6	0.3847	0.2376	0.1121	0.1532	0.0635	0.0489
Average sort	4.6927	3.7515	3.5455	3.3665	3.1430	2.5008
Order	1	2	3	4	5	6

5. Discussions

In this paper, the method of deep learning was used to study the main factors on the effect of the PMPT. Through the testing and training model, the BPNN could better fit the influence of various factors on the $MTQI_{af}$ and obtain the weight of factors. The determination of factors is intended to solve the actual problems of field operation, so it is necessary to reconcile the field operation experience.

When the track quality before tamping is poor, tamping operation can significantly improve the track quality, but when the track quality before tamping is good, tamping operation has little effect on improving the track quality. Therefore, the initial quality of the line has an obvious impact on the tamping effect. The stabilizing operation can rearrange and compact the ballast particles under the sleeper, improve the stability, and restore the bearing capacity of the ballast bed. The different times of inserting picks in the tamping mode affect the ballast arrangement, the compactness of the ballast bed, and the track geometry. In addition, the accuracy of measurement equipment, the operating parameters of the heavy-duty machine, and the tamping machine model also affect the tamping effect. The result obtained by the model accords with the actual situation of the field.

The main factors affecting the effect of the PMPT were clarified, which can scientifically guide the field personnel to strictly control the key links of PMPT. It provides references for the selection of tamping mode, tamping machine model, measurement method, and other operation mode, as well as puts forward some practical suggestions for promoting the application of the PMPT of railway conventional speed lines. It has important practical significance and huge economic benefits to provide a basis for cost reduction and the efficient increase of heavy-duty track maintenance machinery operation.

6. Conclusions

In this paper, through the collection and processing of field data in recent two years, the database of the PMPT operation mode was structured and constructed. The selection index system of the influence of on-site PMPT effects based on big data was established. Based on many multi-source heterogeneous data and track geometry detection data, BPNN technology was used for the first time to select the number of output layers, hidden layer activation functions, and hidden layer neurons that met the characteristics of the data, and it established the main factor model that affected the PMPT operation effect. The accuracy of the neural network model was improved through abnormal data cleaning. The quantitative order of factors influencing the effect of the PMPT were the $MTQI_{bf}$, stabilized track or not, tamping mode, stabilizer vibration frequency, measurement method, and tamping machine model.

Author Contributions: Conceptualization, J.Q.; methodology, J.Q.; formal analysis, P.L.; software J.Q. and P.L.; writing—original draft preparation, P.L.; writing—review and editing, J.Q., P.L., Y.L. and F.X.; visualization, F.X. and Y.L.; funding acquisition, J.Q. All authors have read and agreed to the published version of the manuscript.

Funding: This research was funded by the Technology Research and Development Plan of China National Railway Corporation (CR), grant number (N2021G048), and the Project of the China Academy of Railway Sciences Corporation Limited, grant number (2021YJ216).

Institutional Review Board Statement: Not applicable.

Informed Consent Statement: Not applicable.

Data Availability Statement: Not applicable.

Acknowledgments: The authors thank the 18 Local Railway Administrations for providing the data of the PMPT operation modes. We also thank the Infrastructure Inspection Research Institute of the China Academy of Railway Sciences for providing the track geometry inspection data in this research work.

Conflicts of Interest: The authors declare no conflict of interest.

References

1. Qu, J. Research on Prediction Method of Warranty Period of Track Quality Based on Tamping Modes Using Large Tamping Machine. *J. China Railw. Soc.* **2019**, *41*, 117–122. [CrossRef]
2. Khouy, I.A.; Larsson-Kråik, P.-O.; Nissen, A.; Juntti, U.; Schunnesson, H. Optimisation of track geometry inspection interval. *Proc. IMechE Part F* **2014**, *228*, 546–556. [CrossRef]
3. Martey, E.; Attoh-Okine, N. Modeling tamping recovery of track geometry using the copula-based approach. *Proc. IMechE Part F* **2018**, *232*, 2079–2096. [CrossRef]
4. Chen, Q.; Niu, X.; Zuo, L.; Zhang, T.; Xiao, F.; Liu, Y.; Liu, J. A Railway Track Geometry Measuring Trolley System Based on Aided INS. *Sensors* **2018**, *18*, 538. [CrossRef] [PubMed]
5. Sánchez, A.; Bravo, J.L.; González, A. Estimating the Accuracy of Track-Surveying Trolley Measurements for Railway Maintenance Planning. *J. Surv. Eng.* **2017**, *143*, 05016008. [CrossRef]
6. Luo, Y.; Liu, C.; Yang, X.; He, Y.; Song, T. A New Method for Establishment of Plane Network of Ordinary Conventional-speed Railway Track Control Network and its Accuracy Discussion. *J. Railw. Sci. Eng.* **2014**, *11*, 55–59. [CrossRef]
7. Chen, Q.; Niu, X.; Zhang, Q.; Cheng, Y. Railway Track Irregularity Measuring by GNSS/INS Integration. *Navigation* **2015**, *62*, 83–93. [CrossRef]
8. Nielsen, J.C.; Berggren, E.G.; Hammar, A.; Jansson, F.; Bolmsvik, R. Degradation of railway track geometry–Correlation between track stiffness gradient and differential settlement. *Proc. IMechE Part F* **2020**, *234*, 108–119. [CrossRef]
9. Zhang, M.N.; Wu, Y.B.; Ma, X.C.; Wang, P. Influence of Large Track Maintenance Machine on Sleeper Supporting Stiffness. *Adv. Mat. Res.* **2014**, *919–921*, 1115–1119. [CrossRef]
10. Zhou, T.Y.; Hu, B.; Yan, B.; Peng, Y.X. Study on Motion Parameters of Tamping Operation. *Adv. Mat. Res.* **2013**, *694–697*, 154–157. [CrossRef]

11. Rhayma, N.; Bressolette, P.; Breul, P.; Fogli, M.; Saussine, G. Reliability analysis of maintenance operations for railway tracks. *Reliab. Eng. Syst. Saf.* **2013**, *114*, 12–25. [CrossRef]
12. Wang, X.J.; Geng, X.L. Study on the Effects of Tamping Frequency to the Compaction Degree of Trackbeds. *Appl. Mech. Mater.* **2012**, *178–181*, 1387–1391. [CrossRef]
13. Zhou, T.Y.; Hu, B.; Wang, X.J.; Yan, B. Discrete Element Method Analysis of Mechanical Properties of Railway Ballast during Tamping Process under Different Amplitude. *Appl. Mech. Mater.* **2012**, *233*, 224–227. [CrossRef]
14. Guo, Y.; Markine, V.; Jing, G. Review of ballast track tamping: Mechanism, challenges and solutions. *Constr. Build. Mater.* **2021**, *300*, 123940. [CrossRef]
15. Jadidirendi, K.; Zakeri, J.A.; Teng, H. Field Investigation for Identifying the Effects of Dynamic Track Stabilizing Operation on Track Geometrical Indices. In Proceedings of the 2015 Joint Rail Conference, San Jose, CA, USA, 23–26 March 2015. [CrossRef]
16. Selig, E.; Waters, J. *Track Geotechnology and Substructure Management*; Thomas Telford: London, UK, 1994. Available online: https://www.icevirtuallibrary.com/doi/book/10.1680/tgasm.20139 (accessed on 20 October 2022).
17. Esveld, C. *Modern Railway Track*; MRT-Productions Zaltbommel: Zaltbommel, The Netherlands, 2001.
18. Offenbacher, S.; Koczwara, C.; Landgraf, M.; Marschnig, S. A Methodology Linking Tamping Processes and Railway Track Behaviour. *Appl. Sci.* **2023**, *13*, 2137. [CrossRef]
19. Neuhold, J. *Tamping within Sustainable Track Asset Management*; Verlag der Technischen Universität Graz: Graz, Austria, 2020.
20. Khajehei, H.; Ahmadi, A.; Soleimanmeigouni, I.; Haddadzade, M.; Nissen, A.; Jebelli, M. Prediction of track geometry degradation using artificial neural network: A case study. *Int. J. Rail. Trans.* **2022**, *10*, 24–43. [CrossRef]
21. Sadeghi, J.; Askarinejad, H. Application of neural networks in evaluation of railway track quality condition. *J. Mech. Sci. Technol.* **2012**, *26*, 113–122. [CrossRef]
22. Sresakoolchai, J.; Kaewunruen, S. Track Geometry Prediction Using Three-Dimensional Recurrent Neural Network-Based Models Cross-Functionally Co-Simulated with BIM. *Sensors* **2023**, *23*, 391. [CrossRef] [PubMed]
23. Shafahi, Y.; Masoudi, P.; Hakhamaneshi, R. Track Degradation Prediction Models, Using Markov Chain, Artificial Neural and Neuro-Fuzzy Network. In Proceedings of the 8th World Congress on Railway Research, Seoul, Republic of Korea, 18–22 May 2008; pp. 1–9. Available online: https://www.researchgate.net/publication/268011897 (accessed on 12 March 2023).
24. LeCun, Y.; Bengio, Y.; Hinton, G. Deep learning. *Nature* **2015**, *521*, 436–444. Available online: https://www.nature.com/articles/nature14539 (accessed on 20 October 2022). [CrossRef] [PubMed]

Disclaimer/Publisher's Note: The statements, opinions and data contained in all publications are solely those of the individual author(s) and contributor(s) and not of MDPI and/or the editor(s). MDPI and/or the editor(s) disclaim responsibility for any injury to people or property resulting from any ideas, methods, instructions or products referred to in the content.

MDPI AG
Grosspeteranlage 5
4052 Basel
Switzerland
Tel.: +41 61 683 77 34

Applied Sciences Editorial Office
E-mail: applsci@mdpi.com
www.mdpi.com/journal/applsci

Disclaimer/Publisher's Note: The title and front matter of this reprint are at the discretion of the Guest Editors. The publisher is not responsible for their content or any associated concerns. The statements, opinions and data contained in all individual articles are solely those of the individual Editors and contributors and not of MDPI. MDPI disclaims responsibility for any injury to people or property resulting from any ideas, methods, instructions or products referred to in the content.

www.ingramcontent.com/pod-product-compliance
Lightning Source LLC
Chambersburg PA
CBHW061825210926
51850CB00006B/809
* 9 7 8 3 7 2 5 8 5 6 9 1 6 *